Seed dormancy in grasses

Seed dormancy in grasses

G. M. Simpson
Department of Crop Science and Plant Ecology,
University of Saskatchewan, Saskatoon, Canada

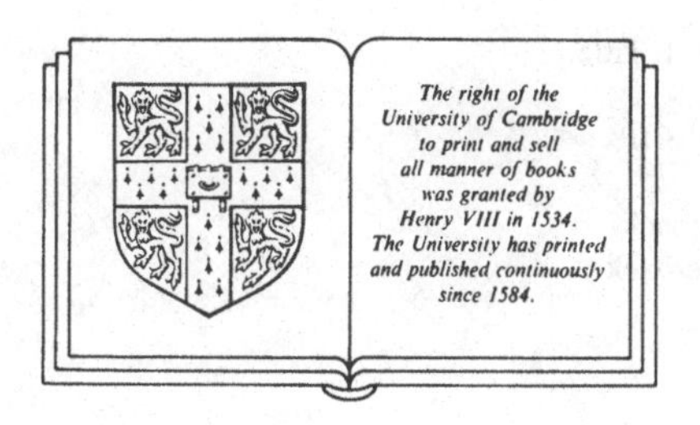

CAMBRIDGE UNIVERSITY PRESS
Cambridge
New York Port Chester
Melbourne Sydney

CAMBRIDGE UNIVERSITY PRESS
Cambridge, New York, Melbourne, Madrid, Cape Town, Singapore, São Paulo

Cambridge University Press
The Edinburgh Building, Cambridge CB2 8RU, UK

Published in the United States of America by Cambridge University Press, New York

www.cambridge.org
Information on this title: www.cambridge.org/9780521372886

First published 1990
This digitally printed version 2007

A catalogue record for this publication is available from the British Library

Library of Congress Cataloguing in Publication data

Simpson, G. M.
Seed dormancy in grasses / G. M. Simpson.
p. cm.
Bibliography: p.
Includes index.
ISBN 0 521 37288 7
1. Grasses–Seeds–Dormancy. I. Title.
QK495.G74S613 1990
584′.90446–dc20 89-17255 CIP

ISBN 978-0-521-37288-6 hardback
ISBN 978-0-521-03930-7 paperback

Contents

Preface

A number of books have been published recently on the subject of germination physiology of seeds. They often have a chapter or two about seed dormancy, either to demonstrate the diversity of mechanisms among seed plants, or to try and simplify the complexity of dormancy mechanisms by establishing general models. A somewhat different approach is used here. Firstly, the subject is confined to seed dormancy in grasses. Secondly, experimental evidence is considered in depth for a single species, the wild oat (*Avena fatua* L.), probably the most widely studied species for understanding seed dormancy in the plant kingdom. The evidence for this member of the family Gramineae is compared with other examples among the Gramineae to reach some general conclusions about the nature of seed dormancy in grasses.

There are several reasons for confining the book to grasses. The grass family is one of the largest (25 tribes and 600 genera) and most diverse in the plant kingdom. From a human nutrition perspective it is the most important family. Grasses are the principal plant life form covering more than 70% of the land surface of the globe and they are of critical importance to the stability of the fragile arid and semi-arid zones. While seed dormancy is of great adaptive significance for survival in nearly all plant species with seeds, it is also the main reason why grass species cause the most serious weed problems in cultivated crops around the globe. It is paradoxical that one of the great achievements of the Neolithic Age was selection against seed dormancy, so that planted crops would germinate quickly and uniformly. Today we are only beginning to understand the nature of dormancy in seeds and how we might use this knowledge to better advantage. Finally, cereal grass seeds in our present age are the main source of diet for the majority of the human race. It is therefore for reasons of great practical, economic and human survival interest that we should fully understand the trait of seed dormancy because it is a primary means for survival of populations, indeed of species, in the plant kingdom.

This book is dedicated to the memory of my colleague for 25 years, the late Prof. J. M. Naylor, who was a pioneer in the field of seed dormancy studies. I am indebted to my students and research colleagues who contributed to our present understanding of seed dormancy in *A. fatua*. The pathway to our understanding has not been along a straight line but something akin to a much compressed helix with much energy spent in going around the circles.

Acknowledgements

Permission for reproduction was granted by Annals of Botany Company and the editors of *Annals of Botany* (Figs. 3.1, 3.13); Association of Applied Biologists and editors of *Annals of Applied Biology* (Fig. 2.8*(a)*); Cambridge University Press, editors of *BioEssays* and Dr. A. J. Trewavas (Fig. 4.1); Crop Science Society of America and editors of *Crop Science* (Fig. 2.6); Editorial Office of *Physiologia Plantarum* (Figs. 3.6, 3.7, 3.12, 3.15); editors of *New Phytologist* (Fig. 3.8); editor of *Phytochemistry* (Table 2.3); editors of *Proceedings of the Association of Official Seed Analysts* (Figs. 2.8*(b)*, 2.8*(c)*); editors of *Seed Science and Technology* (Fig. 3.11); John Wiley and Sons, New York Ltd., (Symbols of Prof. H. T. Odum in Figs. 5.7, 5.8); National Research Council of Canada and editor of *Canadian Journal of Botany* (Figs. 2.1, 2.2*(a)*, 2.10, 3.2, 3.3, 3.14, 3.16); Professor I. N. Morrison (Fig. 2.7*(b)*) and Professor M. V. S. Raju (Fig. 2.2*(b)*). I am particularly grateful to Mr. J. Diduck for help with the figures, to Dr. F. Turel for editorial asistance and to my wife Margarete for patience with me during preparation of the manuscript.

Introduction

The several objectives of this book can be stated in the form of simple questions. What can we conclude about the nature of grass seed dormancy in a single, well-studied, species? Are there commonalities between seed dormancy in this single species and other grass species? Can new conclusions be reached about the nature of the physiological and environmental conditions that establish the state of dormancy in grass seeds? Is it possible to describe new models for seed dormancy that simplify our understanding of grass seed dormancy?

The nature of the first question is in part a semantic problem related to correctly matching the etymological meaning of a word to the reality it attempts to describe. The English word 'dormancy', derived from the French *dormir* (to sleep), itself derived from the Latin *dormire* (to sleep), is defined in the *Concise Oxford Dictionary* as 'lying inactive in sleep'. However, biologists have found that this definition does not encompass observed seed behaviour. Hence the many attempts to divide dormancy into sub-categories that cover the situations in nature where some seeds fail to germinate, whereas others can, in a specific environment (Amen, 1968; Bewley & Black, 1982; Lang *et al.*, 1987).

An adequate description of dormancy must involve at least three major components viz. the seed, the environment and a time element describing changes in state of both the seed and environment. In addition, it is useful to have some measure of incidence of dormancy related to genetic variation in a plant population. The definition of dormancy should be applicable to an individual seed or to groupings up to the population level.

Some years ago I drew attention to the fact that most of the published experimental approaches to understanding the nature of dormancy, as a general phenomenon in biology, have been based on the reductionist approach (Simpson, 1978). This approach emphasises the systematic division of the organism into its constituent parts down through successively lower levels of organisation to the current limits of biological, biophysical and biochemical technology. Using this approach, analysis down to the level of constituent molecules is expected to 'explain' seed dormancy. For example, molecules that inhibit, or promote, germination have been invoked as the *raison d'être* for the presence or absence of dormancy. However, modern systems analysis applied to biology (Koestler, 1967; von Bertalanffy, 1968; Whyte *et al.*, 1969) stresses that the dormancy system includes both the organism and its environment within some specified time period. The whole system is organised into hierarchies of functional and structural levels that are difficult to separate. In a systems description of function the requirements on one level of action are also the constraints on the next level. Thus normal activity at one functional level can only take place when all levels below that level are functioning properly. A functional description of seed dormancy may involve a few or many different levels in the hierarchy depending upon the interest of the observer. A useful description of dormancy may be found in the holistic general view of the system or alternatively at a very detailed sub-level.

The first three chapters show the extent of seed dormancy in the grass family and systematically outline the structural factors within the plant, and the environmental factors encompassing the plant or seed dispersal unit, that together comprise the seed–environment system. Chapters 4 and 5 describe the physiological nature of dormancy, seen through a systems perspective, as well as several approaches to modelling a dormant seed system. A conclusion is reached that there is no single state of seed dormancy common to all grass species. There are as many states of dormancy as there are species, and within species the potential number of states of dormancy increases in direct proportion to the intensity with which observations are carried out on seeds to the lowest levels of the seed–environment system. The view of 'reality' about the state of seed dormancy, or germination, is related as much to the view adopted by the observer as it is to the degree of internal constraint and patterns of determinate behaviour originating from the living system we call a seed.

1

The occurrence of dormancy in the Gramineae

'The grass family is one of the largest and most diverse in the plant kingdom and certainly one of the most important. More than any other, this assemblage of plants feeds man and beast and so clothes the earth that soil may be built and held securely from the forces of erosion. No other group of plants is more essential to the nutrition, well being, or even existence of man.'

J. R. Harlan, 1956
(Theory and Dynamics of Grassland Agriculture)

Gould (1968) has divided the family of grasses (Gramineae) into 25 tribes and 177 important genera. Seed dormancy occurs within 18 of the tribes and has been recorded in one or more species in at least 78 of the genera (Table 1.1). Seed dormancy is one of a number of adaptive traits, such as seed size and shape, that are polymorphic in character. Together these characteristics provide diversity and fitness for both opportunistic settlement and enduring occupation of temporally and spatially diverse habitats (Jain & Marshall, 1967). Within each grass species there can be considerable variation in the degree of polymorphism associated with the expression of seed dormancy. The degree of polymorphism is a function of the form of reproduction (self- or cross-pollination) in each species. Some species rely on genetic diversity and others on phenotypic plasticity for adapting to varying environments through the trait of seed dormancy. Dormancy has been described as an adaptive trait that causes seeds to germinate at a time, or place, favourable to the subsequent survival of the seedling and adult plant (Pelton, 1956).

The grass family occurs in a great diversity of form spread over a wide range of habitats. Grasses survive in even the most stressful extremes of climate (Simpson, 1981) and show great evolutionary plasticity (Harlan, 1956). This plasticity is due to the wide variety of modes of reproduction with aneuploidy and polyploidy. The most common form of reproduction is by cross-fertilization from wind-borne pollen. Nevertheless all methods of fertilization from enforced cross-pollination all the way to obligate self-fertilization, and intermediates, exist among grass species. Because seed dormancy is known to be a genetically inherited trait (Naylor, 1983)

Table 1.1. *List of tribes and genera of the Gramineae (After Gould, 1968) indicating:*
(a) Number of publications about seed dormancy for the genus retrieved in a computer search of world literature – an index of incidence of seed dormancy.
(b) Number of species in the genus listed as weeds (After Holm et al., *1979)*
(c) Geographical distribution of species estimated by average number of countries in which each species is considered a weed problem (After Holm et al., *1979).*

SUBFAMILY I FESTUCOIDEAE

Tribe 1. Festuceae – Bromus (37, 26, 5), Brachypodium (1, 1, 1), Vulpiea (4, 6, 3), Festuca (28, 6, 3), Lolium (37, 9, 12), Leucopoa (–, –, –), Scolochloa (1, –, –), Sclerochloa (–, 1, 1), Catapodium (–, –, –), Puccinellia (1, 4, 1), Poa (77, 7, 15), Briza (–, 3, 10), Phippsia (–, –, –), Coleanthus (–, –, –), Dactylis (20, 2, 20), Cynosurus (3, 2, 7), Lamarckia (–, 1, 7).

Tribe 2. Aveneae – Koeleria (1, 1, 8), Sphenophlis (–, –, –), Trisetum (1, 1, 1), Coryenphorous (–, –, –), Aira (5, 3, 2), Deschampsia (4, 3, 4), Scribneria (–, –, –), Avena (931, 11, 10), Ventanata (–, –, –), Helictotrichon (–, –, –), Arrhenatherum (3, 3, 4), Holcus (2, 4, 4), Dissanthelium (–, –, –), Calamagrostis (2, 5, 1), Ammophila (–, 2, 2), Apera (2, 1, 12), Polypogon (1, 3, 7), Mibora (–, –, –), Cinna (–, –, –), Limondea (–, –, –), Anthoxanthum (2, 2, 6), Hierochloe (–, –, –), Phalaris (18, 9, 13), Alopecurus (10, 12, 7), Phleum (17, 3, 4), Gastridium (–, 1, 6), Lagurus (–, 1, 5), Milium (–, –, –),Beckmannia (–, 1, 1), Agrostis (17, 14, 3).

Tribe 3. Triticeae – Elymus (1, 2, 2), Sitanion (–, –, –), Taeniatherum (–, 1, 1), Hystrix (–, –, –), Hordeum (215, 13, 5), Agropyron (33, 6, 7), Triticum [including Aegilops] (190, 8, 1), Secale (14, 2, 5).

Tribe 4. Meliceae – Melica (–, –, –), Glyceria (1, 3, 3), Catabrosa (–, –, –), Pleuropogon (–, –, –), Schizachne (–, 3, 1).

Tribe 5. Stipeae – Stipa (22, 10, 2), Oryzopsis (29, 2, 4), Piptochaetium (–, –, –).

Tribe 6. Brachyelytreae – Brachyelytrum (–, –, –).

Tribe 7. Diarrheneae – Diarrhena (–, –, –).

Tribe 8. Nardeae – Nardus (1, 1, 3).

Tribe 9. Nonermeae – Nonerma (–, –, –), Parapholis (–, 1, 5).

SUBFAMILY II. PANICOIDEAE

Tribe 10. Paniceae – Digitaria (25, 32, 8), Leptoloma (–, 8, 8), Anthaenantia (–, –, –), Stenotaphrum (–, 3, 3), Brachiaria (10, 19, 7), Axonopus (1, 4, 10), Reimarochloa (–, –, –), Eriochloa (–, 6, 3), Paspalum (14, 37, 7), Paspalidium (–, 3, 4), Panicum (57, 82, 3), Lasiacis (–, 2, 2), Oplismenus (–, 6, 3), Echinochloa (36, 15, 13), Sacciolepis (–, 7, 3), Rhynchelytrum (–, 2, 19), Setaria (38, 26, 10), Pennisetum (19, 14, 7), Cenchrus (5, 15, 5), Amphicarpum (–, 1, 1), Melinus (–, 1, 8), Anthophora (–, –, –).

Table 1.1. *(cont)*.

SUBFAMILY I FESTUCOIDEAE

Tribe 11. Andropogoneae – Imperata (2, 5, 12), Miscanthus (1, 5, 2), Saccharum (12, 5, 7), Erianthus (–, –, –), Sorghum (42, 12, 8), Sorghastrum (7, –, –), Andropogon (13, 27, 3), Arthraxon (–, 1, 4), Mirostegium (–, –, –), Dichanthium (1, 3, 5), Bothriochloa (7, 5, 1), Chrysopogon (1, 3, 7), Hyparrhenia (–, 4, 6), Schizachyrium (13, 3, 1), Eremochloa (1, 1, 1), Trachypogon (–, –, –), Elyonurus (–, –, –), Heteropogon (1, 1, 11), Manisuris (–, –, –), Hackelochloa (–, 1, 7), Tripsacum (–, 1, 5), Zea (23, –, –), Coix (–, 5, 8).

SUBFAMILY III. ERAGROSTOIDEAE

Tribe 12. Eragrosteae – Eragrostis (27, 51, 5), Neeragrostis (–, –, –), Tridens (–, –, –), Triplasis (1, –, –), Erioneuron (–, –, –), Munroa (–, –, –), Vaseyochloa (–, –, –), Redfieldia (–, –, –), Scleropogon (–, 1, 3), Blepharidachne (–, –, –), Calamovilfa (–, 5, 1), Lycurus (1, –, –), Muhlenbergia (2, 5, 1), Sporobolus (4, 18, 4), Blepharoneuron (–, –, –), Crypsis (–, 2, 2).

Tribe 13. Chlorideae – Eleusine (17, 6, 13), Dactyloctenium (2, 2, 27), Leptochloa (–, 8, 8), Trichoneura (–, 2, 5), Gymnopogon (–, –, –), Tripogon (–, –, –), Willkommia (–, –, –), Schedonnardus (–, 1, 1), Cynodon (13, 1, 85), Microchloa (–, –, –), Chloris (12, 14, 6), Trichloris (–, 3, 6), Bouteloua (11, 3, 1), Buchloe (4, 1, 1), Cathestecum (–, –, –), Aegopogon (–, –, –), Tragus (–, 3, 3), Spartina (1, 5, 1), Ctenium (–, –, –), Hilaria (1, –, –).

Tribe 14. Zoysieae – Zoysia (7, 4, 4).

Tribe 15. Aeluropodeae – Distichlis (1, 3, 3), Allolepis (–, –, –), Monanthochloe (–, –, –), Swallenia (–, –, –).

Tribe 16. Unioleae – Uniola (2, –, –).

Tribe 17. Pappophoreae – Pappophorum (–, –, –), Enneapogon (–, –, –), Cottea (–, –, –).

Tribe 18. Orcuttieae – Orcuttia (–, –, –), Neostapfia (–, –, –).

Tribe 19. Aristideae – Aristida (11, 15, 2).

SUBFAMILY IV. BAMBUSOIDEAE

Tribe 20. Bambuseae – Arundinaria (1, 1, 1).

Tribe 21. Phareae – Pharus (–, –, –).

SUBFAMILY V. ORYZOIDEAE

Tribe 22. Oryzeae – Oryza (162, 8, 1), Leersia (1, 3, 11), Zizania (21, 1, 1), Zizaniopsis (–, 2, 2), Luziola (–, 1, 1), Hydrochloa (–, 1, 1).

SUBFAMILY VI. ARUNDINOIDEAE

Tribe 23. Arundineae – Arundo (–, 3, 5), Phragmites (1, 5, 17), Cortaderia (–, –, –), Molinia (–, –, –).

Tribe 24. Danthonieae – Danthonia (5, 1, 1), Sieglingia (–, –, –), Schismus (–, 1, 1).

Tribe 25. Centotheceae – Chasmanthium (–, –, –).

diversity in the mechanisms for achieving seed dormancy can be expected in the grass family because of the diversity in methods of reproduction. One of the major objectives of this book is to determine whether there are just a few, or many, ways of achieving seed dormancy in the grass family.

The grass family can be arbitrarily divided into several categories that reflect their utilitarian value for human needs. Sub-division into wild, forage, cereal and weed grasses in contrast to the taxonomic divisions also gives some indication of the diversity of geographical and agronomic situations in which members of the grass family perform valuable functions.

Seed dormancy has a natural significance for survival of grasses in their undisturbed ecosystem. In addition seed dormancy creates a number of difficulties in the systems of animal and crop husbandry that supply a high proportion of the human diet. Animal husbandry is based on the utilization of both natural grasslands and sown pastures of improved forage grasses. Seed dormancy can be a major problem in the establishment of grasses used for pastures. Crop husbandry based primarily on annual cereal grasses depends on uniform germination hence seed dormancy can cause serious problems at the time of establishment of the crop. Seed dormancy, combined with high seed production, is a primary cause of the persistence of many grass species as weeds competing effectively in the major crops of the globe. Many grasses also have a nuisance value in non-crop localities such as lakes, canals, road sides, railway tracks, airfields and public parks. In these areas of increasing importance linked to the urbanization of a rapidly growing human population seed dormancy plays an important role in the survival of plant species. In recent years the concern for preservation of the genetic diversity of both wild and cultivated plants through the use of gene banks has drawn attention to the difficulty of germinating seeds of many grass species because of persistent seed dormancy (Ellis, Hong & Roberts, 1985b).

In the following sections of Chapter 1 the above somewhat arbitrary categories of wild, forage, cereal, and weed grasses will be assessed for the occurrence and level of significance of seed dormancy occurring in natural and manipulated ecosystems.

1.1 Wild grasses

The term 'wild' is used as a general category for all those species that have not been deliberately modified through selection pressure and breeding methods applied by mankind. 'Wild' grasses comprise the majority of grass species. Even if the top 250 global weeds and the major global crops (about 50 species (Stoskopf, 1985)) were all grasses and these were then added to the total of about 44 cultivated pasture grasses the grand total would be a tiny fraction of the approximately 10000 species of the family Gramineae (Semple, 1970).

It has been pointed out a long time ago (Bews, 1929) that for the majority of grasses there is no economic interest. Thus useful information, other than the simplest morphological description necessary to define taxonomic position, is absent for a majority of grass species. For this reason alone it is difficult to assess the occurrence of dormancy in the 'wild' species. Many of the species found today may represent intermediate stages between the fossil grasses of the bamboo or bamboo-like plants that existed in the Cretaceous period (Semple, 1970) and more modern genera. Seed dormancy is found in bamboos (McClure, 1966; Matumura & Nakajima, 1981) and some species such as *Melocanna baccifera* exhibit vivipary (McClure, 1966). This suggests, in an evolutionary sense, that seed dormancy and vivipary have been important traits in the grass family for a very long period of time.

1.2 Forage grasses

Cultivated forage grasses are mainly selections, starting about the middle of the nineteenth century, of native and introduced grasses with superior productivity. Forage grasses have not yet been substantially altered by plant breeding so that their characteristics are in the main similar to their wild forms. Unlike natural grasslands, modern arable pasture systems need to be frequently re-established as part of a rotational cropping system. Thus seed dormancy in forage grasses prevents successful establishment of a new pasture. Seed dormancy in one of a mixture of grass species can lead to the complete elimination of the species during the establishment phase.

Many native grasses and grassy weeds are utilized as forages and the distinctions between a cultivated forage, a weedy grass, and a native grass disappear in some regions of severe drought when all plants are severely over-grazed. This is particularly true where nomadism still prevails.

Seed dormancy is evidently very common in species belonging to the major grass genera utilized for grazing around the world (Table 1.2). Grasses from the tropical, sub-tropical, and temperate climates can all show seed dormancy. Within genera it is common to find a number of closely related species with seed dormancy. While the sample of approxima-

Table 1.2 *Evidence from the literature of the general occurrence of seed dormancy in forage grasses. The citation is the most recent reference and the number in the column indicates the number of other earlier references in* Bibliography of seed dormancy in grasses *(Simpson, 1987). The common name is given in brackets.*

Sub-family	Tribe	Genus	Species	Reference	No.
FESTUCOIDEA	Festuceae	*Bromus* (Chess grasses)	*-japonicus* (Japanese)	Froud-Williams 1981	7
			-diandrus	Baskin & Baskin, 1981	–
			-catharticus [uniloides] Rescue grass)	Froud-Williams & Chancellor, 1986	3
			-sterilis	Hilton, 1987	7
			-mollis (Soft chess)	Ellis *et al.*, 1985b	–
			-commutatis (Hairy chess)	Froud-Williams & Chancellor, 1986	–
			-inermis (Smooth brome)	Nielsen *et al.*, 1959	3
			-secalinus (Chess brome)	Steinbauer & Grigsby, 1957	1
			-tectorum (Downy brome)	Milby & Johnson, 1987	3
		Festuca (Fescues)		Williams, 1983a	9
			-rubra (Creeping red)	Williams, 1983b	3
			-arundinacea (Tall)	Stoyanova *et al.*, 1984	6
			-pratensis (Meadow)	Linnington *et al.*, 1979	1
			-ovina (Sheep)	Nieser, 1924	–
		Lolium (Ryegrasses)		Williams, 1983b	15
			-perenne (Perennial)	Williams, 1983b	7

		-multiflorum (Italian)	Van Staden & Hendry, 1985	5
		-rigidum (Wimmera)	Gramshaw & Stern, 1977	3
		-temulentum (Darnel)	Cairns & de Villiers, 1986a	–
		-persicum (Persian darnel)	Banting & Gebhardt, 1979	–
	Scolochloa			
		-festucacea	Smith, 1972	–
	Poa	(Bluegrasses)	Naylor & Abdalla, 1982	19
		-pratensis (Kentucky)	Phaneendranath, 1977a	30
		-annua (Annual)	Sgambetti-Araujo, 1978	26
		-trivialis (Rough)	Froud-Williams & Ferris, 1987	7
		-fertilis	Kleine, 1929	–
		-palustris (Fowl)	Rostrup 1897/8	–
		-compressa (Canada)	Andersen, 1947a	9
	Dactylis		Nakamura, 1962	1
		-glomerata (Orchard grass, Cocksfoot)	Probert *et al.*, 1985d	17
	Cynosurus	*-cristatus* (Crested dogstail)	Schonfield & Chancellor, 1983	2
Aveneae	*Aira*	*-flexuosa*	Nelson & McLaggan, 1935	–

Table 1.2. *(cont.)*

Sub-family	Tribe	Genus	Species	Reference	No.
			-*caryophylla* (Silver hair)	Pemadasa & Lovell, 1975	–
			-*praecox*	Roberts, 1986	1
		Deschampsia	-*caespitosa* (Tufted hair)	Roberts, 1986	3
		Arrhenatherum	-*elatius* (Tall oat grass)	Roberts, 1986	2
		Holcus	-*lanatus* (Yorkshire fog)	Schonfield & Chancellor, 1983	1
		Calamagrostis	-*canadensis* (Blue joint)	Conn & Farris, 1987	1
		Anthoxanthum	-*puelli* (Sweet vernal)	Nieser, 1924	–
			-*odoratum* (Sweet vernal)	Schonfield & Chancellor, 1983	–
		Phalaris		Nakamura, 1962	1
			-*arundinacea* (Reed canary)	Berg, 1982	9
			-*minor*	Parasher & Singh, 1984	2
			-*tuberosa* (Harding grass)	Myers, 1963	3
		Alopecurus	-*aequalis* (Shortawn foxtail)	Arai & Chisaka, 1961	3
			-*myosuroides*	Froud-Williams *et al.*, 1984	4
			-*geniculatis* (Water foxtail)	Roberts, 1986	–
			-*pratensis* (Meadow foxtail)	Roberts, 1986	–
		Phleum	-*pratense* (Timothy)	Schonfield & Chancellor, 1983	13
		Agrostis (Bent grasses)		Schmidt, 1969	4

		-*alba* (Red top)	Bass, 1950a	–
		-*tenuis* (Colonial bent)	Toole & Koch, 1977	2
		-*capillaris* (Astoria bent)	Schonfield & Chancellor, 1983	1
		-*palustris* (Creeping bent)	Eggens & Ormrod, 1982	1
		-*gigantea*	Williams, 1978	3
		-(Highland bent)	Pierpont & Jensen, 1958	1
Triticeae	*Elymus* (Wild rye)	-*triticoides* (Beardless)	Gutormson & Wiesner, 1987	–
	Hordeum	-*murinum* (Mouse barley)	Cocks *et al.*, 1976	1
		-*leporinum*	Cocks & Donald, 1973	1
		-*distichum* (2-row)	Ellis & Roberts, 1980	–
		-*pusillum* (Little barley)	Fischer *et al.*, 1982	–
		-*spontaneum* (4-row)	Key, 1987	2
		-*agriocrithon*	Key, 1987	–
		-*jubatum* (Foxtail barley)	Conn & Farris, 1987	2
	Agropyron (Wheat grasses)		Grime *et al.*, 1981	7
		-*cristatum* (Crested)	Maynard, 1963	6
		-*smithii* (Western)	Toole, 1976	5
		-*spicatum* (Bluebunch)	Evans & Tisdale, 1972	–
		-*elongatum* (Tall)	Thornton, 1966a	1
		-*intermedium* (Intermediate)	Kilcher & Lawrence, 1960	–

Table 1.2. *(cont.)*

Sub-family	Tribe	Genus	Species	Reference	No.
			-repens (Couch, Quackgrass)	Conn & Farris, 1987	5
		Aegilops (Goat grasses)	*-ovata*	Datta *et al.*, 1972	1
			-cylindrica (Jointed)	Donald & Zimdahl, 1987	1
			-kotschyii	Wurzburger *et al.*, 1976	10
			-trunncialis (Barb)	Laude, 1956	–
		Secale	*-silvestre*	Plarre, 1975	–
	Meliceae	*Glyceria*	*-maxima* (Manna grass)	Roberts, 1986	–
	Stipeae (Spear, porcupine grasses)	*Stipa*	*-viridula* (Green needle grass)	Fulbright *et al.*, 1983	12
			-leuchotricha (Texas needle G.)	Andersen, 1963	–
			-bigeniculata	Hagon, 1976	–
			-variabilis	Lodge & Whalley, 1981	–
			-poaceae (Feather grass)	Bespalova & Borisova, 1979	–
			-joannis	Ovesnov & Ovesnov, 1972	
			-trichotoma (Nasella tussock)	Joubert & Small, 1982	–
			-stenophylla	Ovesnov & Ovesnov, 1972	–
			-capillata	Ovesnov & Ovesnov, 1972	–

			-neaei	Soriano, 1961	–
		Oryzopsis (Rice grasses)	*-hymenoides* (Indian)	Young *et al.*, 1985	21
			-miliacea (Smilo)	Probert, 1981	5
PANICOIDEAE	*Paniceae* (Millet grasses)	*Digitaria*		Takabayashi & Nakayama, 1981	3
			-sanguinalis (Crab grass)	Miller *et al.*, 1965	5
			-adscendens	Ramakrishnan & Khoshla, 1971	4
			-milanjiana	Hacker *et al.*, 1984	4
			-pentzii	Baskin, *et al.*, 1969	–
			-ciliaris	McKeon *et al.*, 1985	1
			-didactyla	Febles & Harty, 1973	–
		Brachiaria	*-decumbens*	Bewley & Black, 1982	2
			-humidicola	Goedert & Roberts, 1986	–
			-mutica	MacLean & Grof, 1968	–
			-ruziziensis	Renard & Capelle, 1976	1
		Axonopus	*-compressus* (Carpet grass)	Andersen, 1941	–
		Paspalum		Nakamura, 1962	3
			-notatum (Bahia grass)	De Toledo *et al.*, 1981	6
			-dilatatum (Dallis grass)	Johnston & Miller, 1964	2
			-anceps	Mathews, 1947	–
			-distichum (Knot grass)	Huang & Hsiao, 1987	–

Table 1.2. *(cont.)*

Sub-family	Tribe	Genus	Species	Reference	No.
		Panicum [Largest grass family]	*-ramosum* (Brown top millet)	Moore & Hoffman, 1964	4
			-clandestinum (Deetongue)	Chirco *et al.*, 1979	–
			-coloratum (Kleingrass, coloured Guinea grass)	Tischler & Young, 1987	4
			-dichotomiflorum (Fall panic)	Okada & Ochi, 1986	7
			-maximum (Guinea grass)	Harty *et al.*, 1983	14
			-miliaceum	Cavers & Benoit, 1987	2
			-texanum	Egley & Chandler, 1983	–
			-anceps	Mathews, 1947	1
			-antidotale	Koller & Negbi, 1957	–
			-crusgalli frumentaceum	Kim, 1954	1
			-turgidum	Roberts & Ellis, 1982	–
			-phillopogon	Piacco, 1940	–
			-virgatum (Switch grass)	Sautter, 1962	–
			-obtusum (Vine mesquite)	Toole, 1940	–
		Setaria (Foxtail grasses)		Mohamed *et al.*, 1985	5
			-viridis (Green)	Hendricks & Taylorson, 1974	4
			-faberii (Giant)	Taylorson, 1986	4
			-chevalieri	Erasmus & Van Staden, 1983	–

		-lutescens [glauca] (Yellow bristle grass)	Lehle *et al.*, 1981	11
		-italica (Foxtail millet, Bengal grass)	Kim, 1954	1
		-macrostachya (Plains bristle)	Toole, 1940	–
		-anceps	Hawton, 1979	–
	Pennisetum		Pemadasa & Lovell, 1975	2
		-clandestinium (Kikuyu grass)	Cross, 1936	–
		-ciliare (Buffel grass)	Andersen, 1953	–
		-typhoides [glaucum] (Bajra)	Raza, 1977	5
		-polystachyon	Fernandez, 1980	–
		-macrourum (African feather G.)	Harradine, 1980	–
		-pedicellatum (Deenanath grass)	Maiti *et al.*, 1981	2
		-purpureum (Elephant grass)	Oakes, 1959	–
	Cenchrus	*-ciliaris*	Bhupathi *et al.*, 1983	3
		-setigereus	Anonymous, 1962	–
		-longispinus	Twentyma, 1974	–
Andropogoneae	*Imperata*	*-cylindrica* (Cogan grass, satin tail, alang-alang)	Matumura *et al.*, 1983	1
	Sorghum		Marbach & Mayer, 1979	4
		-halapense (Johnson grass)	Tao, 1982	10
		-guineensa	Pont, 1935	–
		-stipoideum	Andrew & Mott, 1983	–
		-intrans	Andrew & Mott, 1983	–
	Sorghastrum	*-nutans* (Indian grass)	Geng & Barnett, 1969	6

Table 1.2. *(cont.)*

Sub-family	Tribe	Genus	Species	Reference	No.
		Andropogon		Stubbendieck & McCully, 1976	–
			-scoparius (Little bluestem)	Prentice, 1981	7
			-gerardii (Big bluestem)	Prentice, 1981	7
			-tener	Eira, 1983	–
			-diastachyus	Goedert, 1984	–
			-gayanus	Goedert, 1984	1
		Dichanthium	*-annulatum* (Bluestem)	Oke, 1952/53	–
		Bothriochloa	*-ischaemum*	Ahring *et al.*, 1975	2
			-intermedia	Ahring *et al.*, 1975	3
			-macra	Watt & Whalley, 1982	2
		Chrysopogon	*-fallax*	Mott, 1978	–
			-latifolius	Mott, 1978	–
		Eremechloa	*-ophiuroides* (Centipede grass)	Delouche, 1961	–
ERAGROSTOIDEAE	Eragrosteae	*Eragrostis* (Love grasses)		Taylorson & Hendricks, 1979a	8
			-trichodes (Sand)	Ahring *et al.*, 1962	1
			-lehmanniana (Lehman)	Haferkamp *et al.*, 1977	3
			-ferruginea	Suzuki, 1987	4
			-abyssinica (Teff)	Katayama & Nakagama, 1972	–
			-ciliaris	Kumari *et al.*, 1987	–
			-tremula	Kumari *et al.*, 1987	–
			-leptostachya	Lodge & Whalley, 1981	–

		-curvula	Voigt, 1973	2
	Sporobolus (Drop seed grasses)		Toole, 1941	–
		-vaginiflorus	Baskin & Baskin, 1967	–
		-elongatus	Lodge & Whalley, 1981	–
Chlorideae	*Eleusine* (Goose grasses)		Popay, 1973	2
		-indica	Hilu & De Wet, 1980	4
		-coracana	Hilu & De Wet, 1980	5
		-compressa	Gupta & Saxena, 1980	–
	Dactyloctenium		Gupta, 1973	–
		-sindicum (Crowfoot grass)	Kumari *et al.*, 1987	–
	Cynodon		Akamine, 1944	2
		-dactylon (Bermuda grass)	Wu, 1980	7
		-plectostachyus (Star grass)	Okigbo, 1964a	1
		-diploidium	Okigbo, 1964b	1
	Chloris		Shimizu *et al.*, 1970	3
		-gayana (Rhodes grass)	Sharir, 1971	–
		-ciliata	Gassner, 1911	3
		-virginata (Feather finger)	Kumari *et al.*, 1987	–
		-truncata (Stargrass)	Lodge & Whalley, 1981	–
	Bouteloua (Grama grasses)		Coukos, 1944	1
		-gracilis (Blue grama)	Thornton & Thornton, 1962	–

Table 1.2. *(cont.)*

Sub-family	Tribe	Genus	Species	Reference	No.
			-*curtipendula* (Side-oats grama)	Major & Wright, 1974	4
			-*eriopoda* (Black grama)	Wright & Baltensperger, 1964	–
		Buchloe	-*dactyloides* (Buffalo grass)	Kneebone, 1960	3
	Danthonieae	*Danthonia*	-*pedicillata* (Silver grass)	Hagon, 1976	–
			-*californica*	Laude, 1949	–
			-*sericea* (Downy oatgrass)	Lindauer, 1972	–
			-*linkii*	Lodge & Whalley, 1981	–
			-*spicata* (Poverty oatgrass)	Toole, 1939	–

tely 200 species is small, in comparison with the total number in the family of grasses, it nevertheless contains a cross-section of the most important genera exposed to grazing by the majority of herbivores on a global scale. It can be argued therefore that seed dormancy in grasses is general rather than exceptional.

To a limited degree the influence of seed dormancy in establishment problems with forages can be estimated by the extent of scientific interest and in turn the number of publications about seed dormancy in each species (Table 1.2). In some cases this is confounded with the degree of weediness. For example *Poa* species are ubiquitous weeds and cause problems in lawns. Both *Sorghum halapense* (Johnson grass) and *Panicum maximum* (Guinea grass) are important forages but also serious weeds.

1.3 Cereal grasses

The cereal grasses that we know today as principal crops were first domesticated through deliberate selection in the regions of western Asia, the eastern Mediterranean, and meso-America about 10 000 years ago (Arnon, 1972; Grigg, 1974). Domesticated cereals of the present time are therefore very different in appearance from their ancestral wild forms and seed dormancy has been to a considerable degree eliminated by plant breeding. Neverthless the eight genera (*Triticum*, *Oryza*, *Zea*, *Sorghum*, *Panicum*, *Hordeum*, *Avena* and *Saccharum*) that contribute the bulk of the carbohydrate for the global human diet each have one or more species that exhibit seed dormancy (Table 1.1). The magnitude of the research on seed dormancy in several of the genera is very great. This reflects both the problem created by seed dormancy in uniformly establishing crops and also the acute weediness of some species closely related to, and found in, major global crops. Only one of the eight above genera (*Zea*) has no important weed species.

Wheat (Triticum *spp.*)

Within the genus *Triticum*, which in modern taxonomic classification includes *Aegilops*, there is evidence of seed dormancy in almost every species and among varieties within species (Harrington & Knowles, 1940; Wverson & Hart, 1961). In most modern cultivars of common bread wheat (*Triticum aestivum*) dormancy of the mature seed only lasts a few days and is lost before the seed is resown (Larson, Harvey & Larson, 1936; Chang, 1943; Ching & Foote, 1961). Nevertheless there are varieties in which dormancy can last for months (Belderok, 1961). Dormancy is a heritable trait (Harrington & Knowles, 1940; Gfeller & Svejda, 1960; Ruszkowski &

Piech, 1969; Piech, 1970; Dyck, Knoll & Czarnecki, 1986) and for some time it was considered that the trait was associated with red seed coat colour (Gfeller & Svejda, 1960). More recent work (Gordon, 1979a) indicates this may be true for some narrow gene pools but in wider pools the association is weaker (Depauw & McCaig, 1983) and dormancy is seen to be associated with a range of heritable traits. The expression of seed dormancy is greatly modified by the environmental conditions during the ripening of seed on the parent plant (Belderok, 1961; George, 1972; Hagemann and Ciha, 1987). After primary dormancy has been lost a secondary form can be re-introduced (Fischnich, Thielebein & Grahl, 1961; Grahl, 1965). There is a relationship between the intensity and duration of expression of primary dormancy in wheat (Grahl, 1984).

An aspect of the lack of dormancy in modern wheats that has received a great deal of attention is the problem of 'pre-harvest sprouting' (Derera, Bhatt & McMaster, 1977; King, 1983). Hydrolysis of starch and protein reserves associated with the premature germination of wheat seeds, while they are still attached to the parent plant, can lead to serious reductions in the milling, baking and nutritional properties of the grain. The premature germination occurs when the humidity is high and temperature is optimal for the germination of cultivars that lack embryo dormancy during seed development (Harrington, 1932; Feekes, 1938; Chang, 1943; Belderok, 1961; Gordon, 1979b). Interest in controlling the pre-harvest sprouting has centred on re-introducing dormancy to the limited extent that it does not later interfere with the normal crop establishment. A series of international conferences devoted entirely to the subject of pre-harvest sprouting has been instituted on a regular basis (King, 1983).

Rice (Oryza *spp.*)

There are two important species of rice used as major crops on a world basis (Purseglove, 1972). *Oryza sativa*, with its two distinct forms called **indica** (long-grained) and **japonica** (short-grained), is grown throughout the tropical and sub-tropical regions of the world. *Oryza glaberrima*, native to Africa, is essentially confined to that continent. Within the species *Oryza sativa* the method of cropping is mainly by irrigated 'paddy' culture where the seed, seedling and vegetative plant are grown in water. An alternative less widespread method called 'dryland' or 'upland' culture is similar to the other major cereals where seed is planted into moist ground and the plant matures under rainfed conditions. The environment for seed germination can thus be very different and distinct cultivars have been developed to suit each condition.

Seeds of most **indica** and other photoperiodic rices have dormancy that can last up to three months (Purseglove, 1972) and this can seriously interfere with the establishment of two or three successive crops in one year, a pattern common in Asia. Seeds of **japonica** and other rices insensitive to photoperiod are less likely to have a period of dormancy. Several reviews of the occurrence of seed dormancy in rice indicate that it causes serious problems in the practice of cropping (Mouton, 1960; Pili, 1968; International Rice Research Institute, 1968).

Seed dormancy occurs among species of *Oryza* (Morishima & Oka, 1959; Misra & Misro, 1970) and among varieties within species (Oka & Tsai, 1955; Takahashi, 1955; Roberts, 1965; Chandrasekariah, Govindappa & Kulkarni, 1976; Jalote & Vaish, 1976; Roy & Gupta, 1976). Variation in the degree of dormancy can occur within a single panicle (Sugawara, 1959). Environmental factors such as temperature (Chaudhary & Ghildyal, 1969; Roberts, 1967), oxygen (Roberts, 1967; Takahashi & Miyoshi, 1985) and moisture (Ghosh, 1962; Roberts, 1967; Sikder, 1967; Sikder, 1983) can modify the expression of dormancy which is a heritable trait (Takahashi & Oka, 1959; Takahashi, 1962; Chang & Yen, 1969; Ikehashi, 1973; Agrawal, 1981; Deore & Solomon, 1982; Tomar, 1984; Rao, 1985).

Seed dormancy has been reported among varieties grown in all the major rice producing areas of the world, e.g. India (Chandraratna, Fernando & Wattegadra, 1952; Agrawal & Nanda, 1969), Japan (Morishima & Oka, 1959; Ikehashi, 1975; Hayashi & Matsuo, 1983; Takahashi & Miyoshi, 1985; Takagi *et al.*, 1986), Viet Nam (Mai-Tran-Ngoc-Tieng & Nguyen-Thi-Ngoc-Lang, 1971), China (Tang & Chiang, 1955; Wu, 1978; Kuo & Chu, 1979, 1983, 1985), Philippines (Ulali, Barker & Dumlao, 1960; International Rice Research Institute, 1968; Veeraraja Urs, 1987), Malaysia (Dore, 1955), Africa (Roberts, 1963a; Misra & Misro, 1970), North America (Mikkelsen & Sinah, 1961; Delouche & Nguyen, 1964; Duke, 1978) and Russia (Kyurdzhiyeva, 1956; Petrasovits, 1958). Dormancy is also found in wild forms of *Oryza* (Takahashi & Oka, 1958, 1959).

As with *Triticum* spp. there is a distinction between the primary form of dormancy found in mature seed (Mouton, 1960; Pili, 1968; Bose, Ghosh & Sircar, 1977) and a secondary form that can be induced in a genetically non-dormant variety (Sikder, 1974).

Pre-harvest sprouting, which is due to absence of dormancy in the later stages of grain maturity, is a serious problem in particular seasons (Chandraratna *et al.*, 1952; Ikeda, 1963; Chen, Chang & Chiu, 1980; Maruyama, 1980; Juliano & Chang, 1987).

Most rice species have tightly adhering hulls (lemma and palea) that remain attached after the seed is shed from the parent plant. They can only be removed with considerable difficulty by a milling process. The hulls have been implicated as causal factors in seed dormancy (Roberts, 1961).

*Maize (corn) (*Zea *spp.)*

Judging by the very small number of research reports in the world literature addressing the subject of seed dormancy in *Zea mays*, it can be assumed that dormancy is not a serious problem in modern maize cultivars. The tetraploid perennial grass teosinte (*Zea perennis*), believed to be the main progenitor of *Zea mays*, does exhibit seed dormancy (Mondrus-Engle, 1981). Dormancy can be expressed in *Zea mays* during the seed development stages and growth inhibitors have been implicated as causal agents (Koshimizu, 1936; Hemberg, 1958; Ermilov, 1961; Ortega-Delagado *et al.*, 1983; Fursov, Kurbakov & Darkanbaev, 1984). The inheritance of this form of dormancy has been studied (Mangelsdorf, 1930; Middendorf, 1938; Pinnell, 1949) and the absence of this dormancy is implicated in pre-harvest sprouting (Mangelsdorf, 1930; Ortega-Delagado *et al.*, 1983; Neill, Horgan & Rees, 1987).

A form of secondary dormancy can occur in seeds submerged under water (Prasad, Gupta & Bajracharya, 1983). Both light (Thanos & Mitrakos, 1979) and low temperature (Pinnell, 1949) have modifying influences on germinability that are distinct from the expression of dormancy.

*Sorghum (*Sorghum *spp.)*

The genus *sorghum* includes a large number of species. There is considerable variation in morphology, methods of reproduction and in agronomic value among the indigenous African types (Doggett, 1970). *Sorghum bicolor* (L.) Moench, a major crop in Africa, has become a very significant crop in India, USA and China in recent years through the incorporation of a dwarfing character and selection for adaptation to a broader climatic range than its centre of origin in N. East Africa (Doggett, 1970). Sorghums grown for cereal grains in Africa and India are annuals adapted to the extremes of drought stress that occur in a semi-arid climate (Simpson, 1981). The sorghums grown in North America are generally grown under irrigation.

Seed dormancy occurs in *Sorghum bicolor* the principal species of sorghum used as a cereal crop (Harrington, 1923; Casey, 1947; Brown *et al.*, 1948; Goodsell, 1957; Wright & Kinch, 1959; Gritton & Atkins,

1963a,b; Clark, 1967; Wilson, 1973). Gritton & Atkins (1963a) studied 33 varieties and all combinations of crosses between eight restorers and six cytoplasmic male-sterile lines and found that dormancy was present in all types for at least one month. There were considerable differences among the varieties. Dormancy also occurs in other closely related sorghums such as *Sorghum halapense*, a perennial which is a serious weed (Harrington, 1916; Tester & McCormick, 1954; Tao, 1982), as well as in wild sorghum species (Mott, 1978).

Seed coats have been implicated as causal agents in the dormancy of sorghums (Harrington & Crocker, 1923; Clark, Collier & Langston, 1968) because of the high content of inhibitory phenolic compounds and impermeability to water. On the other hand dormancy can be under metabolic control (Kamalavalli *et al.*, 1978).

As with other cereal grasses the absence, or comparative absence, of dormancy during the stages of seed maturation can allow pre-harvest sprouting (Clark, Collier & Langston, 1967; Ujiihra, 1982). Alternating temperatures can overcome some forms of primary dormancy (Harrington, 1923; Stanway, 1959). The reduction of germination associated with low temperature is distinct from inherent dormancy of the seed (Robbins & Porter, 1946; Pinthus & Rosenblum, 1961; Gritton & Atkins, 1963b; Huang & Hsiao, 1987) and light probably influences the expression of dormancy (Taylorson, 1975; Huang & Hsiao, 1987).

Until recently, there was not the amount of scientific interest and active research on sorghum species comparable to other major cereal crops. For this reason there is not an extensive literature on seed dormancy in this genus with perhaps the exception of *Sorghum halapense* because it is a serious weed. Nevertheless it is well established that seed dormancy exists and is of agronomic significance in *Sorghum bicolor*.

*Millets (*Panicum *spp.,* Eleusine *spp.,* Setaria *spp.,* Pennisetum *spp.,* Echinochloa *spp.)*

The name millet is applied to a number of cultivated annual grasses included in genera that are not closely related. They are annuals with a short growing season and have small seeds, generally of low quality for human nutrition (Wheeler, 1950). Proso or broomcorn millet (*Panicum miliaceum* is the 'common millet' of Europe grown since prehistoric times. Foxtail millet (*Setaria italica* has been cultivated in China since 2700 BC (Wheeler, 1950). Japanese, or Barnyard millet (*Echinochloa crusgalli frumentacea*) originated in India but is grown extensively in Japan. Pearl millet

(*Pennisetum glaucum* or *typhoides*) has a large number of cultivated races that provide the staple food in the drier parts of Africa and India (Purseglove, 1972).

Many of the millets and near-relatives are useful forage grasses and in some cases serious weeds in various crops. Because the millets include a number of genera, several of which contain a great number of species, it is difficult to generalize about seed dormancy. The approach will be to consider each genus and give some indication of the occurrence and extent of seed dormancy in the crop species and some of the important forage and weed relatives.

Panicum spp. The most important species used for human nutrition is *Panicum miliaceum* (Proso, brown-corn, hog millet) and seed dormancy is well documented (Andersen, 1958, 1961; Moore & Hoffman, 1964; O'Toole & Cavers, 1981; Striegel & Boldt, 1981; Colosi & Cavers, 1983; Cavers & Benoit, 1987). It is probable that seed dormancy is very common in this species because there are both general reports of seed dormancy within the genus (Akamine, 1944; Cullinan, 1947; Krishnaswamy, 1952; Andersen, 1962; Mukherjee & Chatterji, 1970; Johnston, 1972; Shimizu, 1979) and many publications indicating dormancy within certain species. Seed dormancy in *P. maximum* (Guinea grass), an important tropical forage grass, has been studied in considerable depth (Binrad, 1958; Opsomer & Bronckers, 1958; Hanssen & Nicholls, 1965; Febles & Padilla, 1970, 1971; Johnston & Tattersfield, 1971; Harty & Butler, 1975; Pernes *et al.*, 1975; Smith, 1979; Taylorson & Hendricks, 1979a; Gonzalez & Torriente, 1983; Harty *et al.*, 1983). Dormancy has been reported in the following species: *P. dichotomiflorum* (Fall panic) (Taylorson, 1979a; Brecke & Duke, 1980; Taylorson, 1980; Okada, Ochi & Ohta, 1982; Baskin & Baskin, 1983; Okada & Ochi, 1986), *P. coloratum* (Klein grass) (Edwards, 1933; Kijima & Takei, 1971; Butler, Helms & Ogle, 1983; Tischler & Young, 1983), *P. clandestinum* (Deetongue) (Chirco, Goodman & Clark, 1979), *P. anceps* (Mathews, 1947; Anonymous, 1954), *P. antidotale* (Koller & Negbi, 1957), *P. turgidum* (Koller & Roth, 1963; Roberts & Ellis, 1982), *P. virgatum* (Switchgrass) (Sautter, 1962), *P. obtusum* (Vine mesquite) (Toole, 1940a) and *P. phillopogon* (Na & Lee, 1984).

Eleusine spp. Eleusine coracana (Finger millet, Koracan millet, raji) is an important cereal species. Seed dormancy is common (Agrawal & Kaur, 1975; Arora & Bannerjee, 1978; Shimizu, 1979) and there are differences in expression of dormancy among varieties (Shimizu & Mochizuki, 1978).

Both temperature and light affect dormancy (Shimizu & Tajima, 1979).

Goose grass (*E. indica*), a useful forage grass, has seed dormancy (Toole & Toole, 1940; Fulwider & Engel, 1959; Chin & Raja Harum, 1979; Hawton & Drennan, 1980). *E. compressa*, also a forage grass, has dormant seed (Gupta & Saxena, 1980). Other *Eleusine* species have been reported to have seed dormancy (Nakamura, Watanabe & Ichihara, 1960; Nakamura, 1962; Popay, 1973; Hilu & De Wet, 1980).

Setaria spp. Foxtail millet (*Setaria italica*) has been grown for human food in China since the Bronze Age and in this century has become an important forage crop in the USA (Wheeler, 1950). There are numerous reports of seed dormancy (Ito, Kitahara & Kawamura, 1931; Keys, 1949; Kim, 1954; Povilaitis, 1956; Nakamura, 1962; Moore & Fletchall, 1963; Jennings *et al.*, 1968; Kollman & Staniforth, 1972; Hendricks & Taylorson, 1974; Johnston, 1981; Mohamad, Clark & Ong, 1985) that causes problems in crop establishment and makes the species a significant weed. Yellow foxtail (*S. lutescens*), an important weed, has also been well documented for its seed dormancy (Topornina, 1958; Peters & Yokum, 1961; Nieto-Hatem, 1963; Rost, 1972, 1975; Lehle, Staniforth & Stewart, 1978; Norris & Schoner, 1980; Lehle, Stewart & Staniforth, 1981; Lehle, Staniforth & Stewart, 1983) considered to be in part related to the presence of inhibitory compounds (Yokum, Jutras & Peters, 1961). Green foxtail (*S. viridis*) is a particularly bad weed because of the many tiny dormant seeds shed from each plant (Heise, 1941; Born, 1971; Banting, Molberg & Gebhardt, 1973; Kohout & Loudova, 1981). Giant foxtail (*S. faberii*) another weedy member of the family has both primary (King, 1952; Moore & Fletchall, 1963; Stanway, 1971) and secondary (Taylorson, 1982) dormancy. Dormancy has been reported in two minor species of the genus; *S. chevalieri* (Erasmus & Van Staden, 1983) and *S. macrostachya* (Toole, 1940a).

Pennisetum spp. The important cereal species is *Pennisetum typhoides* synonymous also with *P. glaucum* or *P. americanum*. The common names are bullrush, pearl, spiked or cat-tail millet and in India, bajra. As with sorghum, the classification into a single species is probably incorrect (Purseglove, 1972). There is probably a number of distinct races with different geographical origins. Reports indicate seed dormancy in the races classified as *P. typhoides* (Bulgakova, 1951; Adams, 1956; Sandhu & Husain, 1961; Babu & Joshi, 1970; Singh, Datta & Singh, 1971; Raza, 1977; Garcia-Huidroba, Monteith & Squire, 1982). Dormancy has been reported in several other pennisetums: *P. pedicellatum* (Varshney & Baijal, 1978;

Mott, 1980; Maiti, Purkait & Chatterjee, 1981); *P. ciliare* (Buffel grass) (Andersen, 1953) and *P. purpureum* (Elephant grass) (Oakes, 1959).

Echinochloa spp. Echinochloa frumentacea (Roxb.)Link commonly called Japanese barnyard millet is the species used as a cereal. It gets the common name from the fact that it was used as a cereal in Japan although its origin was in India (Purseglove, 1972). It is also an important forage crop in the USA where it is often described as *E. crus-galli frumentacea* (Wheeler, 1950). The species *E. crus-galli* (Barnyard grass) is distinct although probably confused in some publications with *E. crus-galli frumentacea*. The former has become an important weed in addition to its use as a forage crop.

Echinochloa species have prolonged seed dormancy up to many months. There are many reports of seed dormancy in *E. crus-galli* signifying both its importance as a weed and the extent of problems associated with its use as a crop (Ehara & Abe, 1952; Arai & Miyahara, 1962; Nakamura, 1962; Arai & Miyahara, 1963; Harper, 1970; Hayashi & Morifuji, 1972; Hendricks & Taylorson, 1974; Yamasue, Sudo & Ueki, 1977; Rizk, Fayed & El-Deepah, 1978; Takahashi, 1978; Popova, 1979; Jordan, 1981; Vanderzee & Kennedy, 1981; Watanabe, 1981; Barrett & Wilson, 1983; Furya & Kataoka, 1983; Sung *et al.*, 1983, 1987). The control of dormancy is in part associated with the metabolic system (Shimizu & Ueki, 1972; Shimizu, Takahashi & Tajima, 1974; Vanderzee & Kennedy, 1981; Rumpho & Kennedy, 1983; Kennedy, Fox & Siedow, 1987).

Channel millet (*E. turnerana*), a weed, has seed dormancy (Conover & Geiger, 1984a,b). *E. colona* (Jungle rice), a forage that grows in wet places throughout the tropics, also has dormancy (Ramakrishnan, 1960; Ramakrishnan & Khoshla, 1971). Several reports indicate seed dormancy is found among a number of species of *Echinochloa* (Hughes, 1979; Kohout & Loudova, 1981).

*Barley (*Hordeum *spp.)*

Barley (*Hordeum vulgare*) ranks fourth in total production (metric tonnes of grain) of the major world cereal crops (Stoskopf, 1985). Nevertheless it ranks much lower in terms of direct importance in human nutrition because a high proportion of global production is used for making beer and feeding livestock. Although the origin of barley is in a temperate-montane climate, the species is now grown in a number of different climates. As with rice, the mature caryopsis is tightly enclosed in glumes which remain firmly attached when the grain is shed from the parent plant. The glumes have been implicated in seed dormancy of barley (Mitchell, Caldwell & Hampson, 1958; Lenoir, Corbineau & Côme, 1986).

Barley is the most studied of the major cereal grains for reasons of understanding both germination and the nature of seed dormancy. Aside from its use as human food, barley has been used for making beer for several thousand years. The production of the malt for making beer from barley requires uniform germination to bring about hydrolysis of endosperm reserves to low molecular weight sugars used for fermentation to alcohol. Uniform germination on the malting floor can be prevented by the natural dormancy in the mature grain (Larson, Harvey & Larson, 1936; Moormann, 1942; Bishop, 1944; Munn, 1946; Bishop, 1948; Brown *et al.*, 1948; Drake, 1948; Heit, 1948; Harrington, 1949; Pollock, Kirsop & Essser, 1955; Urion & Chapon, 1956; Wellington, 1956; Blanchard, 1957; Brad, Laszlo & Valuta, 1959; Pollhamer, 1960; Belderok, 1962, 1968; Strand, 1965; Corbineau & Côme, 1980; Côme, Lenoir & Corbineau, 1984; Ellis, Hong & Roberts, 1987). In addition, germination can be prevented by a condition called 'water dormancy' which is the imposition of secondary dormancy due to an excess of water during germination (Jansson, 1959; Sims, 1959; Mastovsky & Karel, 1961; Pollock & Pool, 1962; Bloch & Morgan, 1967; Gaber & Roberts, 1969; Dan, 1971; Fischbeck & Reimer, 1971; Narziss *et al.*, 1980; Rauber & Isselstein, 1985). Several of the early classical papers on cereal seed germination were focused on solving the problems in the brewing industry caused by barley seed dormancy (Brown & Morris, 1890; Brown, 1909, 1912; Brown & Tinke, 1916).

Plant breeding techniques have been used in attempts to overcome the natural primary dormancy and a tendency toward secondary dormancy (Harrington & Knowles, 1940; Fischnich, Thielebein & Grahl, 1961; Buraas & Skinnes, 1984; Skinnes, 1984). The naturally occurring plant growth regulator gibberellic acid (GA_3) has been used as a practical measure to overcome dormancy on the malting floor. There is therefore a considerable literature on the mode of action of this and other gibberellins in overcoming dormancy and stimulating the hydrolysis of endosperm reserves (Renard, 1960; Briggs, 1963; Kahre, Kolk & Fritz, 1965; Griffiths, MacWilliam & Reynolds, 1967; MacLeod, 1967b; Khan & Waters, 1969; Palmer, 1970; Rejowski, 1971; Bekendam, 1975; Gaspar, Fazeka & Petho, 1973; Don, 1979).

The embryo is considered to be the main site of dormancy (Bulard, 1960; Wellington, 1964; Grahl, 1969; Obhlidalova & Hradilik, 1973, 1975; Dunwell, 1981). This primary dormancy is initiated during the development of the seed on the parent plant (Fischnich, Grahl & Thielebein, 1962;

Rauber, 1984a,b, 1985) and the degree of expression of the dormancy is influenced by such factors as heat and drought (Essery & Pollock, 1957; Hewett, 1958; Yabuki & Miyagawa, 1958; Grahl, 1970; Nicholls, 1986). There is considerable cultivar variation in the expression of dormancy (Bishop, 1958; Cseresnyes & Buda, 1976; Corbineau & Côme, 1982; Curran & McCarthy, 1986) and dormancy occurs in European (Foral & Bosak, 1971), British (Pollock, 1956), Japanese (Fukuyama, Takahashi & Hayashi, 1973) and Indian (Venugopal & Krishnamurthy, 1972) varieties.

The problem of 'pre-harvest sprouting' due to lack of embryo dormancy during seed maturation can be a problem in some barley varieties in particular seasons (Pammel & King, 1925; Norstog & Klein, 1972; Marais & Kruis, 1983; Buraas & Skinnes, 1985; Black, Butler & Hughes, 1987; Ringlund, 1987). Vivipary has been reported in a cultivar considered to be dormant (Pope & Brown, 1943).

Seed dormancy is common among close relatives of *Hordeum vulgare*. Examples are: *H. leporinum* (Smith, 1968; Cocks, Boyce & Kloot, 1976); *H. pusillum* (Fischer, Stritzke & Ahring, 1982); *H. spontaneum* (Ogawara & Hayashi, 1964; Giles & Lefkowitch, 1984; Key, 1987); *H. agriocrithon* (Key, 1987); *H. jubatum* (Popay, 1975, 1981) and five other relatives (Popay, 1981).

Oat (Avena *spp.*)

The common oat (*Avena sativa*) has been declining in importance as a major source of human diet, mainly because of the improvements in yield and adaptability to a widened range of climates achieved by plant breeding in corn, wheat and rice (Stoskopf, 1985). Nevertheless it still ranks sixth in world cereal grain production ahead of millets and rye. The domestic oat is well adapted to a moist, cool temperate climate.

The level of seed dormancy occurring in *Avena sativa* does not seem to cause the kinds of problems in agriculture and industry found with rice, wheat and barley. Nevertheless seed dormancy is common and has practical significance in establishment of a uniform crop (Hyde, 1935; Johnson, 1935a; Coffman & Stanton, 1940; Bass, 1948; Forward, 1949; Drennan & Berrie, 1961; Corbineau, Lecat & Côme, 1986). In the USA dormancy occurs mainly in the late-maturing varieties (Larson, Harvey & Larson, 1936). The main investigations into the nature of dormancy have been focused on the role of hulls as limiting factors to germination (Peers, 1958; Karl & Rudiger, 1982; Lohaus *et al.*, 1982; Ruediger, 1982; Côme, Lecat & Corbineau, 1987; Lascorz & Drapron, 1987a,b; Lecat, 1987). The genetic control of length of the dormant period has been studied (Nishiyama & Inamori, 1966).

There is an indication that oats can be induced into a state of secondary dormancy by the conditions that cause 'water dormancy' in barley (Walker, 1934). Both primary and secondary dormancy can be overcome by gibberellin (GA) (Kahre, Kolk & Wiberg, 1962; Kahre, Kolk & Fritz, 1965) and inhibitors have been implicated as causal agents in dormancy (Berrie *et al.*, 1979).

Attempts have been made to breed a degree of seed dormancy into *Avena sativa* so that the seed can be autumn-sown, lose dormancy over winter, and germinate as a spring crop before it is normally possible to cultivate the land or sow a crop (Burrows, 1964, 1970; Andrews & Burrows, 1972, 1974). To achieve this, dormancy was incorporated into *Avena sativa* from its close relative *Avena fatua* which has a high degree of seed dormancy.

Rye (Secale cereale)

Rye is the most cold tolerant of the cereal crops and it is important in the extremes of the temperate climate of both the northern and southern hemispheres (Stoskopf, 1985). It was originally a weed of wheat and barley growing on poor sandy soils. Rye is used principally for making ryebread.

Seed dormancy has been reported in rye (Munerati, 1925; Heit, 1948; Kahre, Kolk & Wiberg, 1962; Bekendam, 1975). An assessment of the length of dormancy in rye cultivars has been made (Larson, Harvey & Larson, 1936). There appears to be some genetic linkage between seed dormancy and 'pre-harvest sprouting' and treatment with GA can overcome dormancy (Kahre, Kolk & Fritz, 1965).

*Wild rice (*Zizania aquatica*)*

Wild rice was an important source of food for the indigenous people of North America for several thousand years. In recent times it has assumed increasing importance as a gourmet food. Unlike the other major cereals it is an uncultivated aquatic plant that passes its complete life cycle in shallow slow moving rivers or shallow lakes (Weber & Simpson, 1967). Drying the seed leads to rapid loss of viability (Simpson, 1966a) and the mature seed is normally shed quickly from the panicle directly into water where viability is retained for several years.

The seed may have dormancy that persists for many months (Simpson, 1966a). Mechanical (Oelke & Albrecht, 1978) and chemical (Oelke & Albrecht, 1980) scarification or disruption of the seed coat by ultra-sonic vibration (Halstead & Vicario,1969) can break the dormancy. Unlike other cereals, seed dormancy in *Zizania* cannot be broken by growth regulators such as GA indicating some basic difference in the nature of dormancy in this species.

Sugarcane (Saccharum *spp.*)

Sugarcane ranks first among all crops in total production of dry matter on a global scale and it is the primary source of sugar for the modern human diet (Jones, 1985). Six species of *Saccharum*, of the tribe Andropogoneae, are the main sources of sugarcane (Purseglove, 1972). Almost all commercial cultivars grown today are interspecific hybrids in which flowering has been largely eliminated through breeding, because it terminates sugar production. For this reason the crop is vegetatively propagated by cuttings. Nevertheless seed formation is essential for breeding and wild forms can flower profusely. The seed is minute (1 mm long) and covered with whorls of silky hair adapted for wind dispersal. Sugarcane seeds are thus described within the industry as 'fuzz'.

Ellis, Hong & Roberts (1985b) indicate that with the exception of the wild species *S. aegyptiacum*, which has considerable seed dormancy, there is little dormancy in the modern crop cultivars. Nevertheless there is evidence that after-ripening for six months improves germination in some seeds and seeds generally germinate better in light (Purseglove, 1972) suggesting that the conditions of maturation on the parent plant can influence the degree of dormancy. Lack of dormancy can cause vivipary (Ragavan, 1960).

1.4 Weed grasses

According to Shetty (1976) seven of the worst weeds on a global scale are grasses. Four of these grasses are perennials and the most ubiquitous is *Cynodon dactylon* L. Pers. commonly called Bermuda grass. Bermuda grass is found as a weed in 85 countries (Holm *et al.*, 1979). Two species of the annual grasses of the genus Echinochloa (*Echinochloa colona, E. crus-galli*) each occur in about 70 countries and the wild oat (*Avena fatua* L.) in 58. Some indication of the extent of occurrence of dormancy in grass seeds and its association with success of the species as a weed can be derived from a classification of these two characters among the genera of the 25 grass tribes within the family Gramineae (Table 1.1). Dormancy, or lack of dormancy, within each of the 177 genera of grasses listed by Gould (1968) was characterized by the presence of one or more scientific publications in the world literature focused specifically on seed dormancy. The references were located by a computer search which therefore tended to cover the literature of the last twenty years fairly completely but may have seriously underestimated the occurrence of references in the earlier part of this century. The estimation of the extent of the weed problem can be gauged by (a) the number of species within the genus described as weeds and (b) the extent to which the species are present over a geographical zone of the earth, estimated as the average number of countries in which a single species of the genus occurs. The data are derived from the very detailed listing of the weeds of the world made by Holm *et al.*, in 1979.

Seventy-eight of the 177 genera show seed dormancy among one or more of the species (Table 1.1). Seventy-two of these genera have one or more species that are major weeds. A total of 111 genera can be classed as containing weed species; a statistic that probably reflects the fact that grasses have assumed increasing importance in modern cropping systems because they are not so easily controlled by herbicides as dicotyledenous species. Thirty-seven of the genera appear not to have seed dormancy although they are classed as weeds. In reality a number of these 37 genera and the 66 other genera that are not classed as weeds may well exhibit seed dormancy but for a variety of reasons the condition has never been documented. In addition the computer underestimated the incidence of dormancy because the search could only characterize the condition of dormancy through titles and selected key words. Despite these weaknesses the analysis indicates that approximately one half, at least, of the grass genera have the trait of seed dormancy. A very high proportion (92%) of these genera with seed dormancy exhibit traits of weediness within the world agricultural system.

Among the 72 genera that have successful weed species showing seed dormancy, a high proportion (36%) have more than six species that are weeds: several have as many as 35 and one genus has 82 (Fig. 1.1). By contrast, among the 37 genera that have successful weed species, but apparently no seed dormancy, only a small proportion (5%) have greater than six species. This suggests linkage between the traits of seed dormancy and success as a weed in the agricultural system.

The relationship between seed dormancy in grasses and success as a weed in agriculture becomes more striking if a list of globally serious weeds is examined. Holm *et al.*, (1979) estimate that there is a total of about 250 weeds important in world agriculture. If the top 40 weeds classified as 'serious' (Holm *et al.* used the categories 'serious','principal', 'common' and 'frequent but rank unassigned') are selected by ranking them in the order of number of countries in which they are considered 'serious' weeds, then several important conclusions can be drawn (Table 1.3).

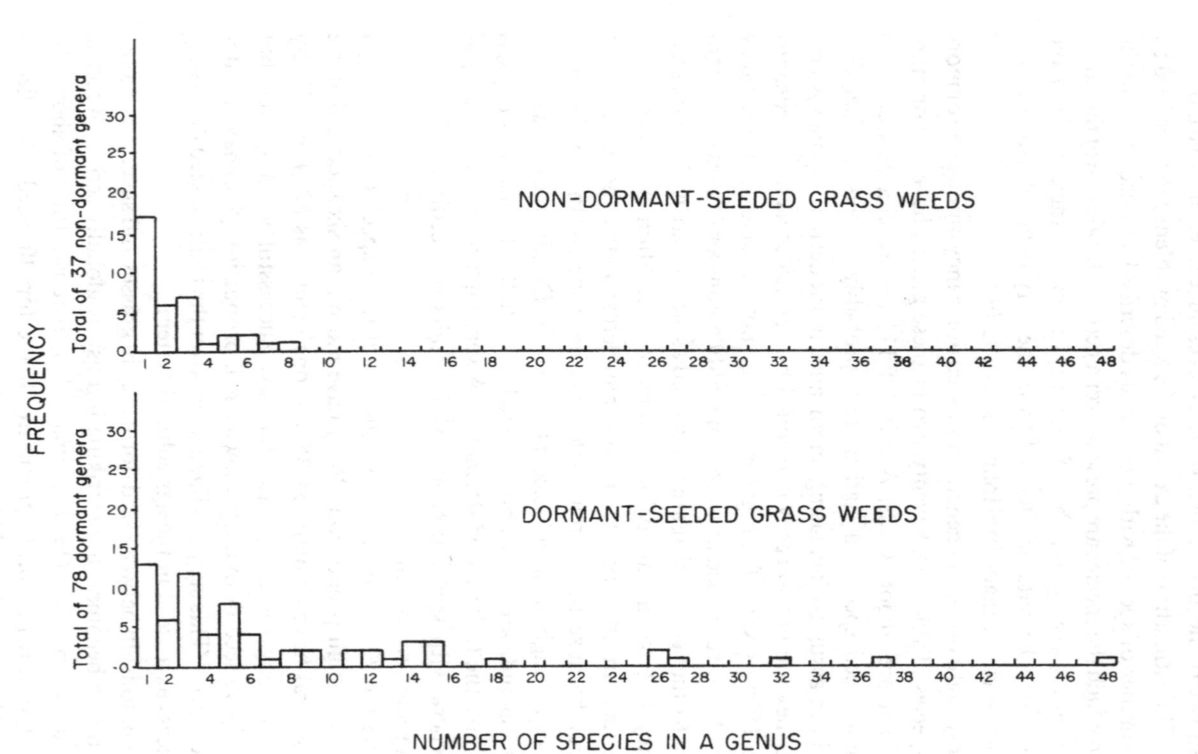

Fig. 1.1. Frequency distributions of the numbers of successful weed species within grass genera showing dormancy, or absence of dormancy, in mature seeds.

Table 1.3. *The forty most serious weeds of the world, found in six or more countries, ranked in order of total number of countries in which each species is considered to be a 'serious weed' (Holm* et al., *1979). The reference includes proof that dormancy is present in the seeds at maturity. Members of the grass family are indicated by (Gr).*

Species	No. countries	Reference
Cyperus rotundus	52	Justice & Whitehead, 1946
Echinochloa crus-galli (Gr.)	32	Harper, 1970
Cynodon dactylon (Gr.)	28	Wu, 1980
Digitaria sanguinalis (Gr)	23	Delouche, 1956
Echinochloa colona (Gr.)	22	Ramakrishnan & Khoshla, 1971
Sorghum halapense (Gr.)	22	Taylorson & McWhorter, 1969
Eleusine indica (Gr.)	20	Toole & Toole, 1940
Cyperus difformis	19	Ahring & Harlan, 1961
Imperata cylindrica (Gr.)	18	Chaudri *et al.*, 1975
Portulaca oleracea	17	Cantoria & Gacutan, 1972
Amaranthus retroflexus	16	Forst & Cavers, 1975
Chenopodium album	15	Chu, Sweet & Ozbun, 1978
Convolvulus arvensis	14	Jordan & Jordan, 1982
Eichhornia crassipes	14	Obeid & Tagelseed, 1976
Poa annua (Gr.)	13	Wu *et al.*, 1987
Datura stramonium	13	Brown & Bridglall, 1987
Paspalum conjugatum (Gr.)	12	Ray & Stewart, 1937
Paspalum distichum (Gr.)	10	Huang & Hsiao, 1987
Panicum maximum (Gr.)	10	Harty *et al.*, 1983
Ischaemium rugosum (Gr.)	10	Holm *et al.*, 1979
Cyperus esculentus	10	Justice & Whitehead, 1946
Fimbristylis miliacea	10	Lawson, 1986
Lantana camara	10	Ellis, Hong & Roberts, 1985b
Polygonum convolvulus	9	Hsiao, 1979
Polygonum lapathifolium	9	Watanabe & Hirokawa, 1979
Raphanus raphanistrum	9	Cheam, 1986
Avena fatua (Gr.)	9	Atwood, 1914
Setaria glauca (Gr.)	8	Peters & Yokum 1961
Panicum repens (Gr.)	8	Ellis, Hong & Roberts, 1985b
Cenchrus echinatus (Gr.)	8	Bhupathi *et al.*, 1987
Ageratum conyzoides	8	Richards, 1966
Amaranthus spinosus	8	Ellis, Hong & Roberts, 1985b
Spergula arvensis	8	Jones & Hall, 1981
Artemisia vulgaris	8	Crescini & Spreafico, 1953
Solanum nigrum	8	Roberts & Lockett, 1978
Mimosa invisa	8	Ellis, Hong & Roberts, 1985b
Sonchus arvensis	8	Fykse, 1974
Polygonum aviculare	8	Courtney, 1968
Rottboellia exaltata	8	Thomas & Allison, 1975
Sinapis arvensis	8	Edwards, 1976

Firstly, without exception, each one of the 41 species has seed dormancy. Secondly, seven of the top ten most widespread species are grasses and two of the others are grass-like sedges (*Cyperus rotundus*, *Cyperus difformis*). Grasses therefore comprise about two fifths (16) of the top 40 weeds indicating that they are disproportionately represented as a family in relation to the rest of the plant kingdom.

If the weeds are ranked in order of the total number of countries in which each species is a weed of some significance then grasses still remain highly ranked (Table 1.4). *Cynodon dactylon* (Bermuda grass), a perennial of cosmopolitan distribution often used as a forage, is the most widespread grass weed; it occurs in 90 countries and is considered a very serious weed in 28 countries. The two other perennial grasses, *Sorghum halapense* and *Imperata cylindrica* are widespread in tropical areas.

Among the annual grasses *Poa annua*, of European origin, is the most cosmopolitan as a weed of crops, lawns and wasteland. *Eleusine indica* is mainly found in tropical regions. The two *Echinochloa* species, *E. crus-galli* and *E. colona* are annuals and the former is the major weed of irrigated rice wherever it is grown. *Avena fatua*, which is of Mediterranean origin, is found particularly wherever the annual temperate cereals – wheat, barley, oats and rye – are grown. *Digitaria sanguinalis* is found in both tropical and temperate regions in both crop and non-crop situations.

The trait of seed dormancy is strongly associated with the persistence of weeds so that it is not surprising that six of the top nine grass weeds are annuals that rely on a period of dormancy in the seed bank to effect population renewal over several seasons. Other traits such as competitive ability at the vegetative stage and seed number will have a strong influence on the success as a weed.

1.5 Dormancy and the persistence of individuals and populations

Survival of an individual grass plant, and in turn the species, depends on avoiding vulnerability during three distinct phases of the life cycle. The persistence of a species as a distinct population in a particular habitat depends on the proportions of surviving individuals in each of these phases. Two of the phases, the vegetative plant and the seed, involve the sporophytic generation and the third involves the gametophytic generation.

In grasses the first and most conspicuous phase, vegetative growth, begins with germination of the embryo. The embryo develops roots, leaves and tillers which compete with surrounding organisms for space, water,

Table 1.4. *Relative ranking of grasses (Gr.) among the forty most globally widespread weeds. Ranking is by total number of countries in which each species is recognised as a weed, without classification into order of economic significance (After Holm et al., 1979).*

Species	No. countries
Cyperus rotundus	91
Cynodon dactylon (Gr.)	90
Portulaca oleracea	78
Solanum nigrum	68
Echinochloa colona (Gr.)	67
Echinochloa crus-galli (Gr.)	65
Eleusine indica (Gr.)	64
Poa annua (Gr.)	62
Chenopodium album	59
Avena fatua (Gr.)	56
Bidens pilosa	56
Digitaria sanguinalis (Gr.)	55
Sida rhombifolia	55
Sorghum halapense (Gr.)	51
Eichhornia crassipes	50
Boerhaavia diffusa	50
Setaria glauca (Gr.)	49
Imperata cylindrica (Gr.)	49
Raphanus raphanistrum	49
Dactyloctenium aegyptium (Gr.)	48
Sonchus arvensis	48
Convolvulus arvensis	48
Rumex acetosella	48
Stellaria media	47
Lantana camara	47
Ageratum conyzoides	47
Amaranthus retroflexus	46
Amaranthus viridis	46
Datura stramonium	46
Phyllanthus niruri	46
Panicum maximum (Gr.)	44
Paspalum distichum (Gr.)	44
Amaranthus spinosus	44
Sinapis arvensis	43
Brassica campestris	43

Table 1.4. *(cont.)*.

Species	No. countries
Senecio vulgaris	42
Polygonum lapathifolium	39
Polygonum convolvulus	38
Galinsoga parviflora	38
Setaria verticillata (Gr.)	36

nutrients and light. Many grasses propagate vegetatively by formation of new shoots arising from buds on stolons and rhizomes. Vegetative propagation is important in colonization of space, in competition against other plants, and for regeneration after some parts of the plant die from environmental stress. The first phase ends with development of an inflorescence. This may occur after one year (annual), two years (biennial) or after a number of years (perennial). The proportions of vegetative to flowering tillers in any one season can be quite variable and in genera such as the bamboos flowering may occur as rarely as once in 60 years and then extend for several years, sometimes without production of fertile seeds (McClure, 1966).

The second phase, the gametophytic generation, begins with formation of monoploid (haploid) gametes. The male haploid gametophytes (pollen) are the dispersal units that can spread through the environment by wind (anemophily), or by insects (entomophily), and they are derived from the parent diploid tissue. Only a tiny proportion of pollen grains succeed to unite with female gametophytes. The female gametophyte develops and remains within the parental floral structure. To achieve this a female megaspore mother cell divides meiotically into four haploid megaspores. Three of the spores die and the fourth develops into the ovoid female gametophyte which contains eight haploid nuclei. In turn only three of these nuclei survive. A pair survive as the potential endosperm nuclei and the single nucleus near the micropyle is the egg nucleus.

At pollination a male pollen grain that has managed to adhere to the sticky surface of the female stigma germinates by forming a tube that grows down through the style of the flower to reach the small opening called the micropyle. One of the two surviving sperm cell nuclei fuses with the egg nucleus to form a diploid zygote. The other nucleus unites with the two haploid endosperm nuclei to form a triploid endosperm; in some cases this can be pentaploid or polyploid. The zygote develops to form the embryo which draws upon the surrounding endosperm tissue for nourishment.

The male and female gametophytic generations are exposed to many environmental hazards, particularly temperature fluctuations and changes in relative humidity that influence shedding and germination of pollen. Conjugation of the male and female elements is further complicated in grasses through specialization of species into those that have enforced cross-fertilization or those with a range of intermediate mechanisms all the way through to enforced self-fertilization (Bews, 1929). Polygamy, the bearing of bisexual (hermaphroditic) and unisexual, usually male florets, in the same spikelet is common in grasses particularly in the highly developed types such as *Chlorideae, Paniceae* and *Andropogoneae*. Monoecism, where separate male and female spikelets are borne on the same plant, is rare in grasses (e.g. *Zea mays* and *Zizania aquatica*). Dioecism, where male and female spikelets occur on different plants, is more widespread among grass genera but also relatively uncommon.

The third phase in the cycle is often called the resting phase that follows the development of the zygote into a mature embryo. The new sporophyte embryo, surrounded by its endosperm tissue, remains enclosed within the remnants of the female parent sporophytic tissue (flower) which together become cut off from the parent plant by formation of an abscission layer. The fruit, or caryopsis as it is called in grasses, consists of the pericarp, usually surrounding and attached to the endosperm (unattached in *Sporobolus* and a few genera of *Eragrosteae*) which in turn surrounds the embryo. This fruit, or grain as it is called in cereals, can separate from the parent plant freely, be enclosed in the lemma and palea tightly, or adhere to one or other of them. The dispersal unit called a 'seed' can therefore be simply a caryopsis or a caryopsis attached to parts of the floral structure. In some species the dispersal unit is comprised of one or more fruits enclosed between several glumes and bracts (a spikelet) or is a cluster of spikelets.

The seed is thus a dehydrated dispersal unit that can be carried by wind, water, predators, grazing animals and human activities to potentially new sites for the species. The seed can also be a survival capsule during periods of drought and severe cold when the vegetative sporophytic phase is not viable. Survival of the individual during the sporophytic seed phase depends on vagaries of the environment, internal mechanisms such as embryo dormancy and impermeable seed coats. A dehydrated caryopsis can survive in a dry environment for many years (Fig. 1.2). Grasses do not survive as long as dicotyledenous species with hard seed coats impermeable

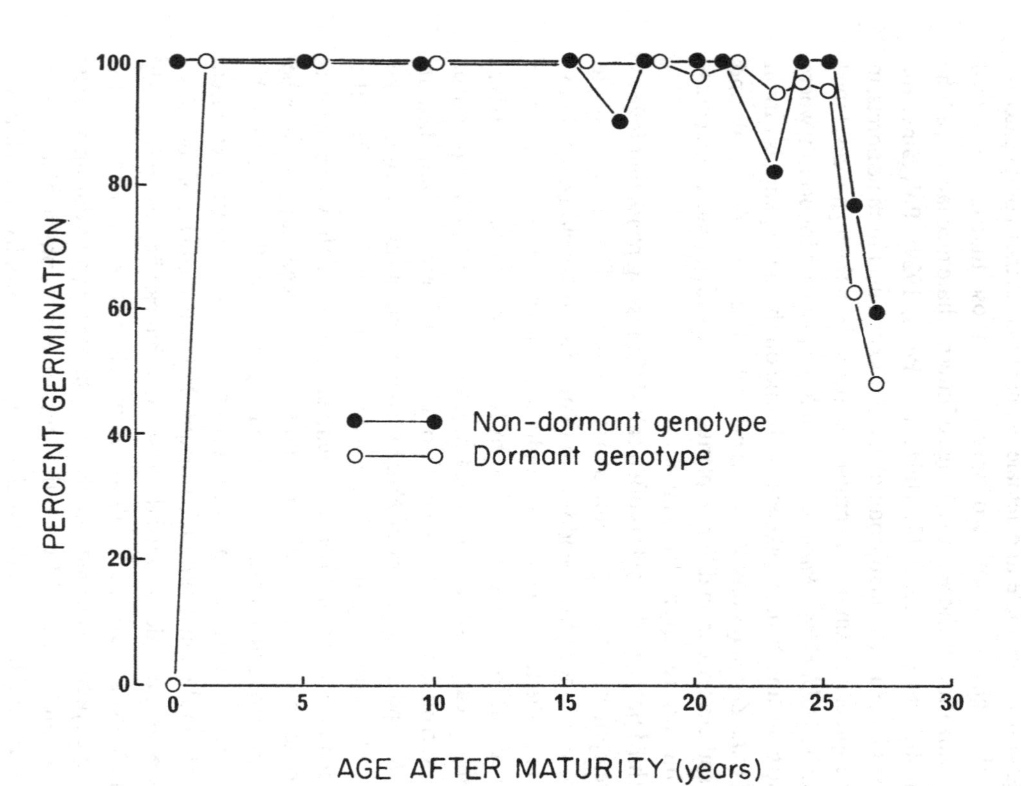

Fig. 1.2. Seed viability in genetically distinct lines of *Avena fatua* characterized by dormant or non-dormant seeds. Seeds were obtained from field-grown plants and stored dry at laboratory temperatures.

to water (Bewley & Black, 1982). Most cereal grasses can remain viable in a dehydrated state at normal temperatures for about 25 or 30 years but survival in a moist soil, without dormancy, is very short, often only for several months. Grass species with small seeds are not generally long-lived in a natural environment unless they have dormancy or seed coats impermeable to water.

Human concern with the viability of seeds has presumably existed since cereal grains were first stored for lengthy periods of time in the Bronze Age. Viability can now be predicted in seeds provided the environmental parameters, principally moisture, temperature and oxygen partial pressure can be estimated (Roberts, 1972). The bulk of the evidence to date indicates that seed dormancy is not of necessity linked to seed viability, which in the long term depends on maintenance of the structural and physiological integrity of the embryo. For example there are indications in rice (Roberts, 1972) and *Avena fatua* (Fig. 1.2) that both dormant and non-dormant cultivars lose their viability at the same rate.

Because dormant seeds have frequently been confused with non-viable seeds in germination tests it has become standard practice to use an ancillary method that can determine whether seeds that fail to germinate are dormant or definitely non-viable (Ellis *et al.*, 1985b). These ancillary methods may take the form of overcoming dormancy by physical or chemical means of scarification of an impermeable seed coat, application of a growth regulator such as GA, or stimulation with a nitrate salt. Many of the publications from the early part of this century failed to distinguish between dormancy and non-viability, principally because there were no satisfactory methods for overcoming dormancy. Methods to indicate that an embryo was alive were often unreliable until the adoption of the tetrazolium test for the presence of respiratory activity (Ellis *et al.*, 1985b).

Seed dormancy is a natural condition imposed within a grass seed at some point during the differentiation of the zygote into a fully mature embryo. The attribute of dormancy confers several adaptive advantages. It prevents the premature germination (vivipary) of the embryo in the presence of water while maturing in its position of attachment to the parent plant. Dormancy also confers a resistance, for some time period, to germination after the seed is shed from the parent plant. This time period permits a number of options for further adaptive advantages such as dispersal by wind, water and organisms; avoidance of temporary conditions that cannot support growth of the vegetative phase after germination (drought, cold, flooding, ingestion, and fire); avoidance of competition from other plants, etc.

From the perspective of a population, or a species, seed dormancy can permit temporal extension of germination in several ways. The dormancy may be limited genetically to some fraction of the population and within this sub-population the individuals can germinate continuously over time ensuring that some proportion will encounter favourable conditions for further growth (Yumoto, Shimamota & Tsuda, 1980). Dormancy may be universal in the population but easily overcome uniformly by some external environmental factor such as a particular temperature, or combination of temperatures (Fishman, Erez & Couvillon, 1987), or light as in the case of buried seeds after cultivation (Roberts, 1972). Complete absence of the dormancy trait also ensures uniform germination of a whole population and this can in certain circumstances confer a competitive advantage.

Further temporal extension of germination can arise from the imposition of secondary dormancy in seeds lying on the surface of the ground, or beneath soil and vegetation. Seasonal changes in temperature, light and water status can lead to both induction and removal of dormancy in the season immediately preceding the period for favourable vegetative growth (Karssen, 1982; Froud-Williams *et al.*, 1984). Similarly, dormancy is induced in the period immediately preceding the unfavourable conditions for vegetative growth. The distinction between winter and summer annuals is thus in part a reflection of adaptation of the vegetative phase to cool (winter) or hot (summer) conditions and appropriate timing of germination through control of dormancy by a plant–environment interaction.

The great majority of grasses pass through their life cycle in relatively undisturbed soil conditions and the seeds lie on, or near the soil surface. The cultivated cereals, which have larger seeds than most other grasses, can be sown at depths of up to 12 cm in soil. Successful weed grasses of cultivated crops and modern cultivated grasses also tend to have large seeds and are buried beneath the soil during cultivation. A small proportion of the seeds of grasses in a natural environment may become buried in soil through falling into the cracks of dry clay soils, by being covered by wind- and water-transported soil, by burrowing insect and other animal transport and through the pressure of hooves of grazing animals. There has been considerable adaptation in grasses to favour persistence of viable seeds with appropriate forms of dormancy in these quite different kinds of conditions. At any point in time the amount of the seed 'bank' distributed between the soil surface, or surface litter, and beneath the soil will be a function of the fecundity of the population (seed production per plant), seed viability, and proportion of viable seeds with some degree of dormancy (Cussans, 1976).

The rate of turnover in the seed bank is largely governed by the mean length of period of dormancy in the population. A distinction between 'transient seed banks' (complete turnover in less than one year) and 'persistent seed banks' (some seed remains in the bank longer than one year) is mainly a measure of the persistence of seed dormancy (Grime, 1981).

Both light and temperature have important influences in maintaining and overcoming seed dormancy, above and within the soil profile. The acute sensitivity to particular light qualities of the plant pigment phytochrome (Frankland, 1981), which is probably present in all grass seeds, together with a temperature sensitive conversion of the active to an inactive form, provides the dual sensitivity to light and temperature that can cue seeds to germinate under very specific environmental conditions (Bewley & Black, 1982). Some grasses respond positively to light and germinate, some respond negatively and become dormant (Pollard, 1982), and others can show a positive or negative response according to stage of after-ripening and the temperature (Hilton & Bitterli, 1983) (see Chapter 3, section 3.2). Thus when soil is cultivated and seeds become exposed to light a general stimulation of germination occurs with some species. Buried seed can be maintained in a dormant state by the inactivity of phytochrome kept in prolonged darkness in conditions that could otherwise promote germination. Prolonged darkness thus leads to light dependence for germination (Roberts, 1972) after other forms of primary dormancy have been lost.

Many grass seeds can be kept in a state of dormancy on the surface of soil, or beneath a vegetation canopy, by the response to a specific ratio of red/far-red light that is relatively independent of light intensity (Taylorson & Borthwick, 1969; Frankland, 1981). The phytochrome pigment that mediates this response can thus determine to a considerable measure the timing of germination in seeds that lie on the soil surface and also to some depth in coarsly structured soils that permit light to penetrate as deep as 10 cm (Wells, 1959).

The persistence of populations of particular species, or sub-species, will depend on the flexibility (genotypic and phenotypic) in adapting to the various selection pressures that occur over time in any niche (Jain, 1982). Seed dormancy provides a useful adaptive trait for grass populations to withstand selection pressures from disease epidemics, overgrazing (Semple, 1970), cultivation (Williams, 1978), herbicides (Jana & Thai, 1987), drought stress (Dogget, 1970), flooding (Conover & Geiger, 1984b) and fire (Meredith, 1955).

1.6 Xerophytes, mesophytes and hydrophytes

Seed dormancy in grasses does not appear to have an association with some particular geographical or climatic zone. Harlan (1956) classified grasses, on the basis of their phylogenetic relationships, into sub-families and tribes that occupy the main ecological zones of the earth. The classification positions the tribes into four major climatic zones: (1) moist tropical – i.e. hydrophytic; (2) tropical and sub-tropical monsoon – i.e. alternately wet and dry; (3) cool temperate, including alpine – i.e. mesophytic; and (4) warm-dry – i.e. xerophytic.

In the first of these climatic zones the three important tribes are *Phareae, Oryzeae* and *Bambuseae*. Examples of seed dormancy can be found in each of the tribes (Table 1.1). In the case of the crop *Oryza sativa*, examples can be found of cultivars that are adapted to a completely aquatic environment or to dryland, yet both types exhibit seed dormancy. A member of the *Oryzeae* adapted to a cool temperate aquatic environment (*Zizania aquatica*) has seed dormancy.

In the second climatic zone two large tribes (*Paniceae, Andropogoneae*) and two minor tribes (*Melinideae, Tripsaceae*) are the important grasses of the monsoon areas. Examples of seed dormancy are common within many species in both the tribes *Paniceae* and *Andropogoneae* (Table 1.1). Examples can be found of closely related species with seed dormancy that are adapted to a completely aquatic or a dryland environment e.g. *Echinochloa* spp. (see earlier section on weeds). It is possible that examples can be found of dormancy in the two minor tribes, but they have not yet been reported.

In the third zone the predominant tribes are *Stipeae, Aveneae, Phalarideae*, and *Hordeae* and they have each evolved from the *Festuceae*. Seed dormancy is found in each of the tribes (Table 1.1). Within this temperate zone, which is relatively mesophytic, examples can also be found of adaptations of species to extremes of cold or moisture, within the same tribe. For example within the genus *Stipa* there are forms adapted to warm desert conditions (Soriano, 1961; Young & Evans, 1980) and semi-arid cool temperate lowlands (Fulbright, Redente & Wilson, 1983) both of which have dormancy.

In the fourth zone the three main tribes are the *Aristideae, Sporoboleae* and *Eragrosteae*. The latter has further evolved into the *Chlorideae* and *Zoyzieae*. Each of the tribes has genera with seed dormancy (Table 1.1) substantiating that seed dormancy is a trait that provides adaptive advantage in every major climatic zone of the earth.

1.7 Terminology and definitions of dormancy

The basis of all discussions about dormancy in plants begins with an observation that growth has been suspended in some particular structure associated with a meristem. This suspension is temporary and the capacity for normal growth is regained. There seems little argument about the general use of such terms as bud, tuber, shoot or seed dormancy. In the specific case of seed dormancy, which is found among a high proportion of seed plant species, the classificatory terminology developed to the present time can be divided into three classes of descriptors that signify (a) the apparent origin of the control of dormancy, for example from structural, environmental or physiological/biochemical causes; (b) the degree, or depth of control; (c) timing of the control. In some cases timing and depth of dormancy become confused. For example, a long period of dormancy may be classified as 'deep' dormancy, in contrast to a short period described as 'shallow' (Nikolaeva, 1969).

Table 1.5 is a re-classification, based solely on the apparent etymological meaning of the words used in the categorization developed by Lang *et al.*, (1987) that supposedly orders dormancy terms into just three classes called endo-, para- and eco- dormancy. Endo-dormancy implies control by physiological factors inside the affected structure, para-dormancy control by physiological factors outside the affected structure, and eco-dormancy control by environmental factors. The three prefixes endo-, para- and eco- fail to give any measure of degree or timing of dormancy. The fact that terms indicating origin, degree and timing of control can occur in each of the categories indicates a lack of comprehensiveness of these classes in categorizing all aspects of dormancy.

Numerous definitions of dormancy have been suggested, disputed, used or rejected in the literature of seed biology. Nevertheless the word 'dormancy' has become accepted, from common usage in biology, as a general descriptor of that state of rest that can appear at some point in the life cycle of almost all living organisms. Other terms have been used from time to time which indicate the essence of the same state, e.g. quiescence, hibernation, anabiosis, sleep, diapause, cryptobiosis, aestivation (Clutter, 1978), but they have not been universally adopted. Some of the terms have been restricted to animals.

The use of the word dormancy is now commonplace in plant biology and little purpose is served in searching for a substitute word. It is regrettable that such a pleasing and dignified term as 'repose' (French – 'l'état de repose'; English - state of rest or repose) used by Lagreze-Fossat (1856) in

Table 1.5. *Three classes of terms that qualify different states of dormancy. Classification is based on the etymological meaning of the terms chosen by the authors. The Table is a re-classification of the division into three classes made by Lang* et al., *(1987) to justify the use of the three terms endo- (En), para- (Pa) and eco- (Oe) dormancy as descriptors of all possible states of dormancy existing in plants.*

Original term	Categorisation by Lang
Class I – Terms signifying the origin of the control of dormancy	
Autogenic dormancy	En
Autonomic dormancy	En
Constitutional dormancy	En
Constitutive dormancy	En
Endogenous dormancy	En
Induced dormancy	En
Innate dormancy	En
Internal dormancy	En
Intrinsic dormancy	En
Organic dormancy	En
Physiodormancy	En
Physiological dormancy	En
Correlative dormancy	Pa
Correlative inhibition	Pa
Conditional dormancy	Oe
Environmental dormancy	Oe
Exogenous dormancy	Oe
External dormancy	Oe
Imposed dormancy	Oe
Class II – Terms signifying depth or degree of control of dormancy	
Deep dormancy	En
Deep physiological dormancy	En
Deep rest	En
Relative dormancy	Pa
Shallow dormancy	Pa
Relative dormancy	Oe
Class III – Terms signifying timing of the control of dormancy	
After-ripening	En
Early dormancy	En
Late dormancy	En
Main rest	En
Middle rest	En
Permanent dormancy	En
Primary dormancy	En
Secondary dormancy	En

Table 1.5. *(cont.)*

Original term	Categorisation by Lang
Spontaneous dormancy	En
True winter dormancy	En
Winter dormancy	En
Winter rest	En
Early rest	Pa
Predormancy	Pa
Preliminary rest	Pa
Summer dormancy	Pa
Temporary dormancy	Pa
After rest	Oe
Post-dormancy	Oe
Post-rest	Oe

En = dormancy regulated by physiological factors inside the affected structure
Pa = dormancy regulated by physiological factors outside the affected structure
Oe = dormancy regulated by environmental factors

one of the first scientific papers published about *Avena fatua* was never retained.

Seed dormancy in grasses is a condition referring to the temporary suspension of the growth phase called germination. Germination can normally be observed by the naked eye as emergence of the coleorhiza through the pericarp. For practical purposes the onset of germination in grasses is recorded at some specific time following the exposure of the caryopsis (with, or without covering structures) to those factors (principally water and optimal temperature) that promote normal germination of a non-dormant seed.

The adjective dormant has sometimes been applied to seeds that are stored in a dry state but this is only a state of suspended growth due to desiccation of the seed following abscission from the parent plant. Dry viable seeds that do not have dormancy germinate readily when hydrated, but dormant seeds will not germinate.

Some of the confusion currently existing in the literature about terminology related to dormancy is caused by the lack of understanding of the true nature of the mechanisms that induce, maintain and terminate the state of dormancy in seeds. Difficulties arise with the use of modifying adjectives that could permit systematic categorization of the many different types of dormancy that are attributed to location (e.g. bud, meristem, seed, embryo, coat), apparent cause (e.g. environmental, genetic, physiological, metabolic, structural), degree (e.g. deep, shallow, semi-) and timing (e.g. innate, primary, secondary, temporary). The notions of 'innate', 'enforced' and 'induced' dormancy (Ellis *et al.*, 1985a) or the terms 'endo-, para- and eco-dormancy' (Lang *et al.*, 1987) may provide some help in categorization, but none of the existing terminologies seem to be able to cover the reality of the great range of interactions between seed morphology, physiological and biochemical functions, and the diverse environments which are always in flux, that together determine the nature of dormancy in any one particular seed.

It is known that the state of dormancy can be induced, or re-introduced, at different periods in the life of a seed. Dormancy is also known to persist for different periods of time: the physiological state of the seed toward the end of the period of dormancy may be different from the state that existed at the beginning. A dormant seed can be moved to other environments and remain dormant until it encounters a specific environment that overcomes the dormancy. Alternatively, exposure to different environments may change the seed so that when it is placed back in the originally non-permissive environment it germinates. The term 'after-ripening' has frequently been used to denote the interaction between seed and environment, taking place over time, that leads to loss of dormancy (Simpson, 1978). This is achieved in nature either by exposure of a stationary seed to a fluctuating environment or by physical transport of the seed to another environment by wind-, water- or animal-transport. Changes in the environment may not be part of the requirement for germination once dormancy has been lost.

Some further confusion in terminology may arise from creating categories that allow factors involved in the induction or maintenance of dormancy to become confused with those factors that terminate dormancy. In a treatise on the nature of dormancy it is essential to widen the time span beyond consideration of only the mature seed to include all stages of the developing seed and its relationship to the parent plant and the often lengthy period, following abscission from the parent plant, that a seed lies in a soil–aerial environment.

Koller (1972) has made the point that dormancy is generally initiated in the zygote at a time when the external environment around the plant is favourable for continued metabolic, synthetic, or morphogenetic activities. The induction of dormancy is most likely controlled endogenously, with the ultimate control located in the tissues of the mother plant, but initiated by specific environmental signals such as photoperiod or cold.

Seed dormancy is a genetically inherited trait (Naylor, 1983). Genetically directed dormancy could thus be an array of the following possibilities:

(a) parental whole-plant influence (parental genotype);
(b) pericarp, testa, aleurone, endosperm influence (interaction of parental and zygote genotypes);
(c) internal embryo influence (zygote genotype);
(d) interactions between a, b and c.

All the above conditions could potentially interact with the different environments and physiological states that occur with the passage of time from anthesis through to separation from the parent plant and later when the seed is either lying on the surface or under soil.

Dormancy is an adaptive trait that optimizes the distribution of germination over time within a population of seeds. It is part of a strategy that ensures survival in an everchanging environment. Variation in the timing of germination of individuals among a population could theoretically be achieved by genetic means in several ways according to the method of sexual reproduction of the species. In completely self-pollinated cereal grasses, such as wheat, barley or rice, a form of dormancy involving a number of alleles controlling different expressions of dormancy, that could each be removed by a different environmental factor, could provide a sequential loss of dormancy that would ensure distribution of germination over time.

In totally or partially cross-pollinated grass species where genes are mixed frequently, a small number of alleles for different types of dormancy could provide a mechanism for spreading germination over time, even in a relatively uniform micro-environment.

Allard has argued (1965) that the great majority of colonizing plants is self-pollinated. Thus species like *Avena fatua* that have limited outcrossing in a predominantly self-pollinated population have a highly flexible genetic system for ensuring survival by such characteristics as seed dormancy. Thus some descriptor that can indicate the genetic basis for the state of dormancy should be valuable in categorizing states of dormancy.

Both dormant and non-dormant embryos appear to develop and mature at the same rate, hence dormancy does not seem at first sight to be a developmental restriction but rather the suspension of the very rapid growth phase called 'germination'. In grasses germination is characterized first by expansion of the cells of the coleorhiza and shortly after by expansion of the cells of the coleoptile and scutellum. It is the visibility of this cell expansion to the naked eye that defines the act of germination.

For the purposes of this treatise the term dormancy will be defined as: *Temporary failure of a viable seed to germinate, after a specified length of time, in a particular set of environmental conditions that later evoke germination when the restrictive state has been terminated by either natural or artificial means.*

1.8 A conceptual approach for understanding seed dormancy

In both a theoretical and practical approach to understanding the nature of dormancy in grass seeds it is desirable to use an analytical framework that is both comprehensive and detailed. The framework should take account of all attributes of the parent plant, including the genetics, morphology and physiology as well as the surrounding environment during the period when dormancy is initiated. The framework should also include the period of maintenance of dormancy, both when the seed is attached to the parent plant and following separation from it. All aspects of the seed and its environment should be examined during the period of termination of dormancy. It has been pointed out earlier (Simpson, 1978) that in reality the main experimental approach used in the past to examine the nature of dormancy in seeds has been the reductionist approach. Reductionism emphasizes the division of the whole into its constituent parts down through successively lower levels of organization to the molecular level. Experimentally it is often only possible to study one factor such as a particular tissue, molecule or environmental parameter at a time. A simple form of explaining seed dormancy may for example invoke some degree of causality between temperature changes and the timing of the induction, maintenance and termination of dormancy (Vegis, 1964). To argue, however, that dormancy can be attributed to any single factor (e.g. a hormone, membrane, seed coat, or temperature) ignores the reality that a seed, like every other biological entity, is an exceedingly complex and dynamic living system situated, at all times, within a complex and ever-changing environment. The parts of the living seed sub-system are inseparable from the environmental sub-system in which they exist. The physical act of dividing up seeds and their surrounding environment for the purposes of investigation only permits us to reconstruct mentally what we imagine to be the total system of the dormant seed in its natural situation. The natural environment for a seed at the time when dormancy is initiated is the parent plant situated in a complex soil–water–air milieu. A new set of environmental relationships, generally on or under soil, will determine both the persistence of the primary dormancy and the conditions favouring germination following maturity and separation of the seed from the parent plant. Primary dormancy can be renewed in a seed after it is shed from the parent plant; this is called secondary dormancy.

The position adopted here for resolving the problem of which is the most appropriate set of terminologies to use in a treatise on the nature of dormancy in grass seeds is the following. As a satisfactory comprehensive set of terms has not yet been universally agreed upon for use in discussions about dormancy in general, or seeds in particular, the terms will be restricted to those defined in Table 1.6. Emphasis will be placed in this section of the chapter on defining the boundaries of the plant–environment system that circumscribe the state of dormancy in a grass seed. The description of this system can provide the nomenclature for categorizing specific aspects of dormancy. Further reference to expansion of the nomenclature will be made in Chapter 4, sect. 4.1 and Chapter 5, sect. 5.5.

If by definition seed dormancy is the temporary postponement of the growth phase known as germination, then a first approach to understanding the **timing** of dormancy requires an assessment of the range of stages in development of a non-dormant grass embryo when germination is potentially possible. The potential range begins at some early stage in the differentiation of the zygote when the shoot and root have developed into distinct structures. The range could extend through all phases of development achieved while the embryo, in its enclosing structures, remains attached to the parent plant. It could extend beyond the time of abscission from the parent tissue where further development (differentiation) is not limited by environmental restrictions such as desiccation. The end of the potential period for expression of dormancy might be assumed to be the termination of embryo viability. Nevertheless, the term 'embryo viability' is not clear cut because in certain cases germination consisting of radicle emergence and coleoptile elongation may occur but further growth and development into a normal seedling do not occur (Haber, 1962).

Experimental investigations with genetically non-dormant strains of grass species, mainly large seeded cereals, indicate that germination is possible in remarkably small and relatively undifferentiated embryos if they are excised from the surrounding tissues and placed in an environment suitable for normal germination of the fully mature and intact non-dormant caryopsis (Lucanus, 1862; Atterberg, 1907; Harlan & Pope, 1922; Gill, 1938; Andrews, 1967). On the other hand germination does not occur, under the same conditions and at most stages of embryo development on the parent plant, with excised embryos from genetic strains characterized by non-germinability (i.e. dormancy of the intact mature caryopsis). It is a general observation that the majority of grass species do not normally

Table 1.6. *Definitions of qualifying terms associated with the state of dormancy in grass seeds.*

Origin of the control	
Genetic	Different expressions of dormancy, or non-dormancy, among a range of genotypically distinct strains within a single species.
Metabolic	An association, within a single seed, between either induction, maintenance or termination of the state of dormancy and a metabolic reaction, or reactions.
Structural	Plant structures external to the potentially dormant zones create limitations influencing either the induction, maintenance or termination of dormancy, e.g. lemma and palea, pericarp, testa.
Environmental	Non-plant factors such as water, temperature, light, gases or soil acting alone or in various combinations to influence the induction, maintenance or termination of dormancy.
Timing of the control	
Primary dormancy	Initiated in the early stages of development of the seed and persisting for different lengths of time according to genotype and environmental conditions.
Secondary dormancy	Re-introduction of a state of dormancy after primary dormancy has been completely, or almost completely terminated. Secondary may, or may not, be qualitatively or quantitatively different from primary dormancy.
After-ripening	Loss of the dormant state over some period of time through the exposure of the seed to a set of environmental conditions following separation, at maturity, from the parent plant.
Degree of control	
Full dormancy	Maximal expression of the genotypic potential involving two or more control mechanisms that in combination limit germination to a narrow range of environmental conditions.
Non-dormancy	Germinability in a wide range of environmental conditions.
Partial- or semi-dormancy	Intermediate stages between full dormancy and non-dormancy.

exhibit premature embryo germination while the intact caryopsis is on the parent plant except under unusual circumstances such as excessive atmospheric moisture where the condition is called 'pre-harvest sprouting' (King, 1983). It is therefore necessary to distinguish in some way between the dormancy characterized by (a) non-germinability of a genetically non-dormant embryo developing in its natural plant tissue environment and (b) non-germinability of a similarly situated embryo from a genetically dormant strain. In the former case excision from the surrounding tissues permits germination, indicating that the parental environment has a role in restricting germination. In part this is due to the increasing desiccation associated with the final stages of seed maturation. In the latter case the parental environment may also have restricted germination but there is an additional condition causing non-germinability present within the embryo and persistent after excision. This persistent non-germinability within an excised embryo of any age, before or after separation from the parent plant, will be designated as **embryonic dormancy** to distinguish it from the more complex expression of dormancy in an intact dispersal unit.

In natural situations the embryos of grass seeds are always situated within surrounding tissues that are different in genetic constitution from the diploid embryo. Typically the endosperm, with its layer of aleurone cells, is triploid or occasionally pentaploid from the combination of two or more sets of maternal genes (Bewley & Black, 1978). The structures that surround the embryo such as pericarp, testa, lemma and palea (hulls), and glumes are parts of the diploid mother plant. Together these structures have a major influence on the germinability of the embryo at all stages of development on the parent plant and following separation of the dispersal unit. Much of the confusion in terminology in the literature associated with characterizing different states of dormancy arises from the experimental difficulties of determining how these genetically and morphologically different structures contribute, individually and through interactions with each other, to the total state of dormancy of a grass seed.

To designate the aspects of dormancy externally imposed on the embryo by these external structures as 'structural', 'environmental', 'positional' etc. (in contrast with 'embryonic dormancy') does not separate the physiological determinants of the two different dormancy states. Embryonic dormancy is a sub-state of the more general condition of dormancy.

If we begin with a simple conceptual framework (Fig. 1.3)
that considers what at first sight seem to be the obvious environmental, structural and positional factors that could be involved in the initiation, maintenance and termination of dormancy of a typical grass seed, two

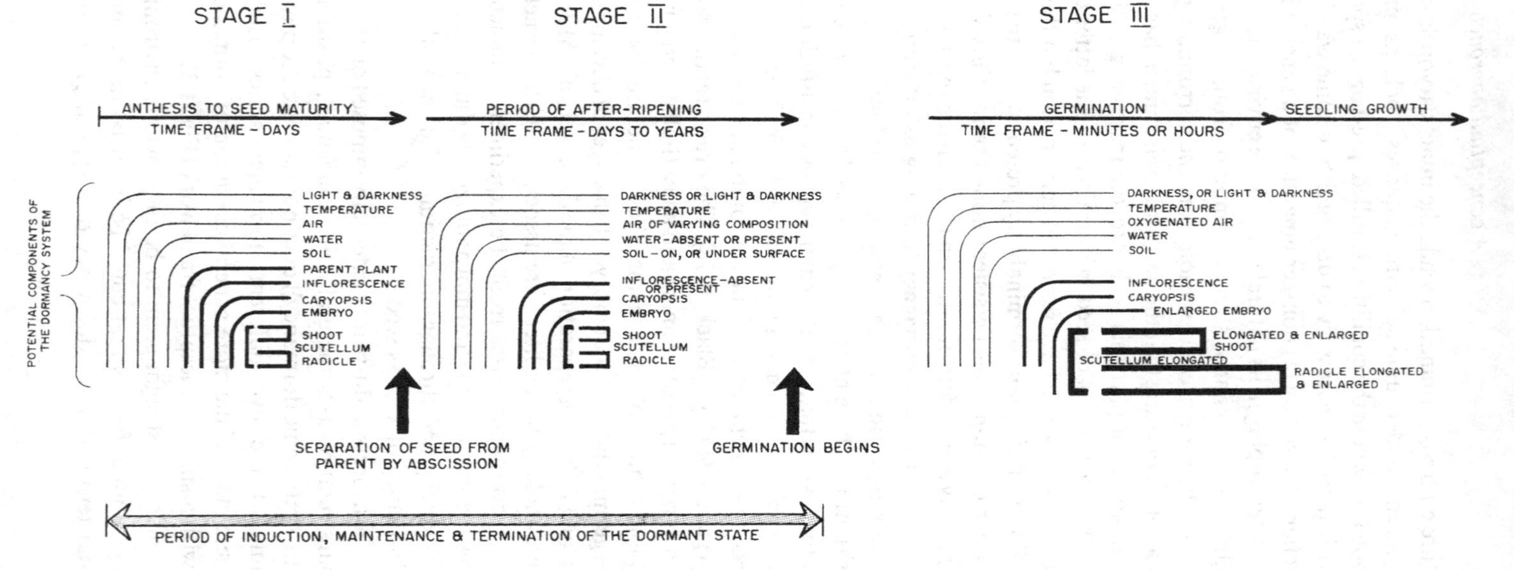

Fig. 1.3. A simple conceptual framework for the factors determining the induction, maintenance and termination of seed dormancy in grasses.

distinct **stages** of dormancy are delineated. The first stage encompasses the time from anthesis to maturity, up to the moment of separation of the seed from the parent plant. The second stage covers the time from separation for an indefinite period when the seed is in a distinctly different environment, or environments. This second stage ends when dormancy is completely lost and normal germination can occur. It is within this second stage that the consequences of seed dormancy, for example poor crop germination or survival of the species through an adaptive advantage, begin in nature.

A physiological explanation of the origin, persistence and termination of dormancy in either of the above stages may be relatively simple or as complex as the currently available experimental methods permit. Modern approaches to physiological analysis have tended to move away from single-factor analysis toward consideration of the major factors, or even all possible factors, that could play a role in a specific physiological condition. This latter approach can be called a 'systems' approach to understand a problematical condition. This approach is considered in some detail below.

Systems analysis applied to biological problems stresses that all parts of a system are organized into a hierarchy of levels that are logically interrelated (Von Bertalanffy, 1968; Whyte *et al.*, 1969). Hierarchies of levels can be constructed for either structural, functional or taxonomic relationships (Fig. 1.4). The requirements on one level act as the constraints on the subordinate level. Thus in a hierarchy of functional relationships the proper

Fig. 1.4. A multilevel hierarchical system (After Whyte *et al.*, 1969).

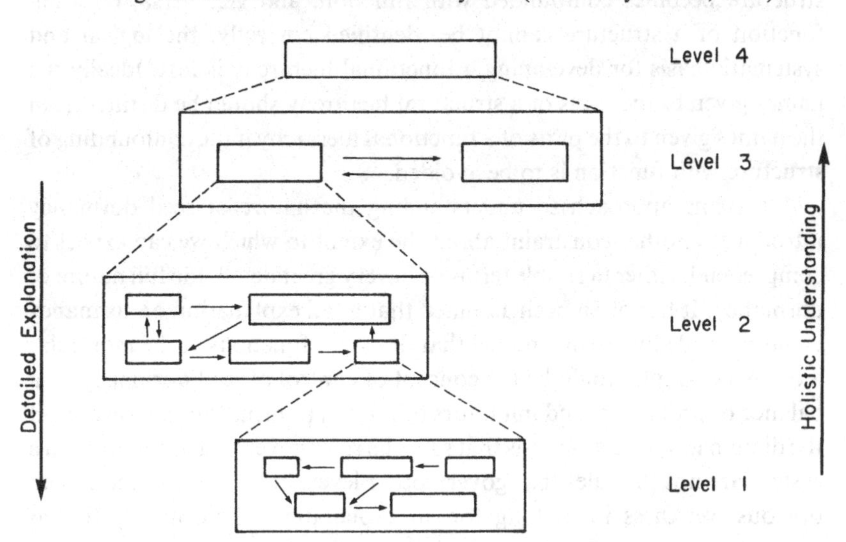

functioning on any one level requires that all the subordinate levels are fully functional. Systems analysis also stresses that the rules that govern functions at one level are not the same rules governing functions at higher or lower levels in the hierarchy. Thus the rules that govern uptake of water into a starch granule will be different from the rules that govern uptake into an intact caryopsis. Defining a functional hierarchy is much more difficult than defining a structural hierarchy. In a structural hierarchy the whole is made up of parts that in themselves have attributes of wholeness and physical autonomy. Starting at the level of an individual cell a structural hierarchy reaches upward through the levels of tissues, organs, organisms, and populations and extends downward through organelles, plastids, macromolecules, molecules, atoms, etc., to the limits of physical identity. The semi-autonomous sub-wholes that are the stable units at any of the levels in a structural hierarchy have been described as **holons** by Koestler (1967). We tend to order structural holons in plants by their anatomical identities defined by boundaries. For example, in grass seeds, anatomical descriptors such as seed, caryopsis, embryo, root, root-meristem, initial cells, and so on are ordered hierarchically into different levels of subordinate structures.

Plant physiology is concerned with explanations of the growth and development of plants over time in the context of environmental change. In the case of the growth and development of a seed it is very difficult to construct logically a functional hierarchy that interrelates the functions determining germination without some reference to structures. When structure becomes confounded with function, and vice versa, or when function of a structure cannot be identified correctly, the logical and systematic basis for developing a functional hierarchy is lost. Ideally the names given to the parts of a structural hierarchy should be distinct from the names given to the parts of a functional hierarchy if the confounding of structure with function is to be avoided.

A systems approach to understanding the nature of seed dormancy introduces another constraint about the extent to which we can expect to comprehend, either in simple terms or in very great detail, the full nature of dormancy. It has often been assumed that a full explanation of dormancy could be made by use of a model that describes functions at the molecular level. An example would be the concept of control of seed dormancy by a balance of promotors and inhibitors of growth (Amen, 1968). However, if the dictum in systems analysis that says the rules governing lower levels in a system are not the rules that govern other levels in the system, there is an obvious weakness in looking for an explanation of dormancy that is couched in terms of only one level of the system; in this case the molecular, or biochemical, level. The most satisfactory description of dormancy will not be associated with great detail at a single low level in the hierarchy but rather will come from the holistic view of the entire system.

For the above reasons it is desirable to analyse the nature of grass seed dormancy with a systematic approach that takes into account the full dimensions of the system that encompasses the induction, maintenance and termination of dormancy. Obvious limitations to this analysis will be the problem of separating structural and functional attributes with appropriate nomenclature and ordering them into structural and functional hierarchies that, viewed holistically, increase our comprehension of the true state of dormancy.

Figure 1.5 describes a hierarchy that orders the structural components of a grass plant situated in its natural environment at some point in time. The ordering is made on the basis of clarifying the level at which the embryonic root of a developing grass seed can be considered in relation to the external environment. The purpose is to interpret the nature of non-germinability (dormancy). In this case the point in time is near seed maturation on the parent plant. This scheme would change with time as the plant structure evolves through growth and development and as the internal and external environments change diurnally and seasonally. There would be a radical change in the appearance of the plant portion of the scheme when the dispersal unit is separated from the parent plant. A continous series of figures could be constructed depicting the plant–environment structural hierarchy for each point in time between anthesis and the point of germination of the seed as it lies on or under the soil. This series would add a dynamic component to the overall view. Categorizing a plant–environment system is clearly a very arbitrary process based primarily on nomenclature derived from studies of plant morphology and the physics of the natural environment. The names generally signify the separateness, or discreteness, of the parts of the system perceived as entities (up the hierarchy) or arrays of fragments (down the hierarchy).

To achieve a physiological understanding of why a radicle and shoot fail to elongate in a situation called 'dormancy' requires firstly the development of a mental image that takes account of the structural position of the root and shoot within the context of the plant–environment system. Figure 1.5 is a graphical representation of such an image. The image cannot be static but must change over time. Secondly, the image should indicate the **functional** relationships between plant parts and the components of the environment. Obviously a single image of the structural hierarchy cannot do this and even

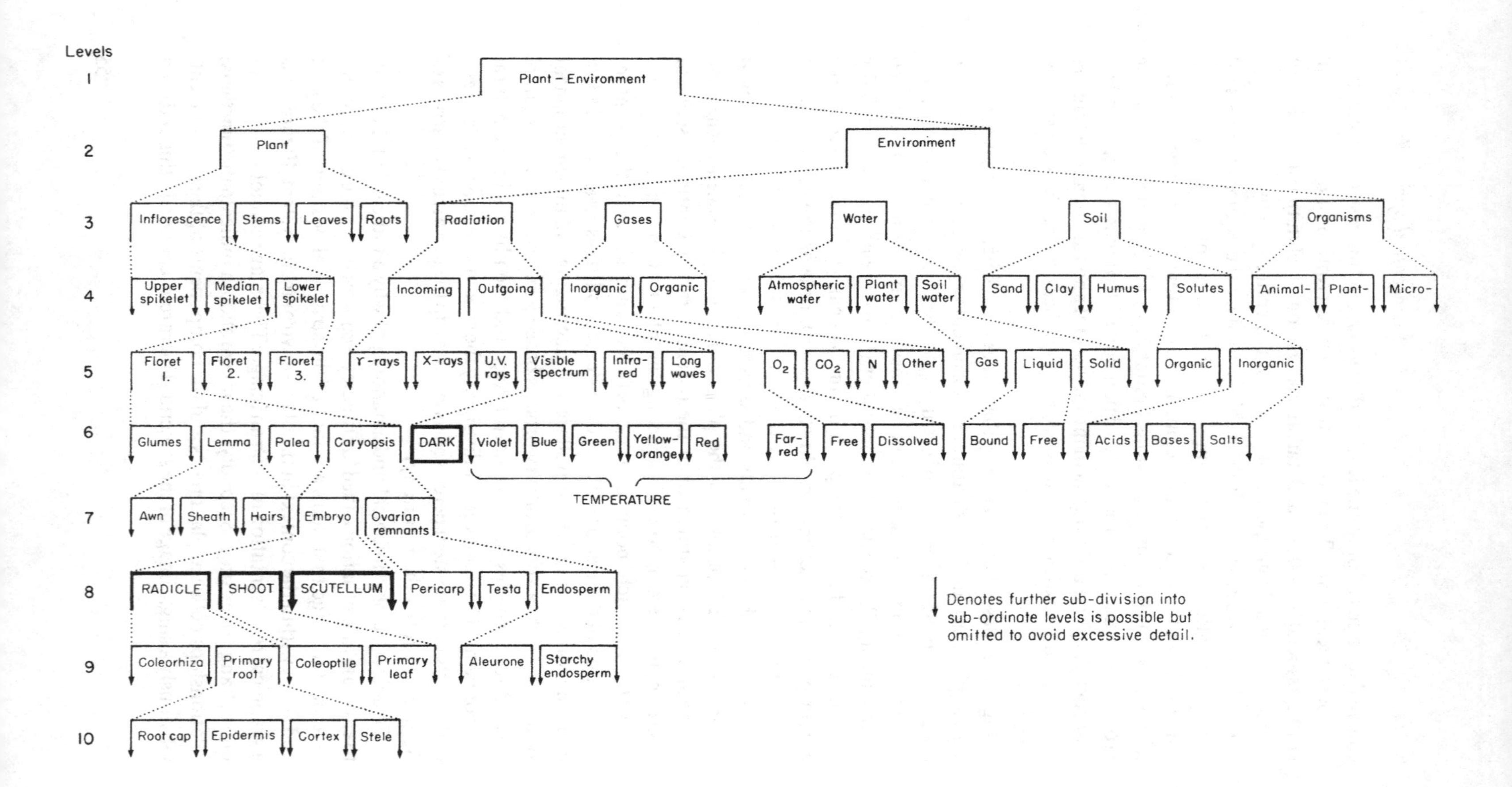

Levels
1
2
3
4
5
6
7
8
9
10
Plant – Environment
Plant
Environment
Inflorescence
Stems
Leaves
Roots
Radiation
Gases
Water
Soil
Organisms
Upper spikelet
Median spikelet
Lower spikelet
Incoming
Outgoing
Inorganic
Organic
Atmospheric water
Plant water
Soil water
Sand
Clay
Humus
Solutes
Animal-
Plant-
Micro-
Floret 1.
Floret 2.
Floret 3.
γ-rays
X-rays
U.V. rays
Visible spectrum
Infra-red
Long waves
O_2
CO_2
N
Other
Gas
Liquid
Solid
Organic
Inorganic
Glumes
Lemma
Palea
Caryopsis
DARK
Violet
Blue
Green
Yellow-orange
Red
Far-red
Free
Dissolved
Bound
Free
Acids
Bases
Salts
TEMPERATURE
Awn
Sheath
Hairs
Embryo
Ovarian remnants
RADICLE
SHOOT
SCUTELLUM
Pericarp
Testa
Endosperm
Coleorhiza
Primary root
Coleoptile
Primary leaf
Aleurone
Starchy endosperm
Root cap
Epidermis
Cortex
Stele
Denotes further sub-division into sub-ordinate levels is possible but omitted to avoid excessive detail.

a continuous series of structural images can only convey a limited understanding of the causal relationships between levels or between parts on a given level. Alternately, a separate image can be created by establishing a hierarchy of **functions** that portray germination (Fig. 1.6). Each element in the hierarchy is a **process** that summarizes the changes over time in a grouping of associated structures. Each process can be summarized in a quantitative way by the use of a mathematical parameter that can measure rate i.e. change in a unit of time. For example, at level two, the process of tissue elongation can be summarized as the rate in centimetres per hour. At level six, changes in respiration can be signified mathematically as micro litres of uptake per minute. Thus the total image conveyed by the hierarchical ordering attempts to convey the causality i.e. the relations between cause and effect that determine germination.

Both the structural and functional hierarchies incorporate names that are common to both systems, and these common denominators tend to be structural rather than functional descriptors. This indicates the difficulty of describing functions without reference to the site, or level in the ordering, where a function can be measured.

A systems approach to understanding dormancy in grass seeds would require firstly a logical ordering of structural levels of the plant environment system and an indication of how the structural elements change over time. Secondly it would require ordering the processes and sub-processes involved in normal germination of a seed into a functional system that could be compared at every level with the system of a temporarily non-ormant seed. Combining the image of the two-dimensional structural hierarchy with the two-dimensional functional hierarchy to form a three-dimensional topological model of either normal germination, or dormancy, may be possible if the sequence of levels in both hierarchies is matched through the use of structurally based common denominators. At the present time the limitations to the development of such a model lie in our inability to measure many of the processes describing the functional interactions that determine germination or dormancy. It is thus not yet possible to complete the image of the functional hierarchy. On the other hand the structural hierarchy can be described much more completely.

The approach that will be used in this treatise on grass seed dormancy

Fig. 1.5. A hierarchical ordering for sub-divisions of a typical grass plant and its environment shortly before seed maturation. The scheme emphasizes the structural positions of the plant components closely affecting seed germination. Sub-divisions of the environment emphasize plant–environment interactions known to influence seed dormancy.

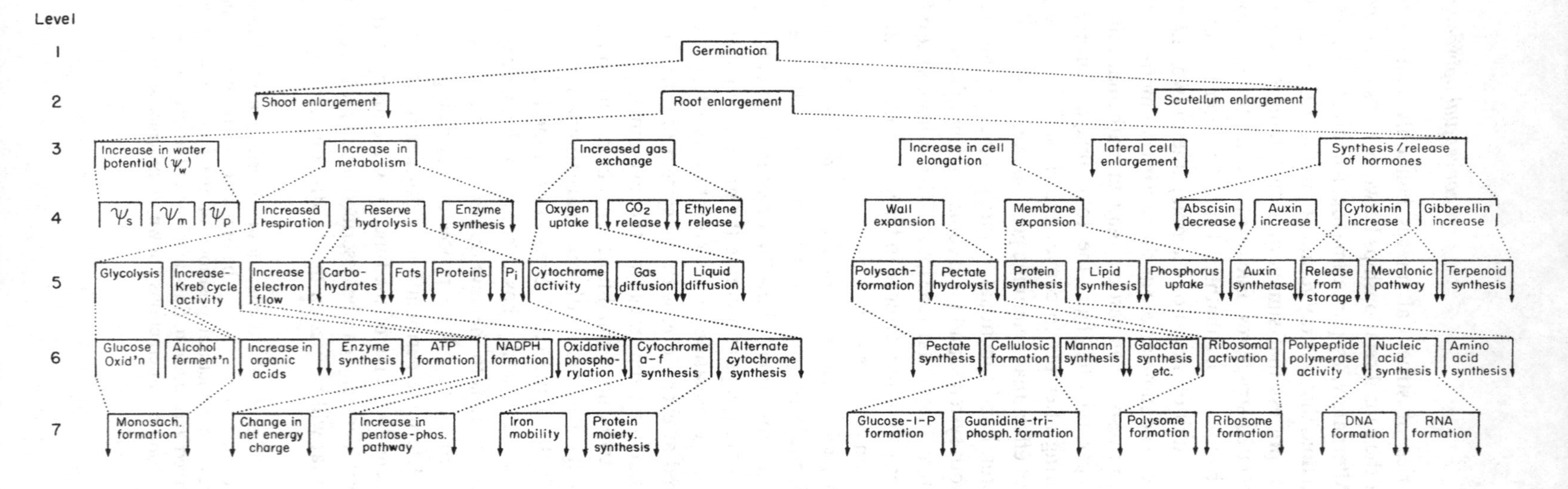

Fig. 1.6. A hierarchy of functions associated with germination of a grass seed. ↓ denotes further potential division into sub-ordinate levels.

represents a compromise between using a full systems approach to describe dormancy, that desirably would reveal the structural and functional images, and a general review of the research information obtained in the past by the various scientists who used a great range of experimental approaches to try and understand the nature of grass seed dormancy. Because the structural image is reasonably well understood and complete it will be described first. The functional relationships will be reviewed systematically but it is not yet possible to complete the image due to the gaps in knowledge about many of the levels.

1.9 *Avena fatua* L., a prototype model of grass seed dormancy

The nature of seed dormancy in *Avena fatua*, the common wild oat, has been studied in considerable detail (Simpson, 1983). More work has been published about seed dormancy in this single species and its role in contributing to the weed problem in cereals than in any other member of the grass family (Table 1.2). The principal reason for this interest is economic. The wild oat ranks tenth in position among global weeds in terms of distribution; it occurs in at least 56 countries (Table 1.4). In terms of 'seriousness' the wild oat ranks at the 27th position (Table 1.3). Nevertheless it is the major weed of wheat (Leggett & Banting, 1958; Jordan, 1977; Sharma, 1979; Wilson, 1979), the world's most widely cultivated crop (Stoskopf, 1985), and of barley, oats, rye and flax. The chief exporters of wheat are the USA, Canada, Australia and Argentina and in each of these countries control of wild oats has proven to be very difficult despite the availability of at least eleven herbicides that give fair measures of control. China, USSR, France and recently the United Kingdom are major producers, and consumers, of wheat and they also have serious problems in controlling wild oats. It is undoubtedly the volume and monetary value of wheat production in these particular countries that makes the wild oat a very costly weed, which in turn has generated a high degree of research interest in this species.

A. fatua will be used as the prototype reference grass for the sections of Chaps. 2–4. These sections are ordered along the lines of the structural and environmental factors that are presently known to influence dormancy in *A. fatua*. In each section the available knowledge for other grasses is compared with existing knowledge for *A. fatua* to bring out similarities and differences.

2

Mutual influences of inflorescence and caryopsis parts on dormancy

2.1 Position of the caryopsis on the parent plant

In the grasses the term 'seed' is commonly used to describe the dispersal unit. However in some cases the dispersal unit may be a spikelet, a floret, or a naked caryopsis. In addition the term 'grain' is frequently substituted for the word caryopsis, particularly among the cereal grasses. In the context of this chapter on structure the term grain will be used as the general descriptor for the fruit. To some extent the terms seed and grain will be used interchangeably, except where clarity in the separation of anatomical structures is of significance.

Example of Avena fatua*:*

The inflorescence in *Avena fatua* is a determinate panicle that matures basipetally but the spikelets mature acropetally (Green & Helgeson, 1957; Raju & Ramaswamy, 1983; Raju, Jones & Ledingham, 1985) (Fig. 2.1). Self-pollination is the rule and the florets are chasmogamous (Raju *et al.*, 1985). Some outcrossing has been observed ranging from one to twelve per cent (Imam & Allard, 1965). Dormancy is present in the grains at an early stage: excised caryopses can germinate as early as three days after anthesis (Morrow & Gealy, 1983) yet by 15 days, dormancy is present and increases up to seed maturation (Thurston, 1957a). The degree of dormancy can vary with position of the spikelet on the panicle (Anghel & Raianu, 1959); grains at the bottom of the panicle are more dormant than those at the top (Schwendiman & Shands, 1943). Small panicles appear to have a higher proportion of dormant seeds than large panicles (Thurston, 1957a). There are usually two flowers, and sometimes three, in a spikelet (Fig. 2.2). The primary floret is larger and better developed than the secondary or tertiary in the genus *Avena* (Raju *et al.*, 1985; Tibelius & Klinck, 1986). The primary grain remains on the inflorescence for a longer period than the secondary or tertiary grains and is also less dormant (Raju, 1983). Many reports indicate that the order of increasing depth of

Fig. 2.1. Panicle of *Avena fatua* (After Raju, Jones & Ledingham, 1985).

Fig. 2.2. Spikelet of *Avena fatua* showing three florets (F) and outer glumes (G). *(a)* at anthesis (After Raju, Jones & Ledingham, 1985). *(b)* at seed maturity (From M. V. S. Raju).

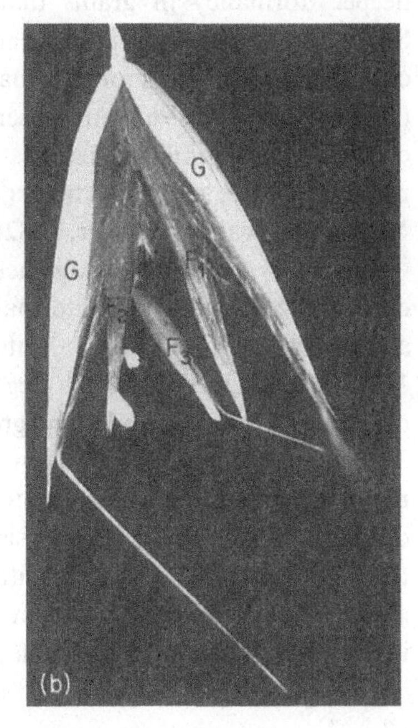

dormancy within a spikelet is tertiary, secondary and primary caryopsis (Johnson, 1935b; Schwendiman & Shands, 1943; Judkins, 1946; Thurston, 1951b, 1952, 1957a; Green & Helgeson, 1957; Anghel & Raianu, 1959; Thurston, 1962a). The explanation is that although the grains mature acropetally within the spikelet so that the primary floret matures first, abscission of each floret is basipetal so that the secondary and tertiary grains are not fully mature when they fall from the plant (Raju & Ramaswamy, 1983). The fully mature primary grain may lack dormancy and this is correlated with the longer period of after-ripening needed for secondary seeds, a device that allows for propagation over two seasons (Johnson, 1935b).

Both genetic and environmental influences on panicle morphology can affect degree of grain dormancy. In natural situations there can be a considerable range in variation of time to panicle emergence. This range can be as great in a localized area as it is for a continental distribution (Somody, Nalewaja & Miller, 1984a). For example, the range in California (46 to 76 days) is as great as the range found within the latitudes of the USA. Time to panicle emergence is a function of photoperiod (Thurston, 1961; Somody *et al.*, 1981, 1984c) and panicle exsertion is completely suppressed in photoperiods of less than 12 h in *A. fatua* (Griffiths, 1961); this leads to enforced self-fertilization. Time until panicle emergence is also determined by temperature. Low temperatures prolong time to emergence and induce deeper dormancy in grains than higher temperatures (Peters, 1982a; Somody *et al.*, 1984a). Time to panicle emergence and grain dormancy are considered to be polymorphic characters in both *A. fatua* and *A. barbata* (Jain & Marshall, 1967) and observations on several very extensive world collections of *A. fatua* indicate that variation in grain dormancy and panicle morphology are associated traits (Toole & Coffman, 1940; Thurston, 1960; Miller, Nalewaja & Mulder, 1982).

A study of genetically distinct lines of *A. fatua* indicates that lines characterized by low grain dormancy have short grain development times and few, heavy grains low in water content (Adkins, Loewen & Symons, 1986, 1987). Lines with high degrees of grain dormancy had long grain development periods and many grains containing a high per cent of water. Thus temperature and photoperiod can interact with the genetic determinants of panicle morphology to exert an influence on the degree of dormancy found in each caryopsis within a panicle. The significant role of the panicle structures in transmitting and enforcing maternal genetic and environmental influences on a developing caryopsis has been demonstrated in *A. sativa* where oil content in the embryo is largely dependent on the maternal characteristics and not on the genotype of the embryo (Brown & Aryeetey, 1973).

Other grasses:

Oryza sativa has a determinate panicle with spikelets borne singly and there is a report of variation in level of caryopsis dormancy among spikelets within a panicle (Sugawara, 1959). *Poa pratensis* has a pyramidal shaped panicle in which there is a considerable variation in germinability among spikelets (Phaneendranath *et al.*, 1978). The degree of dormancy was greatest in immature grains with the highest moisture content. In cereals with spiked inflorescences (Fig. 2.3) there are positional effects on grain dormancy. In *Triticum aestivum*, from Europe, dormancy increased basipetally with the youngest grains showing the most dormancy (Gosling *et al.*, 1981). They noted an influence of the environment during seed maturation on subsequent germination conditions; lengthening the time between anthesis and grain maturity broadened the temperature range for germination of the mature grains. An earlier study (Wellington, 1956), also in a European wheat, showed that dormancy increased basipetally in a spike and was positively correlated with grain moisture. The length of time

Fig. 2.3. An array of spikelets on a spike of wheat (*Triticum aestivum*).

between anthesis and maturity also determined germinability. Upper spikelets were the first to anthese and reach maturity and were less dormant than the lowest spikelets on the spike. On the other hand, a study with American winter wheats indicated that the middle spikelets were less germinable than either the lowest or topmost spikelets (Hardesty & Elliott, 1956). Within spikelets the first and second florets had more dormancy than the third floret, regardless of position on the spike. An investigation of positional effects in European barley (Grahl, Thielebein & Fischnich, 1962) indicated that the bottom and topmost spikelets were more dormant than the middle spikelets.

Examples of heteroblasty have been reported in two species of *Aegilops* that produce dormant and non-dormant seeds within a single spikelet (Datta, Evenari & Gutterman, 1970; Wurzburger & Koller, 1976; Wurzburger, Leshem & Koller, 1976). This heteroblasty is functionally similar to the range of dormancy found in a single spikelet of *A. fatua* in that it serves to ensure germination of an annual plant in several growth seasons. In *Aegilops ovata* the dispersal unit comprises 2–4 spikelets (Datta *et al.*, 1970). The lower two spikelets each contain two caryopses whereas the upper two only contain one. The lowermost caryopsis in each of the lower spikelets is heavier, and has less dormancy, than the upper one. The larger grains also germinate over a wider temperature range than the smaller grains. The hulls of the dispersal unit are believed to contain an inhibitor specific to the smaller grains. In *Aegilops kotschyii* the spikelet has three florets (Wurzburger & Koller, 1976; Wurzburger *et al.*, 1976). The terminal floret is not fertile and one of the other florets has a high capacity for germination beginning shortly after anthesis, whereas the third floret retains dormancy for a long period of time after maturity. The depth of dormancy in this floret was influenced by the environmental conditions to which the parental plant was exposed during seed maturation. Dormancy was greater in grains produced on plants exposed to low temperatures and long days (16 h) than at higher temperatures and short days (8 h). The correlation between dormancy and position of the caryopsis in the inflorescence indicates a strong genetic component determining dormancy, modified by factors in the environment that affect the parent plant during seed maturation.

The great majority of published reports of studies on seed dormancy in grasses (Simpson, 1987) indicates variation in the germinability of the seed samples that generally consist of bulked material from plots or fields. Much of this variation can be attributed to positional differences, particularly in panicular inflorescences, that lead to differences in timing of the onset of anthesis and grain maturation. Collection of seed for commercial purposes is likely to interrupt normal maturation for a proportion of the spikelets so that a difference in germinability may be heightened in comparison to the natural situation. The significant effects of environment and caryopsis position that influence seed dormancy draw attention to the need for reproducible environments in experimental investigations of seed dormancy in grasses.

The paucity of current information about positional effects on seed dormancy in grasses does not permit any generalization other than the probability that dormancy is increased by caryopsis immaturity combined with a high grain moisture content at the time of abscission.

2.2 Lemma, palea, glumes and awns

The structures immediately external to the caryopsis form part of the spikelet, the ultimate unit of the inflorescence in all grasses. There are many reports of the influence of the spikelet on seed dormancy. Although there is considerable diversity among grasses in spikelet structure achieved variously by suppression, multiplication, and fusion of particular parts, it is generally not difficult to identify the key components (Figs. 2.4*(a)* and *(b)*).

Spikelets have sterile subtending glumes that can vary in size from scale-like at the base of the floret to large and completely embracing the spikelet (Clifford, 1986). These glumes may persist on the pedicel at the time of abscission of the grain, or may be shed along with the spikelet. Lemmas, modified bracts, are homologous with subtending glumes and they partially surround the caryopsis. Lemmas are diverse in shape, texture and awn type and can vary in thickness, density, and colour; they are often hairy. Awns may be simple or complex. Generally the lower proximal part of the awn is a twisted column that is hygroscopically sensitive. The twisting of this column in response to changes in relative humidity turns the awl, the upper and distal part of the column, which acts as a lever to propel the dispersal unit from the spikelet and later along the surface of the soil (Raju & Barton, 1984). Lemmas also often bear hooks, or hairs, that have a role in assisting dispersion.

The first true bract on the floral axis is called a palea. It is usually membranous and two-keeled without a midrib. Immediately distal to the palea are lodicules, varying from one to three, which serve to open the flower at anthesis. The floral axis terminates in a gynoecium (pistil) and subtending stamens. The gynoecium becomes the fruit (caryopsis) in the mature spikelet. In some species modified and sterile spikelets form a considerable proportion of the inflorescence.

The dispersal unit may be one or more spikelets within glumes, a single caryopsis enclosed in a lemma and palea or a caryopsis alone. For example, within the *Bambuseae*, *Aveneae*, *Hordeae*, *Chlorideae*, *Agrostideae* and *Phalarideae*, the lemma and palea alone surround the caryopsis because the rachilla disarticulates above the glumes which are thus left on the parent plant (Bews, 1929; Harmer & Lee, 1978). In the *Zoyzieae*, *Oryzeae*, *Phareae*, *Melinideae*, *Paniceae* and *Andropogoneae* the one-seeded spikelets

Fig. 2.4. Schematic floral diagram of a generalized grass floret. *(a)* transverse view. *(b)* sagittal view.

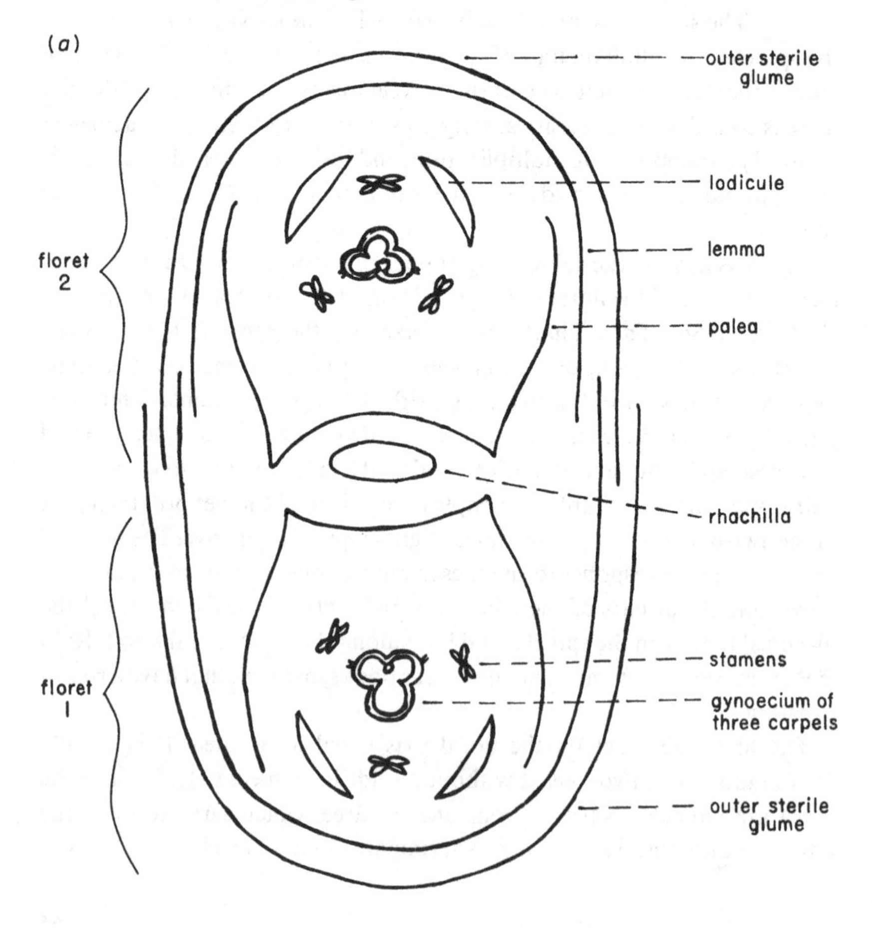

fall off entire because the axis disarticulates below the glumes giving good protection to the caryopsis. In the *Andropogoneae* the whole spikelet is enclosed by a large lower glume that is thick and hard.

In a few species of *Poa* and *Festuca* the dispersal unit is not a seed with storage tissue but a plantlet consisting of a shoot with adventitious roots originating from a single floret (Bews, 1929). This is not true vivipary but rather apogamy where the seed develops without a resting period on the parent plant. These unusual adaptations can provide interesting research material for comparisons with the normal dormant, resting, stages of development in closely allied species.

Example of Avena fatua:

The dispersal unit in *A. fatua* is a floret, so the glumes remain on the parent plant. There is one example in the literature (Richardson, 1979) showing that the basal glumes contribute to dormancy of caryopses within

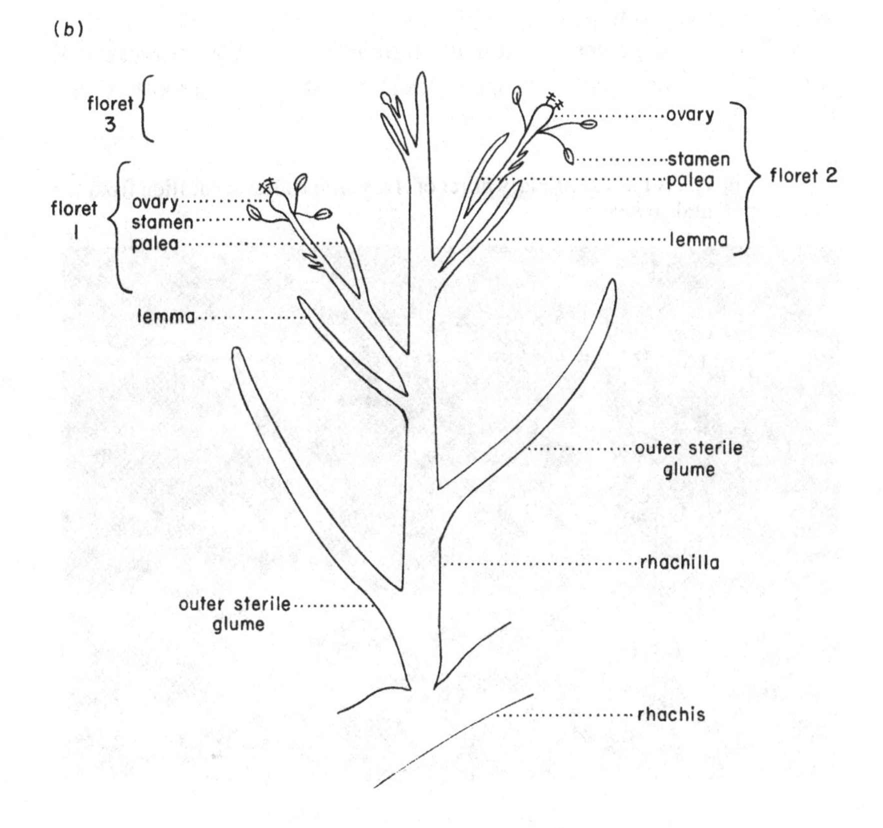

a spikelet: removal of glumes, combined with low night temperature and a short photoperiod, during the period of seed maturation, increased seed dormancy. The first published experiment on dormant wild oat seeds (Lagreze-Fossat, 1856) compared the germination of entire spikelets with individual florets and no differences were noted. The dearth of evidence since that first report suggests that the effect of glumes on dormancy has never been further tested. On the other hand the lemma and palea definitely contribute to reduced germinability (Johnson, 1935b; Kommedahl *et al.*, 1958; Hay, 1960; Thurston, 1962b; Watkins, 1969; Whittington *et al.*, 1970; Banting, 1974; Hsiao & Quick, 1985).

The lemma in *A. fatua* is about twice as thick as the palea (Morrison & Dushnicky, 1982) (Fig. 2.5). It covers the major part of the caryopsis surface and can interfere with water uptake, thus contributing to delayed germination (Haun, 1956; Hsiao & Simpson, 1971). Treatment with partial vacuum to increase water penetration overcomes the interference with rate of water uptake (Kommedahl *et al.*, 1958). Many reports indicate that the lemma and palea together ('hulls') only delay, and do not stop, water uptake to the caryopsis.

The hulls contain germination inhibitors (Black, 1959; Stryckers & Pattou, 1963; Chen, MacTaggart & Elofsen, 1982). Two reports (Hay,

Fig. 2.5. A mature primary floret of *Avena fatua* after separation from the parental spikelet.

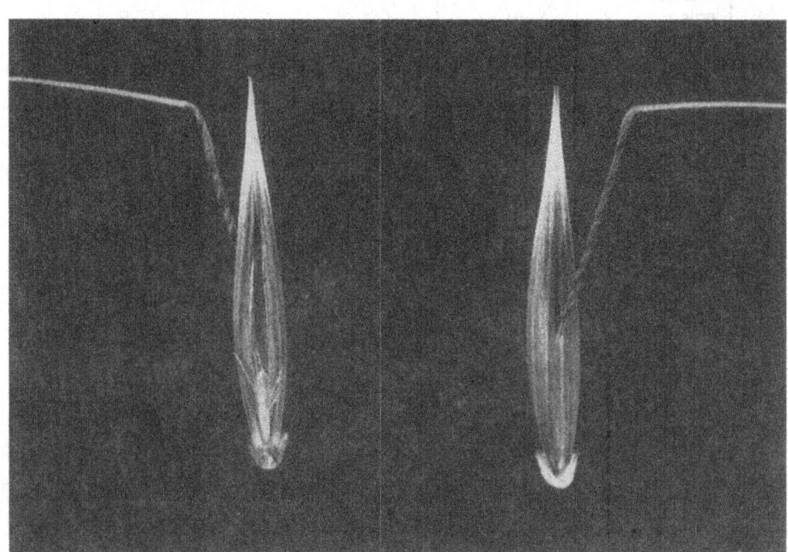

1962; Stryckers & Pattou, 1963) indicate that the inhibitory influence can prevent germination in wild oat caryopses, but others show the influence is effective only on other species (Helgeson & Green, 1957; Koves, 1957; Black, 1959). A triterpenoid glycoside that interferes with root growth has been isolated (Haggquist, Petterson & Liljenberg, 1984). There are several reports of inhibitory substances in the hulls of *A. sativa* (Peers, 1958; Lascorz & Drapron, 1987a,b; Lecat, 1987) and the compound 1,3,4–pentane tricarboxylic acid was identified. Another report suggests a proteinaceous inhibitor (Helgeson & Green, 1957).

Hulls contribute to limitations of gas exchange to and from the caryopsis, that influence metabolic activity in the embryo (Black, 1959; Hay & Cumming, 1959). For example a 40% reduction in germination was noted in caryopses with the hulls on, compared to hulls removed, even in the presence of an atmosphere of 20% oxygen (Hart & Berrie, 1966). In *A. sativa*, metabolic activity in the hulls consumes oxygen and thus deprives the embryo, an effect that is significant only at low temperatures when metabolic activity is low (Corbineau *et al.*, 1986). On the other hand another report about *A. sativa* suggests that low temperature overcomes the inhibitory effect of hulls (Schwendiman & Shands, 1943). Increasing the partial pressure of oxygen with *A. fatua* seeds with hulls increased the per cent germination (Black, 1959).

Lemma characteristics such as colour and hairiness are polymorphic (Thurston, 1957b; Baker & Leighty, 1958; Jain & Marshall, 1967; Rines *et al.*, 1980; Darmency & Aujas, 1987) and appear to be of secondary importance, or even unrelated, to seed dormancy (Von Prante, 1971). However there is one report of grey-lemma types germinating more slowly than black-lemma types (Jain & Rai, 1977). The hulls can interact with light and influence dormancy (Hart & Berrie, 1966, 1967; Whittington *et al.*, 1970; Hsiao & Simpson, 1971) but the primary effect is interference in gas diffusion. If CO_2 concentration is kept at zero when seeds are germinated light inhibits both intact and dehulled seeds (Hart & Berrie, 1966). When CO_2 is added, dehulled seeds are no longer inhibited by light indicating that the hulls modify the rate of diffusion of CO_2 either in, or out, of the fruit.

Lemma awns can have an important influence on the dormancy states induced in the caryopsis (Stinson & Peterson, 1979; Raju, 1983). Secondary and tertiary florets can be prematurely ejected from the maturing spikelet by the twisting forces of awns that rotate in response to changes in atmospheric relative humidity (Raju & Ramaswamy, 1983; Raju, 1984; Raju & Barton, 1984). These ejected immature caryopses are more dormant than those that remain on the parent plant to full maturity.

Mutual influences of inflorescence and caryopsis parts

While dehulling promotes germination and can influence the persistence of at least two separate expressions of dormancy within the caryopsis of *A. fatua* it can only partially contribute to the expression of grain dormancy in this species (Adkins & Simpson, 1988). In recent years with the increase in studies of genetically pure lines of *A. fatua* it has become clear that in many lines removal of the lemma and palea alone is not sufficient to promote germination of dormant florets. There are states of dormancy present within a caryopsis that are independent of any reduction in germinability caused directly by hulls (Kommedahl *et al.*, 1958; Thurston, 1963; Hsiao & Quick, 1985; Adkins & Simpson, 1988). However hulls may modify the expression of these endogenous dormancy states within a caryopsis (Simpson, 1978).

Other grasses:

There is a considerable literature documenting the delay in germination of newly harvested seed caused by the presence of 'hulls', particularly for forage grasses. In the great majority of reports the authors have not distinguished between the true glumes (i.e. whether left on the parent or attached to the spikelet dispersal unit), sterile glumes (lemma and palea of a sterile floret) or the true lemma and palea surrounding a fertile floret. The use of the common term 'hull(s)', or sometimes 'chaff', has arisen from the agricultural practice of threshing cereal seeds to recover the edible grain. Forage grasses are generally small and light. They are frequently cleaned to remove awns, sterile florets and glumes to permit easier sowing. Early observations on the role of the appendages surrounding the caryopsis were thus made from a purely practical perspective. Evaluation of these structures for physiological reasons, and for solving problems associated with gene-bank storage (Ellis, Hong & Roberts, 1985a,b) indicates that part, but not all, of the delayed germination in grasses is associated with them.

The structures surrounding the caryopsis also act to protect the fruit, rather like an extra pericarp, preventing the rapid egress or entry of water (Bews, 1929). The appendages such as awns, hairs and papery glumes aid dispersal by wind and animals. Glumes probably have a significant role in protecting fruits from microbial destruction through the presence of allelopathic compounds. It is well known that dehulled seeds generally have shorter survival in soil than intact seeds (Kilcher, & Lawrence, 1960; Smith, 1971). There is a paucity of literature about the role of the true glumes in seed dormancy. The few reports are for the tribes where one-seeded spikelets fall off entire because the axis disarticulates below the glumes leaving them attached to the dispersal unit.

Sabi panicum (*Panicum maximum*) has a seed which is a caryopsis enclosed in a rigid lemma and palea (Smith, 1971). The single fertile floret is enclosed by the lemma and palea of a sterile floret and in turn these are enclosed by the glumes. Removal of the outer glumes and sterile lemma and palea by acid scarification increased germination from 5 to 40%. The inhibition by these structures was attributed to an inhibitor that could be leached away. When the glumes were removed the grain could be stimulated to germinate by light, or a combination of nitrate and alternating temperatures. Thus two distinct states of dormancy were expressed and it was not until the first state was terminated by removal of the glumes that the second state could also be overcome.

By contrast in *Panicum turgidum*, where the dispersal unit is a single floret surrounded only by a hard shiny lemma and palea, alternating temperatures and light stimulated germination in a similar manner to Sabi panicum, but the surrounding structures had no influence except to limit the rate of water uptake (Koller & Roth, 1963).

In *Brachiaria decumbens* removal of the large sterile glume and of the lemma and palea failed to improve germination in newly matured seed (Whiteman & Mendra, 1982). However in older seed removal of only the glume improved germination from 7.5 to 15%. The further removal of both the lemma and palea increased germination to 97% showing that the glume, lemma and palea together had a major influence on dormancy. The authors concluded that the glume alone had a negligible role due to slight injury to the lemma and palea when the glume was excised. The lemma and palea had no chemical influence in delaying germination. When hulls were placed beside naked caryopses there was no influence on germination. It was concluded that oxygen access to the caryopsis was restricted by the lemma and palea.

The above kinds of evidence suggest that the true glumes alone can have some role in influencing the condition of dormancy in only a few species. *By contrast the lemma and palea have been demonstrated as causal agents in dormancy in almost every study of seed dormancy in grasses.*

Before considering the wealth of experimental evidence available in the literature showing that the lemma and palea have important effects on caryopsis dormancy, it is worthwhile reflecting on the ways in which appendages surrounding a caryopsis might, in theory, influence germinability. It is not possible, experimentally, to observe the effect of the lemma and palea on a single caryopsis because there is no basis for comparison of the hulled and dehulled condition. Estimation of the influence of hulls on germination must, of necessity, be made by comparing populations of hulled and dehulled seeds that reflect uniform genetic backgrounds and similar environmental effects. It is possible that hulls influence caryopsis germinability in at least three ways. They could have a positive effect, a negative effect, or no effect. Only two states of germinability can be recognized experimentally. Germination is either 'on' or 'off'. Thus in a population of caryopses that begin in a completely dormant state and move to a state of non-dormancy after some period of time, there can be six explanations for the outcome of a germination test when hulls surround the caryopsis; eleven explanations after the hulls are removed (Table 2.1).

When hulls are removed from the caryopsis the failure of the latter to germinate could be due to one of the following five possibilities: (1) The caryopsis was inherently germinable but did not germinate because of the previous presence of the hulls. (2) The previous presence of hulls had no influence on germinability but the innate dormancy of the caryopsis prevented germination. (3) There was no germination because of the combination of inherent embryo dormancy with a persistent inhibitory effect from the previous presence of the hulls. (4) The previous presence of the hulls reinforced the inherent caryopsis dormancy but the reinforcement was removed by the removal of hulls. (5) The hulls overcame the inherent caryopsis dormancy while surrounding the caryopsis but this was renewed on removal of the hulls.

Similar hypothetical possibilities can be listed for the role of the coat (pericarp and/or testa) in interacting with embryo dormancy. In both cases possibilities arise that multiply the explanations behind any observation that is only measurable as an 'on' or 'off' state commonly called germination. It is presumably for these reasons that interpretation of causality in experiments with grass seed dormancy has been so difficult. Thus, experiments designed to determine the role of hulls in grass seed dormancy should clarify which of the theoretical possibilities for interaction between a hull and caryopsis actually underlays the observed germination response. For example, it might seem rather unlikely that hulls actually prevent dormancy inherent in the caryopsis. A first assumption is that hulls probably restrict rather than promote germinability. Because the majority of experiments comparing the germination of hulled and hulless grass seeds have indicated that dehulling improves germination, the assumption is generally made that hulls **cause** dormancy. Table 2.1 indicates that it is conceivable that hulls simply **enhance** dormancy or

Table 2.1. *Ways in which hulls could potentially influence germinability of both dormant and non-dormant caryopses. The outcome is recorded as germination (+) or non-germination (−).*

	Physiological effect of hulls			
State of the caryopsis	When surrounding the caryopsis	Germ.	After removal	Germ.
Dormant	No effect	(−)	No effect	(−)
	Enhancement of non-germinability	(−)	Persistent	(−)
			Non-persistent	(−)
	Reduction of non-germinability	(+)	Persistent	(+)
			Non-persistent	(+)
			or	(−)
Non-dormant	No effect	(+)	No effect	(+)
	Enhancement of non-germinability	(−)	Persistent	(−)
			Non-persistent	(+)
	Reduction of non-germinability	(+)	Persistent	(+)
			Non-persistent	(+)

alternatively even overcome, or prevent, dormancy within a caryopsis. The remainder of this section considers evidence that might support the argument for a range of potential roles of the lemma and palea in determining caryopsis dormancy.

It has been suggested that covering structures can exert an influence on caryopsis germinability and dormancy in a number of ways (Bewley & Black, 1982). They could:

(a) interfere with water uptake and/or gas exchange;
(b) supply chemical inhibitors, or prevent the leaching of inhibitors from the caryopsis;
(c) modify the response of the caryopsis to light;
(d) provide a mechanical restraint to an emerging root or shoot;
(e) produce combinations of two, or more, of the above effects.

The above possibilities could be modified in many grass species because in some cases the covering structures adhere so tightly to the pericarp of the caryopsis that they form a restrictive continuum. In others there is a loose arrangement and gas and moisture exchange would be less restricted. In still others, that have naked caryopses following separation from the parent plant, hull effects would be confined to the period of seed maturation on the parent plant.

Mutual influences of inflorescence and caryopsis parts

To fully interpret the influence of hulls on caryopsis germination it is important to consider not just the influence of hulls during the time of germination but also during the period of development of the caryopsis. It is conceivable that the factors described above can exert an influence at every stage of caryopsis development. Some of the changes brought about in the caryopsis could well persist even in the absence of the hulls.

Because lemma and palea are covering structures, the potential range of their influence is similar to that attributed to the pericarp. They undoubtedly reinforce each other and in the intact floret their individual contributions may be difficult to separate.

The literature was searched for experiments in which the lemma and palea were removed by either a surgical operation, or by a process such as gentle rubbing, that did not involve injury to the caryopsis. The many citations in which mechanical or sulphuric acid scarification were employed to remove hulls were omitted on the grounds that in most cases scarification extends also to the pericarp which has a role separate from, but not dissimilar to, the hulls (see section 2.3). The many observations that scarification overcomes some, or all, dormancy in many grass species nevertheless reinforce the general belief that together the hulls and pericarp play an important role in determining dormancy of grass seeds.

The effect on germination of removing the lemma and palea from the floret after seed maturation is summarised for 25 species that normally exhibit a high degree of dormancy at the time of seed maturation (Table 2.2). In the majority of cases the seed was probably after-ripened for some time until part of the population had lost some dormancy. That is, special treatments such as alternating temperatures, or pre-chilling known to overcome hull-imposed dormancy, were not used.

Several general observations can be made:

(a) Removing the lemma and palea gave 100% germination in only one of the 54 comparisons. In the other cases hull removal always left a portion of the population that failed to germinate without some further treatment. When treatments with and without hulls are compared, removal of hulls improved germination from 25 to 54% on average, leaving nearly one half (46%) of the population still dormant.

(b) The comparisons among strains, or cultivars, within a species indicate a great deal of variation in level of dormancy and degree of response to de-hulling. Because of the differences in experimental conditions in most cases, it was not possible to conclude how much of this variation was genetic and how much was environmental.

Table 2.2. *The effect of removing the lemma and palea on embryo germination, in 25 species of grasses. Cultivar and accession signatures are omitted. In some cases germination was tested in both light (L) and darkness (D). In most examples the dispersal units were after-ripened for some time before the germination test. Reference source in brackets.*

Species	Per cent germination: Intact floret	Per cent germination: Naked caryopsis	Dark/Light
Aegilops kotschyii (Wurzburger *et al.*, 1974)	5	84	D
Agropyron repens (Williams, 1968)	3	25	D
Aristida contorta (Mott, 1974)	4	40	L
Bothriochloa ischaemum (Ahring *et al.*, 1975)	15	76	D/L
Bouteloua curtipendula (Major & Wright, 1974)			
Dormant strain	3	90	D
Less dormant strain	59	92	D
Brachiaria decumbens (Whiteman, 1982)	3	63	D/L
Brachiaria ruziziensis (Renard & Capelle, 1976)	30	80	L
Chrysopogon latifolius	11	36	D
(Mott, 1978)	21	43	L
Chrysopogon fallax	14	31	D
(Mott, 1978)	20	66	L
Cynodon dactylon (Ahring & Todd, 1978)	56	67	D/L
Dactylis glomerata (Juntilla, 1977)			
Sub-arctic strains	14	39	L
	11	52	L
Northern European	12	57	L
	55	59	L
	42	42	L
	84	81	L
	72	60	L
	60	52	L
N. American strains	70	85	D/L
(Canode *et al.*, 1963)	55	71	D/L
English strain (Probert *et al.*, 1985c)	31	55	D

Table 2.2. *(cont.)*

Species	Per cent germination		Dark/Light
	Intact floret	Naked caryopsis	
Danthonia californica			
(Laude, 1949)	0	6	D
Dichanthium linkii			
(Lodge & Whalley, 1981)			
Australian strains	22	63	D/L
	41	78	D/L
	22	26	D/L
	26	23	D/L
	9	75	D/L
	2	54	D/L
	0	1	D/L
Dichanthium sericeum	41	78	D/L
(Lodge & Whalley, 1981)			
Echinochloa turnerana	2	100	D
(Conover & Geiger, 1984a)			
Hordeum glaucum	47	62	D
(Popay, 1981)			
Oryza punctata			
(Sano, 1980)			
Diploid-annual	0	16	D
Tetraploid-perennial	20	94	D
Oryza sativa	4	56	D
(Roberts, 1961)	12	60	L
Six cultivars			
(Tseng, 1964)			
A	60	53	D
B	47	49	D
C	22	83	D
D	35	89	D
E	56	95	D
F	23	64	D
Oryzopsis hymenoides	8	23	D
(Young & Evans, 1984)			
Setaria lutescens			
(Rost, 1975)	0	30	D
Sorghum plumosum	7	54	L
(Mott, 1978)			
	7	58	D
Sorghum stipoideum	4	54	L
(Mott, 1978)	2	35	D
Stipa viridula			

Table 2.2. *(cont.)*

Species	Per cent germination Intact floret	Naked caryopsis	Dark/Light
(Fendall & Carter, 1965)			
Dormant strain	8	65	D
Non-dormant strain	64	97	D
Themeda australis	7	33	L
(Mott, 1978)	14	30	D
Zizania aquatica	48	72	D/L
(Cardwell *et al.*, 1978)			

The comparisons among strains of *Dactylis glomerata* indicate very significant differences in response to dehulling that depend on the geographical origin of the strains (Juntilla, 1977). Comparisons among strains of *Oryzopsis hymenoides* grown under uniform conditions (Young & Evans, 1984) indicate that there is genetic polymorphism with respect to hull effects on dormancy and hull colour. The average effect of dehulling was to improve germination from 7 to 17%, and for all the different seed lots of the two different cultivars at least 50% of the seed failed to respond to a combination of dehulling, GA treatment and cool-moist stratification for four weeks.

(c) The effects of hulls on per cent germination are associated with, and only discernible at, the time when a separate state of dormancy of the caryopsis is disappearing within the population of seeds. That is, a fully non-dormant population of seeds with intact florets may exhibit delays of varying periods only up to a few hours, in the time to achieve 100% germination, compared to naked caryopses. However, there is no reduction in the final per cent germination achieved. Thus, hulls appear to interact with a part of the population by either preventing the loss of existing dormancy in the caryopsis, or by reinforcing an existing state of caryopsis dormancy, that is easily lost once hulls are removed.

(d) None of the comparisons indicate a significant reduction in germination due to removal of the hulls.

Traditional explanations for the role of covering structures of seeds, such as the lemma and palea, have been categorized into five possibilities

(Bewley & Black, 1982):

(a) Mechanical restraint on germination.
(b) Modification of a light requirement for the embryo.
(c) Inhibitors that interfere with germination.
(d) Prevention of water uptake.
(e) Impedance of gas exchange.

In not one case among the 54 comparisons made in Table 2.2 was dormancy attributed to mechanical constraint of germination by the hulls. The effects of hulls on preventing germination were evident in both light and dark germination, indicating that the action of hulls does not *per se* depend on light responses in the embryo. Of nine investigations specifically designed to determine whether germination inhibitors were present in the hull, five (Burton, 1969; Wurzburger, Leshem & Koller, 1974; Renard & Capelle, 1976; Landgraff & Juntilla, 1979; Whiteman & Mendra, 1982) found no evidence at all. Two concluded that there might be inhibitors. One of these cases was based on the observation that leaching increased germination (Smith, 1971) and the other that increased germination in the presence of increased oxygen partial pressure could be due to oxidation of an inhibitor (Frank & Larson, 1970). In neither case was there any concrete evidence for an inhibitor. Two investigators reported inhibitors in hulls. In the first, *Dactylis glomerata*, the authors (Fendall & Canode, 1971) found methanol- and ethanol-soluble inhibitors of lettuce seed germination that occurred in the ratio of one third in the hulls and two thirds in the caryopsis. Nevertheless there was no evidence that they influenced germination of *D. glomerata*. In *Bothriochloa* spp., the second case, there was an ether-soluble inhibitor that affected seedling growth, but not germination, of the same species (Ahring, Eastin & Garrison, 1975). *Thus these reports show that the case for involvement of growth inhibitors from hulls in caryopsis dormancy is not yet established.*

On the other hand evidence for the role of hulls in restricting water transport and gas exchange in caryopses is much stronger. In grasses the embryo is situated close to the point of attachment of the floret to the rachilla. In dispersal units that retain the lemma and palea (Fig. 2.5) these bracts press tightly around the region of the embryo, even when the distal region of the floret is open or loose, because of the close proximity of the points of attachment of lemma, palea and enclosed caryopsis. Whether hulls adhere tightly to the pericarp (e.g. rice and barley) or are loose (e.g. wild oat) the embryo region is protected from mechanical injury and to varying degrees from the direct effects of environmental fluctuations of moisture and gases.

The question of the degree of permeability of hull tisssue to water and gases has not been well investigated. It is also surprising that comparatively few studies have ever been made of the anatomy of the lemma and palea in relation to their possible role as barriers to water and gas exchange.

A number of studies with grass seeds indicate that hulls can play a significant role in restricting the availability of oxygen to the embryo in intact florets. In *Aristida contorta*, the hulls adhere tightly to the caryopsis but germination can be stimulated in dormant seeds either by application of oxidizing agents or simply by removal of small sections of the lemma directly adjacent to the embryo (Mott, 1974). Oxygen uptake by a dormant floret is low whereas in dehulled dormant caryopses, or intact non-dormant florets, it is much higher. There is evidence for a similar role of hulls in restricting oxygen availability to the caryopsis in *Brachiaria ruziziensis* (Renard & Capelle, 1976). In both the above species restriction of water uptake to the caryopsis was not a significant factor in the action of the hulls on dormancy. Dissection of a small portion of the lemma adjacent to the embryo stimulated germination of *Aegilops kotschyii* (Wurzburger *et al.*, 1974).

In another species, *Brachiaria decumbens*, there are two sources for the dormancy expressed by caryopses (Whiteman & Mendra, 1982). There is a primary form of dormancy present in the caryopsis at maturity that persists for about 3 months that is overcome by after-ripening. There is also another longer-term dormancy caused by the inhibition of oxygen diffusion to the embryo because of the closely appressed, hard, shiny lemma and palea. Removal of the lemma and palea gave 97% germination, compared with 8% in the intact floret. Soaking in hydrogen peroxide stimulated germination, whereas prolonged soaking in water actually decreased germinability.

In green needlegrass (*Stipa viridula*) about 50% of the dormancy was attributable to the presence of the lemma and palea (Fendall & Carter, 1965). When the lemma was dissected away both water uptake and oxygen uptake increased. In dormant intact florets water uptake was slower than in non-dormant florets. Removal of the hulls improved both rate and total uptake of water in both dormant and non-dormant florets. Oxygen uptake increased in both dormant and non-dormant seeds when the lemma was removed. In intact dormant florets the oxygen uptake was linear and there was no second phase of increased water uptake characteristic of both intact and dehulled non-dormant seeds. When the hulls were removed from dormant florets the oxygen uptake remained linear, but the rate of uptake increased significantly (Fig. 2.6). The authors concluded that the lemma and palea are mechanical barriers to oxygen uptake due to their impermeable nature and the tightness with which they invest the caryopsis.

A separate study on the same species, *Stipa viridula*, compared dehulling, scarification with sodium hypochlorite, and increased oxygen partial pressure as methods for overcoming hull-imposed dormancy (Frank & Larson, 1970). These authors also concluded that the lemma and palea are barriers to gas exchange. Each of the above studies further concluded that the caryopsis had requirements for after-ripening that were distinct from the hull effects, suggesting two separate components of floret dormancy.

Possible mechanisms with which the lemma and palea could restrict availability of oxygen to the caryopsis are suggested by other studies with grasses. There have been very few detailed studies of the anatomy of either the lemma or palea in grasses, in relation to their role in grain dormancy. In one of these rather rare investigations on *Aristida contorta* (Mott & Tynan, 1974), previously demonstrated to have hull-restricted oxygen availability in dormant florets, the authors found that the inner epidermis of the lemma, which completely surrounds the caryopsis and vestigial palea, was marked by a lipid-containing layer. This layer was intact over the whole surface of the lemma in dormant grains but was fractured, and therefore permeable, in

Fig. 2.6. Oxygen uptake of dormant and non-dormant seeds of *Stipa viridula* during the first 70h of soaking at 20°C. Per cent germination after 10 days is shown at the end of each curve (After Fendall & Carter, 1965). Reproduced from *Crop Science* vol. 5, 1965, pages 533–36, by permission of the publisher Crop Science Society of America, Inc.

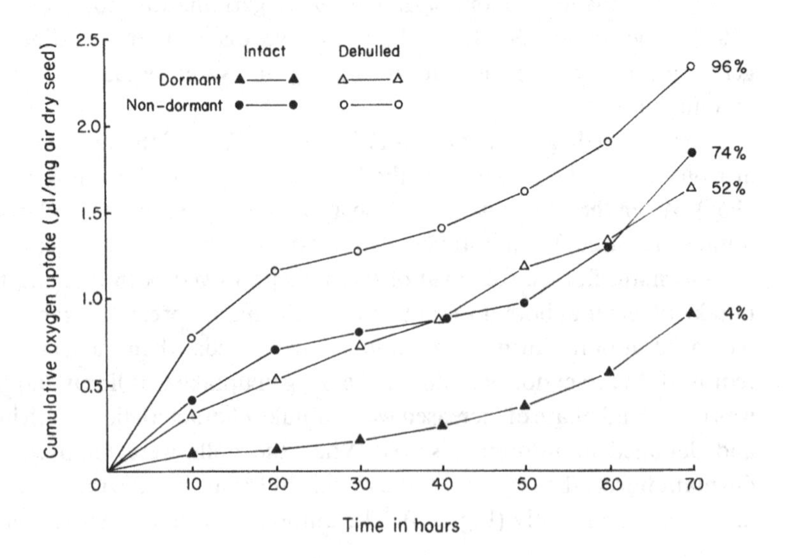

non-dormant grain. The siliceous surface layer that contributes to the density and hardness of the lemmae was similar in dormant and non-dormant florets. The lipid layer retained its integrity when dormant florets were imbibed in water. By contrast the lipid layer in non-dormant florets was very broken with large fractures in the area of the lemma covering the embryo and the proximal region of the lemma. The authors concluded that the lipid layer constitutes a major barrier to oxygen exchange, and therefore uptake by the caryopsis, in dormant florets. Thus regardless of the condition of dormancy or non-dormancy within the embryo, until the lipid layer is broken by the action of alternating temperatures, high temperatures, or mechanical damage, limitations on oxygen exchange, and possibly water exchange, exist in the hulls. In many species the limiting effects of hulls on germination often last longer than the states of dormancy within the caryopsis. Thus the lipid layer, if it is present in other grasses, may be quite persistent. Clearly there is a need for further anatomical investigations of this nature in other grass species to determine whether the lipid layer occurs commonly among grass species.

A second mechanism by which hulls could interfere in oxygen uptake by caryopses has also been suggested by the authors of several investigations on the role of hulls in rice (*Oryza sativa*) grain dormancy (Navasero, Baun & Juliano, 1975; Kuo & Chu, 1982). Although rice hulls at maturity are essentially dead tissue they contain residual peroxidase activity. Following soaking of the intact floret on water the peroxidase activity in the hull competes with the caryopsis for available atmospheric oxygen (Navasero *et al.*, 1975). The hulls slow down the rate of diffusion of oxygen to the embryo. Both after-ripening and heat treatments, which break hull-imposed dormancy in rice, significantly reduced the level of peroxidase in the hulls but not in the dehulled caryopsis or embryo (Table 2.3).

In another study with rice (Kuo & Chu, 1982), stratification and after-ripening at low temperatures (which reduced dormancy) had no influence on the differences in peroxidase activity between dormant and non-dormant grains. Nevertheless, the authors concluded that the hull peroxidase activity could interfere with the availability of oxygen to the embryo.

It is possible that the specific levels of peroxidase activity may not be significant in themselves, but rather that the timing of oxidase activity is critical for preventing germination through interference in oxygen availability. For example, in barley hulls (*Hordeum vulgare*) no differences were found in the levels of peroxidase activity among dormant and non-dormant florets (Lenoir, Corbineau & Côme, 1986). An increase in oxygen partial

Table 2.3. *The influence of after-ripening and heat treatment on peroxidase activity in rice (*Oryza sativa*) florets harvested at maturity. (After Navasero* et al, *1975: with permission of the Editor of* Phytochemistry.*)*

	Peroxidase activity (oxygen uptake (nmol/hr))			
	Intact floret	Hulls	Caryopsis	Embryo
Dormant floret	20.9	13.4	24.8	11.5
After-ripened floret	14.6	7.1	29.3	11.0
Heat-treated floret (4 wk at 28–30°C)	13.6	8.9	23.0	10.7

pressure did not break dormancy. Nevertheless in dormant seeds oxygen uptake occurred immediately on imbibition of the seeds on water, whereas there was a delay of nearly 10 h in non-dormant seeds. This lag in activity could be related to a time-lag in water uptake caused by the presence of a lipid layer of the type described by Mott & Tynan, 1974. Thus competition for oxygen in the early phase of imbibition of water may be important in determining the nature of metabolic activity. In turn this activity may enhance an existing state of embryo dormancy, or induce a form of secondary dormancy (see later section in this chapter).

Lipoxygenase activity has been reported in the hulls of both dormant and non-dormant oats (*Avena sativa*) (Lascorz & Drapron, 1987b). This provides further support for the presence of enzymatic activity in tissues previously considered to be inert. The activity persisted for many weeks. Nevertheless the evidence from oats does not indicate that hulls in dormant florets starve the caryopsis for oxygen because the levels of lipoxygenase were similar in hulls of both dormant and non-dormant caryopses. Obviously further studies are needed to determine whether embryonic tissues in dormant caryopses have a higher demand for oxygen at the initiation of germination than non-dormant embryos. If this were so then any competition for oxygen by surrounding tissues could persist until oxygenase activity disappeared with after-ripening.

All of the above studies reinforce the conclusion that any influence of the hulls on dormancy is expressed through their influence on embryonic activity. If the embryo is already in a state of dormancy at maturity, the effects of hull limitations cannot be detected until this embryonic dormancy is lost. The question of whether the hulls cause a prolongation of this original embryo dormancy, or introduction of a secondary dormancy, will be answered in the later sections of this chapter.

2.3 The grain coat – pericarp and testa

The caryopsis, the only type of fruit found in the Gramineae, is a dry monospermic indehiscent fruit and the pericarp varies in thickness and degree of adherence to the seed-coat (Sendulsky, Filgueiras & Burman, 1986). Generally the pericarp is thin and rarely fleshy but in some bamboo genera (*Melocanna, Dinochloa, Ochlandra, Olmeca* and *Melocalamus*) there is a thick fleshy pericarp which produces a 'berry-like' caryopsis. For example, in *Melanocanna bambusoides* the cells of the pericarp have thickened membranes packed with food material and the true endosperm is a mere relic, and the scutellum is very large, filling the entire cavity within the pericarp. This exceptional fruit may reach 13 cm in length and 8 cm in breadth (Arber, 1934).

In grasses the ovary wall is nearly always partially resorbed and only two cell layers remain forming the pericarp. This fuses with the testa of the seed (Bor, 1960). In a few grasses it is possible to separate pericarp and testa: in *Eleusine* there is only a thin pellicle representing the outermost layer of the pericarp and in *Crypis* and *Sporobolus* it is transformed into mucilage. The surface of the pericarp of most grasses is smooth and glossy. The pericarp and testa together are considered to have an important role in affecting the emergence of embryos from dormancy (Werker, 1981).

Example of Avena fatua:

In *A. fatua* the grain coat forms the surface of the caryopsis (Fig. 2.7*(a)*) and is composed of a single or double layer comprising the thin-walled pericarp that presses tightly against the underlying testa (Morrison & Dushnicky, 1982; Cairns & de Villiers, 1986a) (Fig. 2.7*(b)*). The testa has an inner and an outer cuticle that are present over the entire caryopsis except in the area adjacent to the embryo. In 1906 Crocker showed that pricking the outer layers of the caryopsis promoted germination of dormant seeds. Later Atwood (1914), in a classical and frequently cited paper, concluded that stimulation of germination by piercing the seed-coat was not related to promotion of water uptake but rather to an increase in availability of oxygen for the embryo. His conclusion was based on the observations that increased germination by excision of the embryos, or from increases in oxygen partial pressure, and increased respiration of embryos following pricking of the coat, were all consequences of increased availability of oxygen.

Since these early observations, there have been at least 25 more reports indicating that pricking the seed coat, particularly adjacent to the embryo, promotes germination of dormant seeds. The effectiveness of the treatment depends on the genotype of the material. Genetically distinct lines with considerable dormancy do not respond as effectively, or as rapidly, as less dormant lines (Thurston, 1963; Foley, 1987; Raju, Hsiao & McIntyre, 1988). In genetic lines that have deep embryo dormancy, piercing the seed-coat alone is not sufficient to break dormancy (Simpson, 1978; Cairns & de Villiers, 1986a; Foley, 1987); in these strains excised embryos still exhibit dormancy.

Chemical scarification of the seed coat can simulate the effects of piercing. For example, sodium hypochlorite breaks both primary (Hsiao, 1979; Hsiao & Quick, 1984) and secondary dormancy (Hsiao & Quick, 1985) more effectively than a powerful oxidizing agent like hydrogen peroxide. The authors concluded that the effect was due to membrane breakage leading to enhanced permeability to oxygen. Treatment with ammonia gas (Cairns & de Villiers, 1986a) or aluminum phosphide (Cairns & de Villiers, 1980) promoted germination. A principle effect of these

Fig. 2.7. A caryopsis of *Avena fatua*. *(a)* Entire caryopsis. *(b)* Cross-section of integuments (After Morrison & Dushnicky, 1982).

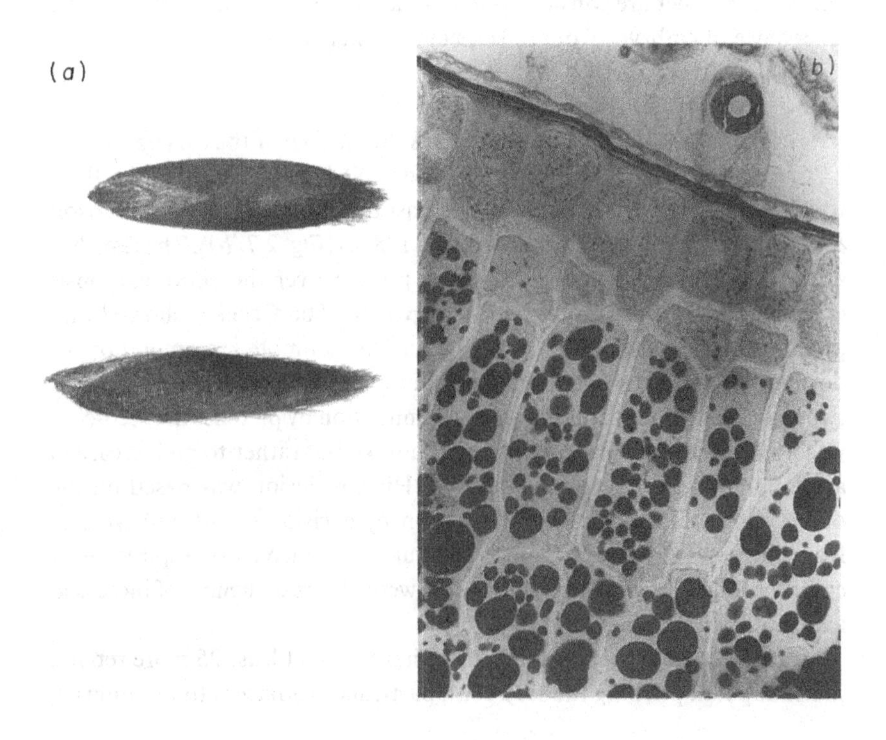

treatments was an increase in 'leakiness' of ions indicating membrane damage. Pricking the seed coat adjacent to the embryo was as effective as breaking the dormancy with potassium nitrate (Johnson, 1935b). Scarification with a simple mechanical abrasion also promotes germination of dormant seeds (Watkins, 1969).

Most of the observed increases in germination following damage to the seed coat have been attributed to increased permeability to oxygen and in turn enhanced availability of oxygen to the embryo. However, none of the experiments were refined enough to prove these points definitively. For example, there have not been any reports on the specific permeability to oxygen of seed-coat tissues in either dormant or non-dormant grains.

In recent years an alternative explanation has emerged based on the relative permeability to water of seed-coats in dormant and non-dormant caryopses. This, despite earlier conclusions by several authors (Atwood, 1914; Quick & Hsiao, 1983) that the seed coats and aleurone layers are not in themselves barriers to the uptake of water.

The basis for believing that at least part of the germination promotion following piercing of the grain coat is related to increased availability of water for the embryo is as follows. When non-dormant caryopses are imbibed on water the first response of embryonic tissues is an expansion of the root, followed later by an expansion of the scutellum (Raju, Hsiao & McIntyre, 1986). When the seed coat of a dormant caryopsis is pierced in a position close to the embryo the same sequence of tissue expansion occurs. However, when the coat is pierced at some distance from the embryo but immediately above the scutellum the order of tissue expansion is reversed: the scutellum elongates and papillae develop between 48 and 72 h prior to any root expansion. The authors concluded that the localized wounding increases permeability to water in that region only and the delay in movement to other regions imposes restrictions on expansion of other tissues. They infer that an intact seed coat prevents the development of a sufficiently negative water potential to support expansion growth of the embryo. The effect of after-ripening, which breaks dormancy, was similar to piercing the seed-coat immediately adjacent to the embryo axis; the root emerged well in advance of any expansion of the scutellum. Embryos excised from these dormant imbibed grains take up further water within 30 min and germinate after 18 h.

Recent work in the same laboratory with genetically distinct lines (Raju *et al.*, 1988) indicates that the ability of the embryo to develop the necessary solute potential to take-up water is a factor in addition to seed coat permeability. Comparison of these lines in order of their germinability indicated that even after excision of the embryo, when restriction of water availability is circumvented, the order of germination events is still the same as with primary dormancy.

Other grasses:

There are many reports from the literature that the intact caryopsis, freed of the restricting influence of the lemma and palea, still shows varying degrees of dormancy according to genotype, environment during maturation, and period of after-ripening after maturity (Table 2.2). Scarification of intact florets, either by mechanical abrasion or sulphuric acid, has been widely reported to overcome dormancy in grass seeds. However prolonged scarification can remove both the lemma and palea restrictions and break the integrity of the grain coat. In many publications about the influence of scarification on grass seed dormancy it is impossible to distinguish between these hull and coat effects. For this reason in the following discussion only those papers that describe the effects of scarification, pricking, piercing and cutting of **intact, naked, caryopses** will be considered in relation to the role of the caryopsis coat in dormancy.

There are many examples of the stimulatory effect on seed germination of disrupting the integrity of the caryopsis coat (Table 2.4). Several general conclusions can be drawn from this Table. In only one case (*Poa pratensis*) was puncturing the coat deleterious to germination. In all other examples damage to the coat produced a significant increase in germination. The magnitude of the response varied according to the condition of the seed at the time of treatment and with the origin of the accession or cultivar. In many cases the intact naked caryopsis was completely dormant before puncturing, in other cases the removal of the hulls before puncturing had clearly promoted germination. Only 30% of the puncturing treatments achieved a germination greater than 90%; in the remainder the embryo still retained some degree of dormancy. *Thus, it can be concluded that coat-imposed dormancy may be a general phenomenon in grasses. There can be considerable genetic and environmental variation in expression and the effect eventually disappears following a period of after-ripening.*

Explanations for the promotive effect of damaging the grain coat, among the papers cited in Table 2.4, fall into several groupings. In many cases there appeared to be negligible differences in the rate and total amount of water taken up by the caryopsis and the conclusion was made that improvement in gas exchange, rather than water status, was the reason for enhanced germination. In other cases it was argued that improvement of the water status of the embryo was the key reason for improvement. There were very

Table 2.4. *The influence of puncturing the pericarp and testa on overcoming dormancy in 31 grass species. The conditions for germination are ommitted. In the majority of examples the tissues were cut with a needle or scalpel. In a few cases scarification was by mechanical abrasion or with chemicals. Cultivar and accession signatures are ommitted. Reference source in brackets.*

Species	Accessions	Per cent germination	
		Control	Punctured
Agropyron elongatum			
(Thornton, 1966a)			
(111 samples over 4 yrs)		75	90
Agropyron smithii			
(Kinch, 1963)	A	44	63
(Bass, 1955)	B	16	19
	C	56	91
(Delouche & Bass, 1954)			
	D	5	72
	E	16	72
	F	11	90
	G	30	86
	H	33	92
Bouteloua curtipendula		4	45
(Major & Wright, 1974)			
Buchloe dactyloides		49	73
(Thornton, 1966b)			
Chloris orthonothon		0	40
(Solange *et al.*, 1983)			
Dactylis glomerata			
(Chippindale, 1933)	A	40	53
(Probert *et al.*, 1985c)	B	55	98
Danthonia californica			
(Laude, 1949)	A	0	81
	B	0	71
	C	0	21
Digitaria milanjiana		49	92
(Baskin *et al.*, 1969)			
Digitaria pentzii		49	91
(Baskin *et al.*, 1969)			
Digitaria sanguinalis		0	84
(Giangfagna & Pridham, 1951)			
Digitaria adscendens			
(Shimizu, 1959)		0	60
Distichlis spicata			
(Amen, 1970)		0	75

Table 2.4. *(cont.)*

Species	Accessions	Per cent germination	
		Control	Punctured
Echinochloa crus-galli			
(Shimizu *et al.*, 1974)		0	55
Echinochloa turnerana			
(Conover & Geiger, 1984a)		28	100
Eragrostis lehmanniana			
(Haferkamp *et al.*, 1977)	A	6	61
(Wright, 1973)	B	0	40
Hordeum vulgare		35	100
(Dunwell, 1981a)			
Oryza sativa			
(Delouche & Nguyen, 1964)	A	6	72
(Roberts, 1961)	B	12	96
Panicum anceps			
(Mathews, 1947)		0	56
Paspalum notatum			
(Mathews, 1947)	A	11	29
(Burton, 1939)	B	0	32
Pennisetum typhoides		21	82
(Burton, 1969)			
Phalaris arundinacea		10	86
(Juntilla *et al.*, 1978)			
Poa pratensis			
(Delouche, 1958)		83	65
Setaria lutescens			
(Rost, 1975)		30	70
Sorghum bicolor			
(Gritton & Atkins, 1963a)	A	25	74
	B	42	86
(Ahring, 1963)	C	90	95
	D	79	98
	E	44	99
Sorghum halapense			
(Harrington & Crocker, 1923)			
	A	0	65
	B	0	44
(Huang & Hsiao, 1987)	C	26	76
Sorghum vulgare			
	cv.sudanense	0	98
(Harrington & Crocker, 1923			
	cv.saccharatum	42	99
(Gaber *et al.*, 1974)			
Stipa viridula			
(Av. of six accessions)		45	78

Table 2.4. *(cont.)*

Species	Accessions	Per cent germination	
		Control	Punctured
(Wiesner & Kinch, 1964)			
(Fendall & Carter, 1965)		8	63
Triticum aestivum			
(Harrington, 1923)	A	56	100
	B	64	100
	C	0	16
Uniola paniculata			
(Westra & Loomis, 1966)		0	100
Zizania aquatica			
(Simpson, 1966a)	A	4	92
(Cardwell *et al.*, 1978)	B	0	55

few cases of evidence for the leaching away of inhibitors from the seed-coat. It is noteworthy that in only one or two examples, attempts were made to investigate the anatomy of the pericarp and testa and then to relate these intact and damaged structures to the germinability of the embryo. It is from these latter investigations, coupled with others in which the caryopsis was pierced at selected positions in relation to the embryo, that some insight can be gained into the role of the grain coats in dormancy.

Schroeder (1911) discovered that there is a semi-permeable membrane situated in the coat of the wheat caryopsis that only permits the entry of solutes and water in the area adjacent to the embryo. In barley, cuticular membranes are barriers to the penetration of water through the coat (Collins, 1918). In *Lolium perenne* there are membranes that give a cuticular staining reaction indicating lipids (Brown, 1931). Solute penetration from outside the coat can only penetrate as far as the membranes which are attached to the outer surface of the caryopsis (Fig. 2.8*(a)*). A heavily suberized outer integumentary layer occurs in *Agropyron elongatum* (Thornton, 1966a) and also in *Buchloe dactyloides* (Thornton, 1966b). In *Sorghum halapense* the outer integument is absent but the heavily suberized inner integument is fused with the pericarp to form an impermeable membranous layer investing the whole caryopsis (Harrington & Crocker, 1923). This layer is extremely resistant to the action of both sodium hypochlorite and chromic acid. Penetration of the membrane by solutes occurs largely at the proximal end of the caryopsis adjacent to the embryo. The thickest portions of the membrane are over the hilar and micropylar

Fig. 2.8. Transverse sections of seed coats from several grasses. *(a) Lolium perenne* (After Brown, 1931). (b) *Agropyron elongatum* (After Thornton, 1966a). *(c) Buchloe dactyloides* (After Thornton, 1966b).

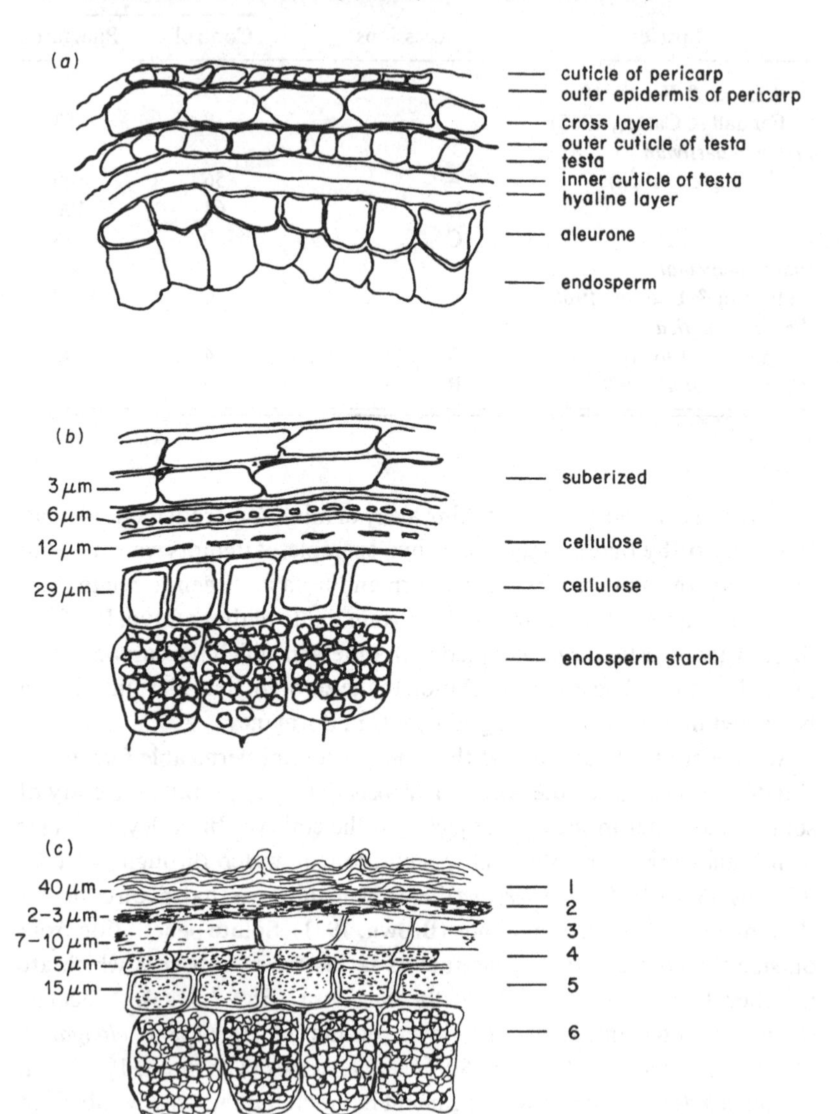

regions and in the particular case of *S. halapense* at the point of radicle emergence. From this rather detailed study Harrington hypothesized that the fused pericarp and integument layer, impregnated with lipid, decreases permeability to solutes and physically limits the imbibitional swelling of embryos so that their water content is insufficient for initiation of germination.

Other observations on water uptake into both *Lolium perenne* (Brown, 1931) and *Hordeum vulgare* (Collins, 1918) indicate that the earliest absorption of water occurs in the basal region, penetrating first through the furrow on the ventral surface and then spreading diagonally to the dorsal surface adjacent to the scutellum and then the embryo. It was noted in *Lolium* that where fungal hyphae penetrated the coat in distal parts of the caryopsis, water penetration could occur, but otherwise when the distal region was exposed to water, penetration to the proximal part of the caryopsis was slight due to the impermeability. In this same study it was concluded that absorption of water by dry caryopses with intact membranes can only begin with water penetrating the micropyle. This water softens the adjacent membrane which then swells and thickens from 10 to 30 μm in thickness because of its mucilaginous nature. In turn this swelling stretches the membrane on the dorsal surface, over the embryo, further increasing permeability to water.

Experiments in which the seed coat is pricked or cut at positions proximal or distal to the embryo indicate that germination is poor unless water initially penetrates close to the embryo indicating the poor permeation of water through either the starchy endosperm or along the aleurone cells (Table 2.5).

Pricking the distal end of the caryopsis in *Sorghum halapense* induced the initiation of growth and enzyme secretion in the scutellum before the embryo germinated (Harrington & Crocker, 1923). A similar reversal was noted in *Avena fatua* (Raju *et al.*, 1986). These experiments indicate that both the timing and specific location of water entry into the caryopsis can determine the sequence of tissue responses. Unless the coat is easily permeable to water over most of the surface, or alternatively is preferentially permeable adjacent to the embryo, water uptake by the embryo is insufficient to trigger normal germination. It is possible that the water potential of the tissues surrounding the embryo is more negative than the embryo axis. Unless the solute potential of the embryo is altered, or liquid water is directly available to the surface of the embryo, it cannot derive enough water from the surrounding tissues by the process of simple diffusion. Puncturing the coat directly adjacent to the embryo is one way of

Table 2.5. *The influence of coat puncturing site on dormancy of a grass caryopsis. Numbers indicate per cent germination. Reference source in brackets.*

Species	Puncture site with respect to embryo		
	Distal	Proximal	Directly over
Avena fatua (Raju, Hsiao & McIntyre, 1986)	—	20	100
Oryza sativa (Delouche & Nguyen, 1964)	20	84	98
Oryza sativa (Roberts, 1961)	25	45	53
Zizania aquatica (Cardwell, Oelke & Elliott, 1978)	0	25	55

by-passing the process of diffusion from other competitive tissues so that embryo water uptake is then in direct proportion to its combined solute and matric potentials.

Repeated wetting and drying of caryopses has frequently been used as a method of accelerating the natural after-ripening of dormant grass-seeds (Ellis *et al.*, 1985b). The permeability of seed-coat membranes is probably affected by this treatment through the formation of cracks and zones of increased permeability. In *Dactylis glomerata*, previously soaked and dried seeds take up nearly 70% more water in the first three hours of imbibition than non-soaked seeds (Chippindale, 1933). This increase in initial rate of permeability increased the per cent germination from 42 to 65 (averaged over six accessions) but the total uptake of water, on a dry weight basis, was similar by the time of germination.

The conclusion has often been drawn, in many of the experiments in which cutting the seed coat improved germination, that increased gas exchange and not water availability was the cause of the improvement. In most cases the conclusion was not based on demonstration of increased gas exchange but rather on the observation that there was little difference in water uptake between seeds with intact or damaged coats. Nevertheless, it is possible that in many cases the initial rates of water uptake were quite different in seeds with intact membranes compared to seeds with damaged permeable membranes. For example, old seed of *Echinochloa turnerana* (Conover & Geiger, 1984a) imbibed water much faster than freshly harvested seed. Because the impermeable membrane zone immediately adjacent to the embryo is usually the thinnest compared to either the distal, ventral or hilar zones (Harrington & Crocker, 1923), it is probable that any factors such as after-ripening, scarification, or wetting and drying, preferentially improve the permeability of water to the embryo compared with the tissues of the endosperm or scutellum.

Among the 31 species investigated for the role of coats in dormancy (Table 2.4) only four reports indicated specific effects of enrichment of oxygen in the germination environment. In each case there is an alternative to the conclusion that stimulation of germination was the direct result of improved availability of oxygen to the embryo when coats were damaged. In *Dactylis glomerata* (Juntilla, Landgraff & Nilsen, 1978) increased oxygen partial pressure was claimed to improve germination but no data were provided. The authors also indicated that soaking control seeds for four days in aerated water improved germination to the same extent as supplying oxygen. This suggests a water relationship rather than an oxygen effect. In other experiments with *D. glomerata* (Probert, Smith & Birch, 1985c) enrichment of the atmosphere with oxygen improved germination of intact caryopses from 64 to 76%, compared with air. Nevertheless this was less than the 98% achieved by cutting the coat, implying that an effect on water uptake was more significant than gas exchange. In *Digitaria sanguinalis* (Gianfagna & Pridham, 1951), increasing oxygen partial pressure to 100% had no influence on germination of intact grains. Heavily scarified grain germinated 100% in atmospheres containing oxygen ranges between 40 to 80% but there was no air control to make comparisons with. The authors noted that non-dormant grains took up fractionally more water in 24 h than dormant ones. Thus the improvement in water relations, rather than gas exchange, following damage to the coat could be the explanation. In *Eragrostis lehmanniana* (Haferkamp, Jordan & Matsuda, 1977) increasing oxygen concentration from 10 to 21% actually reduced the germination from 24 to 17% in intact caryopses and made no difference in scarified caryopses (75%). The authors noted that old grains imbibed water faster than new ones and uptake of water was faster in scarified than non-scarified grains. Again, the explanation could be a water relationship and not a gas-exchange improvement from cutting the coat.

Clearly there is a need for new critical evaluations of the structure of grain coats in grasses and their role in the control of water and gas exchange for the embryo. The possibility of inhibitory substances within the coat, or presence in other parts of the caryopsis because of the integrity of the coat, also needs further investigation.

2.4 Embryo and scutellum

The embryo in most grasses is small and with the exception of some bamboos is usually less than half the length of the caryopsis. The most visible structure of the embryo is the scutellum, a shield-like organ, which is generally achlorophyllous during development (Arber, 1934) (Fig. 2.9). Scutella have been classified into four classes on the basis of shape, structure and function (Negbi, 1984): (a) typical in the strictest sense, where the only form of growth during germination is the development of every epithelial cell into a separate papilla involved in the secretion of hydrolases, such as amylases, gibberellin and other hormonal factors. These factors

Fig. 2.9. An embryo of *Avena fatua* (After Raju, Hsiao & McIntyre, 1986). *(a)* Diagram of a sagittal section of a mature embryo. *(b)* Sagittal section of an embryo 36h after commencement of germination. *As*, adaxial scale; *Co*, coleorhiza; *Ep*, epiblast; *Pv*, provascular tissue; *Ra*, radicle; *Rc*, root cap; *Sc*, scutellum.

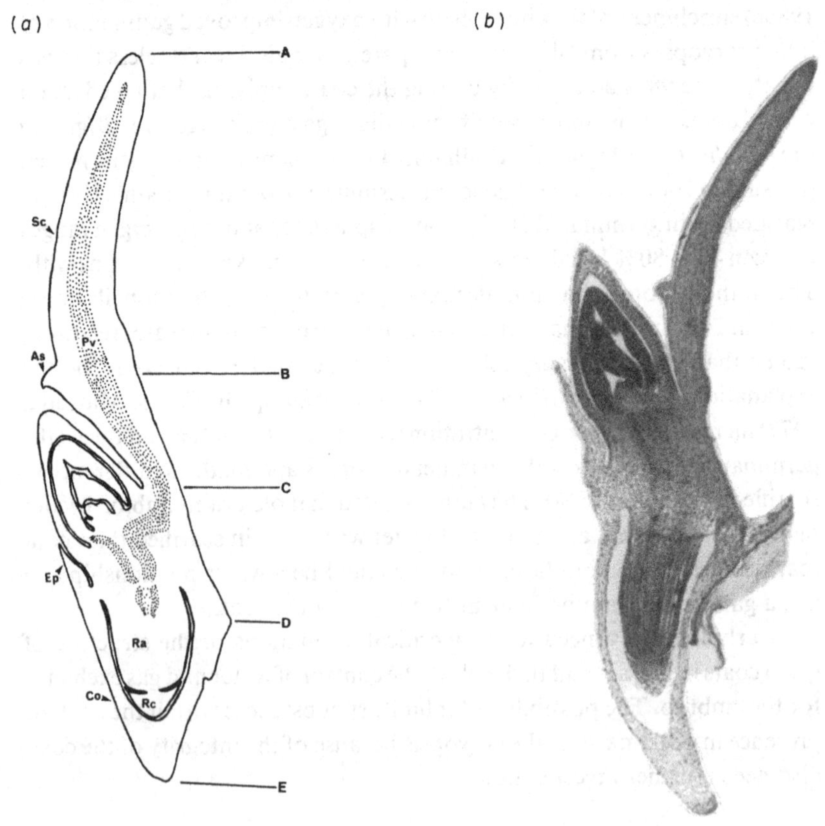

also activate the aleurone layer to mobilize endosperm reserves which are then absorbed by the scutellum. (b) The type characterized by *Avena* and found in several other genera, where the tip of the scutellum elongates during germination to reach the distal end of the endosperm sac and develops papillae over its whole surface. (c) The *Zizania* type where the scutellar tip extends to the distal end of the caryopsis during development but not further during germination. Only the abaxial surface develops papillae. (d) The storage organ type, characterized by *Melanocanna*, replacing typical endosperm which is vestigial. The scutellum constitutes a major part of the embryo.

The function of the scutellum is to act as a digestive organ for utilization of the endosperm reserves and it is connected to the embryo axis by large vascular strands (Fig. 2.9). The role of the scutellum in tranforming monosaccharides into sucrose that is then transported to the germinating embryo has been well described in wheat (Edelman, Shibko & Keys, 1959) and rice and oats (Danjo & Inosaka, 1960).

The developing and mature grass embryo has been described for *Zea*, *Avena* and *Triticum* (Avery, 1930). The grass embryo develops rapidly and may reach maturity in about one third of the time taken for full caryopsis maturation (Anton de Triquel, 1986), which is typically about 30 days. In grasses characterized by dormancy the embryo may be dormant at any time following the beginning of development (Simpson, 1978). There is a clear distinction between ability to continue development and germinability of the embryo. The latter is an interruption of the former, and development continues even in the presence of dormancy. Evidence of non-dormancy in a developing embryo is demonstrated by 'precocious' germination of the type characterized as 'pre-harvest sprouting' in the presence of moisture or a high humidity. True embryo dormancy can be demonstrated by excision of the embryo from the surrounding structures that contribute to part of the expression of dormancy (see sections 2.2 and 2.3). Excised embryos from non-dormant caryopses will germinate normally when placed on water, and at the same temperatures that germinate intact caryopses.

Example of Avena fatua:

The first isolation of embryos from surrounding tissues (Atwood, 1914) indicated a low proportion of dormancy (13%). The experiments were made with a genetically heterogeneous population. Excising the embryos improved germination from 41 to 87% compared with intact caryopses. More recently, following the realization that dormancy can be a genetically inherited trait, a number of reports indicate the existence of true embryo dormancy that is dormancy quite distinct from that imposed by covering structures.

Embryos excised from freshly mature caryopses of a dormant, genetically pure line failed to germinate on water (Naylor & Simpson, 1961a,b). During an after-ripening period of up to three years excised embryos showed increases in both rate of germination and proportion of germinable embryos. Thus embryos recover the capacity to germinate following a period of after-ripening. Excised embryos from very dormant strains fail to germinate on a liquid medium containing sugars, amino-acids and vitamins that supports normal and rapid germination of embryos from non-dormant (after-ripened) caryopses of the same strain (Simpson, 1965). Excised embryos of genetically non-dormant strains germinate normally on the same liquid medium. When GA is added to the medium dormant embryos germinate normally (Fig. 2.10). The general conclusion reached in this study was that excised dormant embryos lack the ability to synthesize a GA-like substance that in turn promotes the synthesis of enzymes needed for the utilization of sugars, amino-acids, etc., derived from the degradation of endosperm reserves. The synthesis of this GA-like activator was

Fig. 2.10. Breaking the dormancy of excised embryos of *Avena fatua* by addition of gibberellic acid to the culture medium (After Simpson, 1965). *(a)* Nutrient medium. *(b)* Nutrient medium plus gibberellin.

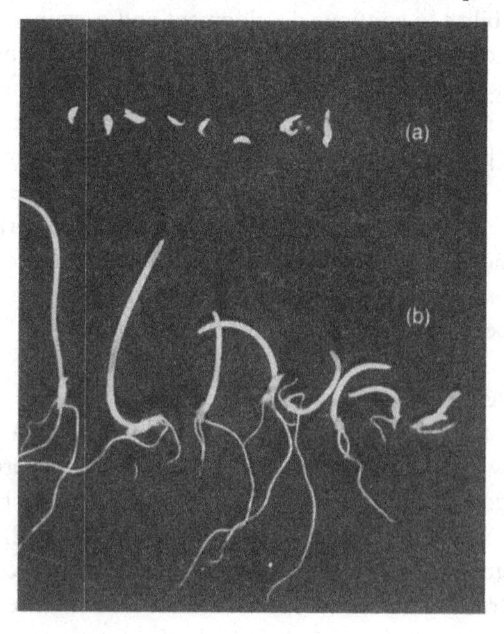

inhibited by CCC ((2–chloroethyl)trimethylammonium chloride) in non-dormant embryos (Simpson, 1966). The ability to produce GA-like substance(s) increased during after-ripening up to a period of about 18 months, a time when dormancy is normally lost in intact caryopses. At this time excised embryos were no longer dormant. Excised developing embryos of *A. fatua* show dormancy at all stages of maturity after anthesis (Zade, 1909; Atwood, 1914; Lute, 1938; Andrews & Simpson, 1969). Embryos of *A. ludoviciana* are similarly dormant from anthesis to maturity (Quail & Carter, 1969; Morgan & Berrie, 1970).

The mechanism of dormancy in embryos is complex and not due simply to the absence of a single GA-like hormone (Simpson, 1978). The suggestion has been made that dormant embryos are physiologically and metabolically 'quiescent'. However comparisons between dormant and non-dormant embryos, both from the same population and in genetically pure lines differing in degrees of dormancy, indicate that in many respects the physiological activity in both conditions is similar. It appears that some subtle differences contribute to the prevention of germination in dormant lines. Some of these similarities and differences are outlined in the remainder of this section.

Protein synthesis can occur in imbibed dormant embryos (Chen & Varner, 1970). Comparison of the levels of protein synthesis among dormant and non-dormant embryos indicate that in one specific strain (AN265), when dormancy is broken with GA_3, there are no significant differences in soluble protein or in SDS (sodium dodecyl sulphate) soluble protein between treated and untreated embryos (Keng & Foley, 1987). On the other hand in another strain the same kind of GA treatment did induce an enhancement in biosynthesis of protein and RNA in the embryos before they germinated (Chen & Chang, 1972). The levels of 3'-nucleotidase and ability to synthesize protein were lower in dormant embryos, between 24 and 48 h after imbibition, compared with non-dormant embryos. GA overcame this difference (Simpson, 1965).

Direct comparisons of dormant and non-dormant strains indicate that protein and phospholipid turnover is similar in the membranes of imbibed embryos of both types (Cuming & Osborne, 1978a,b; Osborne & Cuming, 1979). This suggests that dormant seeds can repair and maintain membrane integrity rapidly following imbibition of water. Both dormant and non-dormant embryos show synthesis of RNA within minutes following imbibition of water (Anonymous, 1983). Nuclear DNA is also repaired and there is continuous DNA synthesis. Nevertheless, only the non-dormant embryos achieve full replication that leads to cell division and germination.

The hypothesis that available energy, such as ATP, is too low for germination in dormant embryos (Zorner, 1981) has not been confirmed. In other studies (Adkins & Ross, 1981a,b; Jain *et al.*, 1983) increases in ATP occurred in dormant embryos without germination, indicating that changes in ATP alone are not sufficient to break dormancy.

Dormant excised embryos require a carbohydrate source for germination (Simpson, 1965). Both sugar accumulation from the endosperm and its utilization by dormant embryos are restricted. Both of these restrictions can be overcome by GA. Several attempts have been made to determine where the block to utilization of sugar occurs within the metabolic system of embryos. An initial hypothesis was that glucose oxidation via the pentose phosphate pathway (PPP) was essential to overcome embryo dormancy (Roberts, 1973). However, a series of investigations (Simmonds & Simpson, 1971, 1972; Upadhyaya, Simpson & Naylor, 1981: Fuerst *et al.*, 1983) eventually led to the conclusion that there are no significant differences with respect to the activity of this pathway before germination occurs, among strains differing in degree of dormancy.

Investigations into the various aspects of the oxidation of sugar within embryos has generated further information on the nature of the metabolic differences between dormant and non-dormant embryos. Inhibitors of cytochrome oxidase such as cyanide, azide, carbon monoxide, methylene blue and hydroxylamine can break dormancy in a number of grasses (Roberts, 1969). These same inhibitors can prevent the germination of non-dormant embryos. A series of investigations into these contrasting effects of azide and cyanide were made with *A. fatua* (Upadhyaya, Naylor & Simpson, 1982a, 1983). First indications were that stimulation of germination with these two inhibitors involved an increase in activity of the alternate pathway of electron transfer to oxygen as the final stage of the catabolism of sugar. However, later investigation (Tilsner & Upadhyaya, 1987) indicated that the effects of azide and cyanide are different, taking effect at different stages of the after-ripening process. Cyanide stimulates germination in seeds with very little after-ripening. On the other hand azide, at the same relative concentrations and over the same treatment duration, has no effect until seeds have been after-ripened for several months when they still show some dormancy. The duration of the after-ripening period before azide breaks dormancy varies with the conditions under which the seed batch was grown. The response to azide is a heritable trait (Jana, Upadhyaya & Acharya, 1988). Recent studies on the action of an inhibitor of the alternate pathway, salicylhydroxamic acid (SHAM), suggest that the actions of both azide and cyanide are not via a stimulation of the alternate pathway but rather through stimulation of a 'residual' form of respiration that is insensitive to both SHAM and cyanide (Upadhyaya, 1986; Tilsner & Upadhyaya, 1987). The stimulatory effects of azide on germination occur in seeds with secondary dormancy indicating similarities between the metabolism of seeds with primary or secondary dormancy (Tilsner & Upadhyaya, 1985).

Other studies with a different group of germination promoters tested on intact caryopses indicate that: (a) several organic acids (citric, succinic, fumaric, malic, pyruvic, lactic and malonic) can induce germination in partially after-ripened genetically distinct lines (Adkins, Simpson & Naylor, 1985; Adkins, Symons & Simpson, 1987). The action of these acids in overcoming dormancy seems related to the simple effect of lowering pH. The neutral salts of the same acids are without effect. The values of pH that overcome dormancy also stimulate oxygen uptake. The uptake is not inhibited by the inhibitor SHAM indicating again that the alternate pathway of electron transport is not involved in this breaking of dormancy. (b) Sodium nitrate, a substance commonly used for breaking seed dormancy in many grass species, had a similar effect to that of both organic acids and azide (Adkins, Simpson & Naylor, 1984a,b). The sensitivity to nitrate increased with the length of the period of after-ripening indicating that some other inhibitory block must be removed first before the stimulatory effect of nitrate can be seen. The less persistent block and the nitrate-sensitive block were both overcome by ethanol (Adkins, Simpson & Naylor, 1984c). The stimulatory effect of ethanol occurred well before germination (Adkins, Naylor & Simpson, 1984). Ethanol promoted both oxygen uptake and germination, without stimulation of the alternate pathway, by directly affecting the activity of the Krebs cycle. (c) The actions of ethanol, nitrate and organic acids seem to involve direct effects on the activity of the Krebs cycle, which in turn lead to stimulation of germination.

The effects of these particular dormancy breaking compounds were compared in some genetically different pure lines exhibiting a gradient in persistence of dormancy in mature seed (Adkins & Simpson, 1988). Two overlapping, but quite distinct, phases of dormancy were discernible among the different lines. Each phase lasted for a different length of time during the period of after-ripening. The least persistent phase occurred during the early stages of after-ripening. During this period only GA and ethanol could break dormancy; azide and nitrate were ineffective. This state varied in persistence according to the genotype of the parent. The line with the most persistent dormancy required approximately 180 days of after-ripening before the insensitivity to azide and nitrate was lost.

The second state of dormancy, characterized by sensitivity to both azide and nitrate, was about 40 days in duration in each line, following the ending of the first phase. The second phase of dormancy can also be overcome by ethanol and GA. There was a third, very short-lived, phase evident in the most dormant lines. This latter phase was present at grain maturity, lasted for only a few days and could be overcome by GA.

The duration of the two most persistent states of dormancy was influenced not only by genotype but also by the environment during the period of seed development on the parent plant and later during seed storage. The duration of both phases was increased if caryopses matured under low (15°C) temperatures and decreased under high (25°C). Similarly the two states increased in duration when seeds were after-ripened under low (4°C) and decreased under high (45°C) temperatures. Dehulling the caryopses prior to after-ripening also reduced persistence of both states. These observations on the plasticity of embryo dormancy correlate well with the known effects of field environments on seed dormancy reported not only for *A. fatua* but also for many other grasses, that is that low temperatures during seed maturation increase seed dormancy compared to high temperatures; high temperatures speed up the after-ripening processes and low temperatures prolong dormancy in mature seeds.

It has been suggested that the growth inhibitor abscisic acid (ABA) may have a role in controlling dormancy in grasses (Walton, 1980–81). ABA can speed up the abscission of florets of *A. fatua* (Anonymous, 1983), but it is either absent, or present only in very low amounts in hulls (Zemenova, 1975; Chen, MacTaggart & Elofsen, 1982; Ruediger, 1982). ABA has been identified in the caryopses of both *A. fatua* and *A. ludoviciana* (Quail, 1968). However, there is no conclusive evidence yet for a role in embryo dormancy. ABA applied exogenously can inhibit the germination of non-dormant embryos (Andrews & Burrows, 1972; Holm & Miller, 1972) and also reverses the GA-induced stimulation of germination in dormant *A. ludoviciana* (Quail, 1968). However, these effects may only be related to the suppression of root elongation and not direct effects on dormancy. ABA has significant effects in inhibiting amylolysis in endosperm-aleurone tissues of many grasses. In protoplasts derived from the aleurone of *A. fatua*, ABA can prevent the synthesis of α-amylase. However, amylase secretion in the endosperm is essentially a post-germinal event (Drennan & Berrie, 1961) in *Avena* species.

There is no clear indication yet that the scutellum plays a significant role in embryo dormancy. Nevertheless the scutellum is influenced by the condition of dormancy in the embryo axis. The ab- and ad-axial cells that develop into papillae in the scutellum of non-dormant embryos remain undeveloped in dormant imbibed embryos (Rao & Raju, 1985). Wounding that promotes germination in dormant embryos also promotes formation and growth of these papillae (Raju *et al.*, 1986). An early report suggested that the ability of the scutellum to form sucrose from low molecular weight sugars, such as glucose and maltose formed from starch hydrolysis in the endosperm, is restricted in dormant embryos (Chen & Varner, 1969). However, the application of sucrose does not break dormancy unless GA is present. The production of low molecular weight sugars from endosperm reserves is a post-germinal event in *A. fatua* (Drennan & Berrie, 1961). Application of GA to break embryo dormancy stimulates the growth of the scutellum (Rao & Raju, 1985). The scutellum of both dormant and non-dormant excised embryos can secrete considerable quantities of α-amylase (Simpson & Naylor, 1963).

Other grasses:

Many species that lose some dormancy when the caryopsis coat is broken have residual dormancy that can be attributed to the embryo (Table 2.6). This dormancy is eventually lost with further after-ripening. The most convincing proof of embryo dormancy is to excise the embryo and demonstrate that germination does not occur on a medium, and at a temperature, that supports normal germination of a non-dormant embryo. The embryo of most forage grasses is very small and difficult to separate physically from the enveloping structures of the caryopsis. For this reason, information about true embryo dormancy is meagre and mainly derived from the large grains of cereals.

Experiments with the large-grained cereals can be broadly classified into two categories. The first category covers those situations where the primary purpose of the experiment was to determine whether excised embryos show dormancy. The second category covers those experiments where the purpose in excising the embryo, usually at different stages of development in the floret, was to find an *in vitro* method of culturing the embryo through the stages of germination and seedling growth to become an autonomous plant. A review by Belderok (1961) of the first category indicated very few reports of true embryo dormancy in cereals. Table 2.6 summarises a review of both categories covering many of the most important cereal grasses. In the majority of cases, within each species, there is little indication that excised embryos are dormant, either on water or more complex media.

Table 2.6. *Evidence for the germination, or dormancy, of embryos excised from the caryopsis then transferred to water or a nutrient solution. Examples include both immature and mature caryopses. Reference source in brackets.*

Species	Germination	Dormancy
Avena sativa	(Von Guttenberg & Wiedow, 1952)	—
Hordeum vulgare	(Merry, 1942; Kent & Brink, 1947; Konzak, Randolph & Jensen, 1951; Ziebur & Brink, 1951; Schooler, 1960; Wellington & Durham, 1961; Briggs, 1962; Chang, 1963; Radley, 1967; Taylorson, 1970; Cameron-Mills & Duffus, 1977)	(Norstog & Klein, 1972 b; Dunwell, 1981 a,b)
Oryza sativa	(Amemiya, Akemine & Toriyama, 1956 a,b; Sircar & Lahiri, 1956; Nakajima & Moroshima, 1958; Bouharmont, 1961; Juliano, 1980)	(Amemiya, Akemine & Toriyama, 1956 a,b)
Panicum maximum	—	(Smith, 1971)
Secale cereale	(Purvis, 1948)	—
Setaria lutescens	(Lehle, Staniforth & Stewart, 1978)	—
Triticum aestivum	(Moormann, 1938/9; Augsten, 1956; Fitzgerald, 1959; Shuster & Gifford, 1962; McCrate *et al.*, 1982; Symons *et al.*, 1983)	(Schleip, 1938; Moorman, 1938/9; Gaspar *et al.*, 1977; McCrate *et al.*, 1982)
Triticum durum	—	(Grilli *et al.*, 1986)
Zea mays	(Gain & Jungelso, 1915; Andronescu, 1919; Haagen-Smit *et al.*, 1945; Pieczur, 1952; Dure, 1960)	—
Zizania aquatica	(LaRue & Avery, 1938)	—

Nevertheless, there are isolated examples from each of the species indicating that true embryo dormancy, although apparently uncommon, can be found among some strains or cultivars.

Given the common observations of dormancy in the intact caryopsis in each of these species it could be inferred that this dormancy is strictly an effect of the covering structures and not in the embryo. Alternatively, it might mean that in most cases the grains were after-ripened before the excision of embryos so that true embryo dormancy was already lost. However, a large number of the experiments were actually made with embryos excised at various stages in the development of the floret from anthesis to maturity. It would be therefore reasonable to expect to find some evidence of embryo dormancy during this period, if it were present, based on the demonstrations of embryo dormancy in *A. fatua* (section 2.4). Thus the predominance of non-dormancy in most embryos of cereals can be attributed to two causes. The first is, that there are clearly genetic differences among cultivars within each species that reflect gradations in, as well as absence of, dormancy. Secondly, it can be surmised that the process of domestication of cereals during the last several thousand years has placed heavy selection pressure on removing embryo dormancy as a trait in mature grains. Continued selection pressure for grains that germinate quickly and uniformly at sowing time would tend to remove this dormancy. Nevertheless, the observations reported in the section of this chapter on the role of hulls and coat in grain dormancy indicate that most cereals still retain some dormancy. This suggests that selection pressure during domestication has been most successful in removing the embryo dormancy that is more often present in wild plants.

Several interesting observations can be made on the studies of dormancy in cereal embryos. The first is related to the influence of hydration and dehydration on the development of dormancy during the development of the caryopsis. In a study of 4000 cultivars of barley grown in the USA, the three most dormant (winter) cultivars were exposed to a treatment of high moisture around the developing floret (Pope & Brown, 1943). To do this, the lateral flowers around the chosen floret were dissected away at a point in time between 7 and 10 days after anthesis. A wick of moist filter paper was inserted in contact with the exposed embryo end of the caryopsis. Within 5 days, treated embryos germinated and began the seedling growth characteristic of pre-harvest sprouting. The viviparous embryos of the 'dormant' cultivars were thus no more dormant than any of the other cultivars. The authors concluded that the seed coats were freely permeable to water. Thus dormancy in these barley cultivars must arise in the final stages of maturation of the caryopsis, when tissues are undergoing desiccation and maturation of the endosperm reserves.

In another comparison of 18 wheat cultivars showing a range of resistance to pre-harvest sprouting and presence of post-harvest dormancy, the excised embryos from each cultivar germinated equally well on water

(McCrate *et al.*,1982). However, they responded quite differently to the presence of a water-soluble inhibitor present in equal amounts in the caryopses of all cultivars. The degree of inhibition of germination in the excised embryos was in proportion to the degree of resistance to pre-harvest sprouting in the intact caryopses. The inhibitor did not decline with after-ripening of the caryopsis, but the response of the embryo to the inhibitor declined with after-ripening. GA could counteract this inhibitor. This particular study indicates that the non-germinability of an intact caryopsis can be due to factors other than seed coat permeability that interact with otherwise non-dormant embryos.

The capacity of immature embryos to either germinate or continue further development when excised is partly dependent on the nutritional status of the medium. For example, excised embryos of *Oryza sativa* did not germinate until 9 days after anthesis, when development was nearly complete (Amemiya, 1956a,b). Continued normal development did not take place unless sugars were present in the medium. Similarly excised *Hordeum vulgare* embryos continued to develop in the presence of sugars and other factors from either the endosperm or lemma and palea (Kent & Brink, 1947; Ziebur & Brink, 1951; LaCroix, Naylor & Larter, 1962). If these factors were withheld the embryos germinated. Nutritional factors from the endosperm contributed to the narrowing of the range of permissible temperatures for germination in intact barley caryopses (Dunwell, 1981a). When the embryo was excised it germinated over a wide range of temperatures on water. In another study, precocious germination of otherwise non-dormant barley embryos was suppressed by low availability of oxygen, high temperatures, light and high sucrose concentrations (Norstog & Klein, 1972a,b). The first three of these factors are commonly important in the initiation of secondary dormancy in many grasses. In the same study embryos excised from caryopses from plants grown in cool conditions were more dormant than those in warm conditions. None of the factors that suppressed germination appeared to interfere with embryogenesis. The general conclusion was reached that precocious germination is simply the absence of dormancy determined by the environment in the early stages of embryo development.

The relatively small amount of information about dormancy in excised cereal embryos still supports several general conclusions. *Embryo dormancy is found among several cultivars of the important cereal crops. The absence of dormancy in an excised embryo does not necessarily indicate an absence of dormancy when the embryo is* in situ *in the caryopsis.* Interactions of the embryo with aleurone, endosperm and soluble factors derived from various structures, as well as osmotic status, can lead to the expression of embryo dormancy in an intact caryopsis.

Further understanding of the nature of embryo dormancy in grasses must be inferred from observations made on semi-intact or intact caryopses. However, many of these observations are difficult to interpret because of the varied contributions to the total expression of embryo dormancy made by the structures external to the embryo. In most attempts to understand dormancy in grass seeds manipulation of environmental factors has been much more significant than manipulation of seed structures. These environmental influences are considered in chapter 3. Further discussion of innate embryo dormancy is therefore found in chapters 3, 4 and 5.

It is not clear yet whether the scutellum of grasses generally has any significant role in seed dormancy. The scutellum has been studied for the purpose of understanding its role in the secretion of amylases to degrade the starch reserves of the endosperm, or to indirectly stimulate the aleurone tissue to do the same. Such studies in *Zea mays* (Dure, 1960), *Hordeum vulgare* (MacLeod & Palmer, 1966; MacLeod, 1969), *Oryza sativa* (Miyata *et al.*, 1981; Mitsui *et al.*, 1985) and *Avena sativa* (Zee, Chan & Ma, 1984) have emphasized the post-germinal events. The utilization of endosperm reserves by the scutellum has been considered a post-germination event in a supposedly non-growing tissue (Bewley & Black, 1982). Nevertheless, the scutellum is an integral part of the dormant embryo, linked to it by prominent vascular strands, and it seems likely that there is some form of interaction between the tissues prior to germination. The observations on cutting the pericarp (section 2.3) show that the scutellum can be made to expand and function either prior to, or following, embryo expansion according to the point of entry of water. Thus activation of the scutellum is not necessarily a post-germinal event. Competition between the scutellum and the rest of the embryo for water is in itself a modification of the germination potential of an embryo.

The volume of the scutellum is usually at least as great, and generally greater than that of the embryo axis. Many of the recorded metabolic events that seem related to the disappearance of dormancy were measured on the entire embryo axis and scutellum. It is thus not clear what relative contribution was made by each tissue to these events. It is possible that the scutellum is an important source of immediately available energy, derived from either starch or fat reserves, necessary for the initiation of membrane repair and protein and RNA turnover that occur rapidly after imbibition of water in both dormant and non-dormant caryopses.

2.5 Interrelation between the embryo and maternal storage tissue

In grasses the diploid embryo, with its prominent scutellum, is positioned at the proximal end of the caryopsis. At maturity the scutellum may be extended for various proportions of the total length of the caryopsis (section 2.4). The remainder of the cavity of the mature fruit is filled with the predominantly starchy endosperm. This is surrounded by an aleurone layer that subtends the integumentary layers. Both the endosperm and aleurone tissues have a common origin from the fusion of the second male nucleus with the two polar nuclei of the embryo sac, one of which is the sister nucleus of the egg. Arber (1934) has humorously described the tissue as a 'spoiled second embryo that is unique among living things because in a sense it has three parents but no descendants'!

The aleurone layer of mature, moistened, caryopses secretes hydrolytic enzymes, for example amylases, cellulases, proteinases, etc., that degrade the starchy endosperm to low molecular weight sugars and amino acids that can be easily taken up by the scutellum, particularly in the post-germination stages of early seedling growth (Briggs, 1973; Akazawa & Miyata, 1982; Jacobsen, 1983). It is generally considered that GA originating from the embryo axis is the trigger for initiating enzyme activity in the aleurone tissue (MacLeod, 1967a,b).

Many of the products of endosperm hydrolysis are undoubtedly used for the support of post-germination seedling growth. For this reason there has been a tendency to disregard the endosperm as having a significant role in the control of dormancy. Nevertheless it is a tissue that surrounds much of the embryo and is intimately connected with the coat that plays an important part in the maintenance and release of dormancy. In a caryopsis with long-term seed dormancy, under field conditions, alternate wetting and drying as well as fluctuating temperatures are presumably a hazard to the preservation of endosperm reserves. Some form of linkage between the timing of persistence of dormancy and preservation of endosperm reserves would seem essential to long-term dormancy and grain survival. For example, if hydrolytic activity begins in the endosperm on the first occasion the grain is exposed to moisture the reserves could be lost before the embryo loses dormancy. In addition, invasion of the caryopsis by pathogens is enhanced when non-utilized sugars increase solute potential and then leak from the swollen grain. There are mechanisms for the preservation of endosperm integrity. When seeds of *Avena fatua* were submerged in canal water to depths of 120 cm the embryos lost viability after three months but the endosperm reserves remained intact (Bruns, 1965).

Interrelation between embryo and maternal storage tissue

The relationship between embryo and endosperm in the expression of dormancy has been investigated by biologists in several quite different ways. One approach has been to explore the genetic determinants, particularly the maternal contribution, by crossing genetic lines that show different degrees, or the presence or absence of dormancy. Reciprocal crosses can give some insight into the potential influence of maternal tissue because the zygote-embryo of grasses develops and matures within the parental tissue. The diploid embryo is also positioned within a triploid tissue in which two of the three sets of genomes are contributed by the female parent. Thus the maternal contribution to dormancy might be expected to be significant. The other approach has been made through the study of hormonal–metabolic–physiological relationships. This section considers these several approaches to understanding the links between embryo and endosperm that relate to dormancy.

Example of Avena fatua:

(a) The genetic approach Within the genus *Avena* we find expressed a wide range of degrees of dormancy (Nishiyama & Inamori, 1966; Jorgenson, O'Sullivan & Vanden Born, 1974). Some species such as *A. abyssinica* apparently have no dormancy at all (Nishiyama & Inamori, 1966). A range of degrees of dormancy also exists within each of the groups of diploid, tetraploid and hexaploid species. Within the hexaploids the domesticated species *A. sativa* has the shortest period of dormancy and *A. fatua* the longest (Nishiyama & Inamori, 1966). Within the single species *A. fatua*, which is essentially self-pollinated (Imam & Allard, 1965), there is an array of genetically distinct ecotypes exhibiting a gradation from non-dormancy through to persistent long-term dormancy (Naylor & Jana, 1976). Early in this century it was observed that occasional out-crossing can occur between wild and domestic oats (Zade, 1912). There was for some time a controversy about the origin of these so-called 'fatuoid' oats because some biologists believed that they originated as mutations. Others believed they were simply the outcome of a natural crossing between the species. Phenotypically fatuoids appear to be like the wild oat but the physiological behaviour is more like the domestic oat. When the fatuoid arose in a field where the variety of domestic oat showed natural dormancy, the fatuoids also showed seed dormancy (Coffman & Stanton, 1940). When dormancy was absent as a character in the domestic oat, the fatuoid was non-dormant.

When specific crosses were made between *A. sativa* and *A. fatua*, delayed germination was found to be linked to the fatua-type (Garber & Quisenberry, 1923). In these crosses every seed that failed to germinate because of dormancy became soft and partially decayed during the 13–day germination test, indicating the dissolution of the endosperm. Another early genetic study of *A. sativa* concluded that dormancy was due to two genes, one of which determined the nature of the caryopsis coat and the other the nature of the embryo (Moormann, 1942). The coat genes were considered to be strictly maternal with segregation taking place in the later generations. When both genes were present dormancy was very persistent. From a series of controlled reciprocal crosses between *A. fatua* (dormant) and *A. sativa* (non-dormant) it was concluded, after observing three generations, that germinability is dominant over non-germinability (Johnson, 1935a,c). Germinability was inherited on the basis of three factors of more or less similar germinative potencies, one of which was related specifically to grain type. The authors inferred that only the embryos with six dominant alleles are germinable at maturity. Following a period of after-ripening, a progressively smaller number of dominant allelomorphs is germinable. At a certain stage embryos with the triple recessive (Aa,AA,AaAa) will be non-germinable; later, AA and AaAa types can germinate and finally the triple recessive (*A. fatua*) types will germinate. Dormoats have been produced from crosses between *A. fatua* and *A. sativa* with the general character of the *sativa* species but with a degree of dormancy that can persist through a winter after the seeds are sown in the autumn (Andrews & Burrows, 1972). Excised embryos of dormoats are non-dormant indicating that the endosperm and coat contribute to the main expression of dormancy.

Crosses have been made within the species *A. fatua* between pure lines differing in degrees of dormancy expressed as duration of dormancy (Jana, Acharya & Naylor, 1979). Observation of the progeny showed that the parental lines differed for at least three genes that control the germinability of grains during after-ripening. In the crosses between dormant and non-dormant parents, F_1 grains matured on the parent plants fell into two groups. Grains could either germinate within the first 3 weeks or remain dormant for longer than about 12 weeks. Grains from both these early and late germinating parents, including the reciprocal crosses, did not differ significantly in the timing of dormancy. This indicates an important maternal influence on dormancy in the original crosses.

Other indications of maternal influence in the modification of the expression of dormancy comes from a study of seven crosses of pure lines differing in degree of dormancy (Jana & Thai, 1987). Crosses were made in two ways: dormant by non-dormant and dormant by dormant. After selfing the hybrids, reciprocal back-crosses were made for several generations. The time-course of germination was then compared in each of the parental and derived progeny, and back-crossed lines. Three interacting loci related to dormancy were found. One locus was related to the sensitivity of the seed to dormancy loss in the presence of sodium azide. The two most dormant lines differed in their response to sodium azide (AN-51 was more responsive than M-73). When these two particular lines were crossed, F_1 seeds were only responsive to azide if AN-51 was the maternal parent suggesting a modifying influence of the non-embryonic tissue on expression of dormancy.

The susceptibility of the endosperm to hydrolysis by the triggering effect of hormones secreted by the embryo is different in genetic lines showing different degrees of dormancy (Upadhyaya, Hsiao & Bonsor, 1986). Several substituted phthalimides, that can mimic the triggering effect of naturally-occurring GA's secreted by the embryo, were applied to the endosperm tissues of grains from a range of lines with different degrees of dormancy. Some lines failed to respond to a particular phthalimide that triggered amylase release in other lines.

The hypothesis that natural selection would inevitably lead to co-adaptation of both embryo dormancy and dependence for amylolytic activity of the endosperm on the triggering effect of a GA-like hormone from the embryo after loss of dormancy, has been tested in several ways. In the first experiments six pure lines of both dormant and non-dormant genetically pure lines were selected at random from a large number of lines (Upadhyaya, Naylor & Simpson, 1982b). Autonomous (i.e. independent of the presence of the embryo) α-amylase production was compared in endosperm segments after a period of incubation in conditions similar to those for normal germination of a caryopsis. None of the endosperm segments from the six dormant lines produced α-amylase unless GA was supplied exogenously. On the other hand, among the six non-dormant lines, endosperm segments of four of the lines produced substantial amounts of α-amylase and reducing sugars indicating a degree of autonomy from hormonal control by the embryo. This led the authors to conclude that the coupling together of the traits of embryo dormancy and rigorous dependence of the endosperm on GA from the embryo must be due to multiple alleles and not pleiotropy.

The hypothesis was also tested in another way (Jong, 1984). Endosperm segments from lines characterized by 'autonomous' endosperm behaviour were shown to have the ability to synthesize both amylases and GA. After-ripening reduced the length of the lag phase in the development of this endosperm autonomy in both dormant and non-dormant lines. Both low

(4°C) and high (28°C) temperatures suppressed the development of autonomy for amylase formation in non-dormant lines but not in dormant lines. Reciproal crosses were made between parents having either a dormant embryo combined with a non-autonomous endosperm, or a non-dormant embryo combined with an autonomous endosperm. In the F_2 seeds a strong association was found between dormancy and non-autonomy of the endosperm and *vice versa*, indicating a physical linkage between the genetic loci controlling the two phenotypes.

(b) The physiological approach The relationship between embryo and endosperm, determining germinability, can be altered by changing the environment of the parent plant during seed maturation. When maturing panicles were excised from the parent plant and transferred to a medium high in sucrose or maltose, both the embryo and the endosperm became insensitive to the stimulatory effect of GA_3 (Cairns, 1982; Cairns & de Villiers, 1982). The coupled sensitivity could not be induced by low concentrations of the same sugars, or equimolar high concentrations of glucose or fructose. The authors concluded that the effect was not related to osmolarity but to a direct nutritional influence. Further evidence of the linked nature of endosperm stability and embryo dormancy is provided by the way dormant caryopses respond to ammonia gas. Ammonia breaks dormancy of the embryo and also stimulates α-amylase activity five-fold in isolated endosperm segments of the same seeds (Cairns & de Villiers, 1986b).

ABA, identified in the caryopses of *A. fatua* (Quail, 1968; Zemanova, 1975), has been considered as a possible candidate for linking embryo dormancy to maintenance of endosperm stability. ABA can inhibit the development of α-amylase activity in endosperm (Buller, Parker & Grant Reid, 1976; Buller & Grant Reid, 1977) and also negates the GA-stimulated loss of dormancy during the early stages of after-ripening in the embryo (Quail, 1968).

Other grasses:

(a) The genetic approach There have been many reports of genetic variation within grass species for the character of persistent dormancy (see earlier sections). However, attempts to understand the genetic determinants of dormancy in grasses have been confined to just a few cereals. In most studies of crosses between dormant and non-dormant parents the objective has been to determine the number of genes involved and the heritability of dormancy.

A general survey of the genus *Oryza* (Morishima & Oka, 1959), including perennial and annual forms, indicated that germinability appears to be associated with the morphological characteristics of the low-dormancy *O. sativa* type. From another study with a series of crosses between strains showing a range of expressions of dormancy, observed to the F_3 generation, it was concluded that grain dormancy has a complex inheritance depending on multiple gene action (Chang & Yen, 1969). Heritability estimates based on the F_4/F_3 regressions indicated that the environment during ripening has a substantial effect on the expression of dormancy. From crosses among six varieties the conclusion was reached that in one cross the dormancy was due to a single gene, in another to two major genes (Tomar, 1984). The only study to locate the site of gene action for dormancy in *O. sativa* (Takahashi, 1962) concluded that the embryo, endosperm and caryopsis coats make separate contributions to the overall expression of dormancy. Coat impermeability to water was shown to be recessive and long-term dormancy was dominant over short-term dormancy.

In barley, dormancy has a high heritability and is determined by several recessive genes (Buraas & Skinnes, 1984). Wheat also has a high degree of heritability (Gfeller & Svejda, 1960). Observations of the developmental pattern of wheat dormancy in four cultivars indicated that only one cultivar (Pembina) showed a linkage of embryo dormancy and α-amylase activity. In this particular study there was no relationship apparent between the presence of ABA and dormancy. Studies of crosses between wheat cultivars, observed to the fourth generation, suggest that dormancy is controlled by two additive genes (Ruszkowski & Piech, 1969).

Studies with sorghum show that late-flowering cultivars have dormancy (Gritton & Atkins, 1963), but in rice there is no indication of a relationship between flowering time and dormancy (Chang & Yen, 1969).

None of the above studies shed much light on the kind of interrelations between the embryo and endosperm in these species that have been demonstrated in *A. fatua*. Clearly there is considerable room for more investigation of dormancy using the genetic approach.

(b) The physiological approach Barley endosperm has been the subject of many investigations since the early discovery that α-amylase secretion can be activated in this tissue by a factor from the embryo (Yomo, 1958). The factor, since shown to be a GA, also breaks dormancy of the embryo. There is a correlation between the state of embryo dormancy and the susceptibility of barley aleurone to gibberellin. After-ripening, heating, and increase

in the partial pressure of oxygen during soaking in water all break barley embryo dormancy. The same factors also increase the GA (Crabb, 1971). Barley endosperm normally has a suppressing effect on germination of a developing embryo, at the same time that it is promoting development and differentiation (Ziebur & Brink, 1951). This is true also of wheat (Fitzgerald, 1959), maize (Dure, 1960) and several species of *Digitaria* (Baskin, Schank & West, 1969).

In *Setaria lutescens*, the excised embryo from a dormant caryopsis germinates normally on a liquid medium containing a simple carbohydrate source such as glucose, fructose, maltose or sucrose (Lehle, Staniforth & Stewart, 1978). However, in the intact caryopsis the endosperm tissue is not susceptible to degradation by α-amylase unless it is first treated with protease. In this way the endosperm controls the germinability of the embryo. Both the embryo and endosperm of dormant *Sorghum bicolor* are insensitive to GA, kinetin and IAA. This indicates that the pattern of endosperm dependence on GA secretion from the embryo, typical of barley, is not the same in this species. Similarly *Distichlus spicata* does not respond to GA or kinetin during the early stages of after-ripening (Amen, 1970).

Correlative control of both embryo germination and endosperm activity, effected by an inhibitory substance originating in the glumes and hulls, has been demonstrated in *Aegilops kotschyii* (Wurzburger & Koller, 1973, 1976). The smaller of the two caryopses present in a single spikelet fails to germinate whereas the larger germinates. The inhibitory substance found in the hulls can prevent the activity of GA_3 in initiating amylase activity in barley endosperm, as well as inhibit embryo germination in the smaller of the caryopses of *Aegilops*.

The endosperm of *Uniola paniculata* has a leachable inhibitor that controls germinability of the embryo, normally non-dormant when excised (Westra & Loomis, 1966). Thus the endosperm is the major determinant of dormancy in this case.

There are now several reports implicating the growth inhibitor ABA as a potential correlative inhibitor of both embryo and endosperm hydrolysis in grasses. Interrelations between the activities of both GA and ABA on embryo activity and hydrolytic activity in endosperm, have been demonstrated in *Hordeum vulgare* (Dashek, Singh & Walton, 1977; Jacobsen, 1983; Mapelli, Lombardi & Rocchi, 1984), *Triticum aestivum* (King, 1976; Safaralieva, Alekperova & Mekhtizade, 1977; Leshem, 1978; Varty *et al.*, 1983; Walker-Simmons, 1987; Walker-Simmons & Sesing, 1987), *Oryza sativa* (Gupta, Bhandal & Malik 1985; Hayashi, 1987), *Secale cereale*

(Dathe, Schneider & Sembdner, 1978; Weidner *et al.*, 1984), *Zizania aquatica* (Albrecht, Oelke & Brenner, 1979) and *Oryzopsis hymenoides* (McDonald & Khan, 1977). The questions of exactly how the ABA is produced by the maternal tissue, or whether different tissues have different sensitivities to ABA, have not yet been resolved. These studies do, however, indicate that the relation between the embryo and endosperm of non-dormant caryopses, typified by the many elegant experiments with barley, may not occur in all other non-dormant grass species. It seems certain that an explanation of the interrelation between the embryo and endosperm of dormant grass seeds is much more complicated than the simple idea that endosperm hydrolysis is only a post-germination event.

3

Environmental influences on seed dormancy

3.1 Water – liquid and vapour phases

Water has a primary role in sustaining tissue activity and viability of both the parent plant and the developing caryopsis. The natural process of seed maturation involves a dehydration of the caryopsis during the final stages that culminates in abscission. Water stress on the intact plant can interact with temperature and influence both the rate of development and level of dormancy attained by the caryopsis. In its new environment, after abscission, the dormant seed may be exposed to a range of varying environmental factors, particularly moisture and temperature. For example, if the seed remains on the soil surface it will be exposed to diurnal cycles of desiccation and hydration according to the level of radiation from the sun and increased relative humidity at night. Alternatively seeds buried in soil can be exposed for long periods of time to excessive moisture or dryness. Variations in available moisture influence the persistence of dormancy. Excess of water can induce secondary dormancy in many grass species. Some optimum amount of water and suitable temperature will ultimately determine the time for germination. Seed coat structures can modify any of the above moisture variations by limiting ingress of water to particular rates or amounts (Chapter 2).

Interactions of moisture with temperature, controlling dormancy and germination, can occur in several ways. Temperature affects the relative humidity of the atmosphere. In turn this determines the ease, or difficulty, with which a seed hydrates by the hygroscopic action of the structural and reserve materials. Germination can occur at high relative humidities, by seeds directly absorbing water vapour (Simpson, 1978). On the other hand uptake of water in the liquid phase can be very rapid due to the high matric potential of the seed. The first phase of imbibition of water, usually over in a few minutes when a seed has an optimal supply of liquid water, is followed by a much slower phase of temperature-dependent cell activity. This second phase may sustain dormancy or lead, after hours or days, to germination with a later phase of water uptake associated with seedling growth.

The following section covers the variations in moisture that influence the initiation, maintenance and termination of dormancy.

Example of Avena fatua:

Water stress on the parent plant during seed maturation reduces the level of seed dormancy (Peters, 1978; Naylor, Sawhney & Jana, 1980; Peters, 1982a,b; McIntyre & Hsiao, 1983). It can also increase the α-amylase activity of mature seeds as much as four times that of non-stressed plants (Peters, 1982b). The degree of reduction in dormancy from water stress varies among genetically distinct lines in proportion to the level of primary dormancy (Peters, 1985). Lines with a high degree of dormancy also have a high seed water content at maturity when compared with non-dormant lines. After abscission of seed from the parent plant the persistence of dormancy is markedly affected by conditions of soil moisture. For example, in soil with higher than 50% moisture secondary dormancy was induced (Kiewnick, 1964; Schonfeld & Chancellor, 1983). At high humidities on the soil surface, or within soil at field capacity, seeds autolyse and rot indicating that although the embryo is dormant the endosperm disintegrates (Kiewnick, 1964; Sharma, McBeath & Vanden Born, 1976). Because light soils dry faster than heavy soils dormancy is lost fastest in the former. Optimum germination of non-dormant seed occurs between 15–17% moisture on a seed dry weight basis. The second phase of water uptake that leads to germination is initiated at around 11% (Justice & Bass, 1978). Seed water contents higher than this lead to secondary dormancy that persists (Hay & Cumming, 1959; Barralis, 1965). These observations correlate with field studies that indicate optimal germination occurs between 50–70% field capacity (Koch, 1968a; Sharma, McBeath & Vanden Born, 1976).

Alternate wetting and drying decreases germination by inducing secondary dormancy (Kommedahl, DeVay & Christensen, 1958). The site of this induced dormancy is within the caryopsis because the level of dormancy induced is similar in hulled and dehulled florets. If partly after-ripened seeds are given a single soaking in water, followed by incubation in a humid atmosphere for two weeks, secondary dormancy is induced: six weeks of dry storage removes the induced dormancy (Andrews & Burrows, 1974). The induction of secondary dormancy occurs near the optimal temperature for normal germination (e.g. 21°C) but not at low temperatures, indicating involvement of metabolism in water-induced secondary dormancy (Thurston, 1961). Soaking seeds in water for as little as 15 min induced secondary dormancy (Hsiao, 1979) at a time when adenosine triphosphate (ATP) was already detectable (Jain, Quick & Hsiao, 1983).

A moist atmosphere combined with high temperature hastens after-ripening in stored seeds (Quail, 1968). Dormancy is lost within 18–36 months when seeds are stored dry in a laboratory environment, but under the variable moisture of the soil seeds may remain dormant as long as seven years (Simpson, 1978).

Embryos seem to have a requirement for exposure to the liquid phase of water to be able to terminate dormancy (Black, 1959). This may be related to the inability of the embryo to develop a sufficiently negative water potential (Andrews, 1967). When dormant embryos are excised from fully soaked caryopses they immediately take up further water indicating that the surrounding tissues of the caryopsis prevent the embryo from reaching its full water content (McIntyre & Hsiao, 1985).

Other grasses:

There are surprisingly few studies of the influence of water status of the parent plant on dormancy of the maturing floret. Water deficit increases the level of dormancy in *Hordeum vulgare* when it occurs close to the time of anthesis (Aspinall, 1965), but if it occurs at the late stages of maturation dormancy is reduced. In *Triticum aestivum* post-harvest dormancy can be increased or decreased at any particular moisture level of the environment during the final week of maturation, according to whether temperature is increasing or decreasing (Grahl, 1975). Studies with barley and wheat in Europe indicate that predicting the level of post-harvest dormancy simply from meteorological data is very difficult (Belderok, 1961; Grahl, 1975). Nevertheless it is well known that cereals with low potential for dormancy exhibit 'pre-harvest sprouting' during the later stages of maturation when the atmospheric moisture level is high (Belderok, 1961; Grahl, 1975; Black *et al.*, 1987). Desiccation of the developing caryopsis decreases dormancy in wheat (Symons *et al.*, 1983; Nicholls, 1986) and barley (Nicholls, 1986). Conversely barley harvested with a high moisture content has considerable dormancy (Phaneendranath, Duell & Fund, 1978) similar to *A. fatua*, decribed in the previous section.

A lot of interest has been focused on the influence of seed and atmospheric moisture on post-harvest dormancy, particularly with forage crops, to find ways of improving germinability at planting time. Moisture in the storage environment of seeds after harvest, as well as specific treatments such as pre-soaking and alternate wetting and drying, have major effects on the persistence of dormancy (Table 3.1). In many species dormancy is

Table 3.1. *The influence of change in moisture status on the germinability of mature, dormant grass seeds. Changes in dormancy are indicated by: A = induced, B = sustained, C = reduced. Reference source in brackets*

Species	Treatment	Dormancy
Agropyron pauciflorum	Alternate wet and dry.	C
(Griswold, 1936)		
Aristida armata	Dry storage at high temp.	C
(Brown, 1982)	Stratification at 4 °C.	C
Avena sativa	Hot water soak plus drying.	C
(Guzevskii, 1947)		
Bromus anomalus	Rapid drying.	C
(Griswold, 1936)		
Dactylis glomerata	Soak at 5–6 °C and dry.	C
(Haight & Grabe, 1972)		
(Chippindale, 1933)	Presoak and dry.	C
Echinochloa crus-galli	Dry at 23 °C or higher.	C
(Sung *et al.*, 1987)		
	Prick seed after water content is lower than 18%.	C
	High-temperature drying, 40–50 °C.	C
	Low temperature of 5 °C.	C
(Harper, 1970)	Pre-dry 7 days at 35 °C.	C
Hordeum nodosum	Alternate wet and dry.	C
(Griswold, 1936)		
Hordeum vulgare	Hot water soak then dry.	C
(Guzevskii, 1947)		
(Heit, 1945)	Reduce water content of seed to less than 12.7%.	C
	Heat seed to 39 °C without loss of water.	C
Oryza sativa	Pre-soak and dry.	C
(Juliano, 1980)		
(Roberts, 1962)	Store dry at 3 °C.	B
(Ellis *et al.*, 1983)	Pre-soak at 3 °C.	C
(Dighe & Patil, 1985)	Pre-soak in hot water.	C
(Sikder, 1973)	High humidity night, sun in day.	B
	Low humidity night, sun in day.	C
	Dry on parent plant a long time.	C
(Yasue, 1973)	Soak at high temp. to get 30% water content.	B
	Air dry then give high temp.	C
Oryzopsis hymenoides	Moisten at 3 °C.	C
(Wang *et al.*, 1986)		
Panicum maximum	Moisten at low temp.	B
(Harty & Butler, 1975)		
Panicum turgidum	Desiccate prior to germination.	C

Table 3.1. *(cont.)*

Species	Treatment	Dormancy
(Koller & Roth, 1963)	High temp. of 23–26 °C.	C
Phalaris arundinacea (Juntilla *et al.*, 1978)	Moisten at low temp.	C
(Landgraff & Juntilla, 1979)	Moisten at low temp.	C
(Griffeth & Harrison, 1954)	High seed moisture plus high temp. just before maturity.	A
Poa annua (Phaneendranath, 1977a)	(Genetic variation in response).	C
Poa capillata (Kearns & Toole, 1938)	Moisten at low temp.	B
Poa secunda (Griswold, 1936)	Alternate wet and dry.	A
Setaria faberii Taylorson, 1986)	Water stress with PEG (0.3 MPa).	C
	Slow hydration in humid atmos.	C
	Dry after partial hydration.	C
Sorghum bicolor (Huang & Hsiao, 1987)	Store dry.	C
(Inoue *et al.*, 1970)	Store dry at 40 °C.	C
	Moisten at low temp.	C
(Wright & Kinch, 1959)	Artificial drying at 100 °F of prematurely harvested seed with water content of 45%.	B
	Store dry.	C
Stipa columbiana (Griswold, 1936)	Alternate wet and dry.	A
Stipa lettermani (Griswold, 1936)	Alternate wet and dry.	C
Stipa viridula (Rogler, 1960)	Moisten at low temp.	C
Triticum aestivum (Grahl, 1965)	High humidity plus high temp.	A
(Guzevskii, 1947)	Hot water soak plus drying.	C
(Yasue, 1973)	Desiccation at high temp.	A

reduced, or eliminated, when the mature floret or naked caryopsis is simply desiccated. Repeated desiccations by means of alternate wetting and drying can speed up loss of dormancy. In many cases the physical effect of desiccation is to crack, puncture or scarify the pericarp and testa (Sung *et al.*, 1987). In other cases (e.g. *Sorghum*, *Oryza*, *Poa* and *Panicum maximum*) the moistening stage of the wetting and drying cycle may sustain the original dormancy, or even induce secondary dormancy, at either high or low temperatures. This may reflect a direct effect on embryonic dormancy distinct from the effects of desiccation. Genetic variation in the response to pre-germination soaking or desiccation has been noted with several species (Rogler, 1960; Phaneendranath, 1977b; Juliano, 1980; Ellis *et al.*, 1983). For example, within *Oryza sativa* no single pre-germination treatment, involving change in water, was able to improve germination in all accessions (Ellis *et al.*, 1983).

In many cases the experimental description of the pre-soaking treatment does not indicate whether seeds were dried before the germination conditions were given, or whether they were transferred directly to a higher germination temperature. In those cases where the pre-germination soaking was at a low temperature and seeds were simply transferred to the higher optimal temperature for germination it would seem that desiccation is not involved in the breaking of dormancy. The effects of temperature obviously interact with and complicate interpretations of the influence of moisture on dormancy.

Temperature effects are considered in the next section of this chapter.

3.2 The effects of radiation on dormancy – heat/cold, light, microwaves, magnetism, photoperiodism

Radiation from the sun is the source of energy causing the distribution of heat, water and organic substances on the earth. There are large regional differences in incident radiation depending on latitude, altitude and cloud frequency (Larcher, 1983). The tropical regions may receive as much as 50% more radiation than temperate regions. The important bands of wavelengths influencing grass seeds lying on the soil surface are: (a) heat (wavelengths longer than 3000 nm) that is eventually radiated back to space. (b) Light (wavelengths approximately 380–800 nm) that if absorbed is eventually transformed into longer wavelengths (heat) and transmitted back to space.

Buried seeds receive attenuated amounts of heat and light according to depth and transmissivity of the soil. Soil temperature lags behind air temperature diurnally and seasonally due to the high storage capacity of the former. In winter, soil temperatures may exceed the daily maximum for air: in summer they may be less because of the buffering effect of mass.

Seeds use heat radiation as a source of energy to initiate and sustain germination. In addition, a very high proportion of all plant species shows some degree of dependency on light radiation for the timing of germination (Koller, 1972). Heat and light are major determinants of both the induction and termination of dormancy. Two regions of the light spectrum have separate influences on germination and dormancy: red to near infra-red (660–800 nm) and blue to ultra-violet (290–520 nm). The ratio of incident red/far-red light governs the physiological activity of a receptor pigment (phytochrome) in grass seeds. Phytochrome in its hydrated state can either promote or inhibit germination by its interaction with temperature and length of night in a timing mechanism. Thus factors such as latitude, season, and both wavelength and intensity attenuation under canopies of vegetation, influence the activity of phytochrome which in turn influences dormancy. Phytochrome plays a major role in controlling the seasonal emergence of seedlings from seed banks both above and below the soil surface (Kendrick & Frankland, 1983). The blue region of the light spectrum can also influence the phytochrome pigment and have other major effects on plant growth (Bewley & Black, 1982). The ultra-violet portion of the spectrum has photo-destructive effects on plant tissues, particularly nucleic acids and proteins involved in germination (Larcher, 1983). Microwave and nuclear radiations have subtle and longterm ageing and genetic effects on seeds that survive long periods in either a dry or moist state of dormancy (Osborne, 1980).

Because the literature about the role of phytochrome in seeds is very extensive the reader is referred to recent treatises on the subject (e.g. Kendrick & Frankland, 1983). Emphasis in this section is placed on determining the extent to which phytochrome activity is involved in the dormancy of grass seeds. There is a considerable literature indicating the influence of temperature on grass seed germination. Much of it, from the early part of the century, documents the many attempts to find practical means of overcoming dormancy by manipulating temperature either prior to or during germination. A majority of these studies gives little insight into the physiological changes induced in the dormant seed by these treatments. Because temperature influences the free energy of water, growth, and metabolism, it is difficult to separate the primary physiological effects of temperature from interactive effects on dormancy and germination.

This section surveys the role of radiation in its influence on the induction of dormancy during the development of the seed on the parent plant. It also considers the way temperature and light terminate the state of dormancy. Discussion of the roles of temperature and light in the normal process of germination of a non-dormant grass seed will be limited.

Example of A. fatua:

(a) Temperature The temperature changes induced by cycles of increasing and decreasing incident radiation from the sun influence dormancy in at

least four ways. Firstly, induction of dormancy in the seed as it develops on the parent plant is very sensitive to temperature. Secondly, persistence of the primary state of dormancy is strongly influenced by temperature. Thirdly, reinforcement of the primary state of dormancy, or induction of a secondary state of dormancy, may occur in response to temperature changes at any stage of the post-abscission period of the seed in some genotypes. Fourthly, temperature is a major determinant in the triggering of germination after the seed has imbibed water.

In any discussion of each of the above responses to temperature it must be emphasized that the species *A. fatua* in its natural environment occurs in polymorphic populations exhibiting considerable genetic diversity in response to temperature and other environmental factors. Thus within a population of seeds taken at random from nature, the response to a specific germination temperature is an average response. To say, therefore, that the optimal temperature for germination of *A. fatua* is 22°C (Koch, 1974) simply means that a majority of the population will germinate at that temperature. Alternately it also means that the optimum temperature for some seeds is above, or below, this average optimal value. Until the selection of distinctly different genotypes into pure-breeding lines (Marshall & Jain, 1968; Whittington *et al.*, 1970; Naylor & Jana, 1976; Seely, 1976; Peters, 1982a,b) it was not possible to clarify the role of temperature, or for that matter any environmental factor, because of the confounding effects of variable and even opposite responses among individuals of a population. In the above sense, any of the generalizations made about temperature, or any other environmental factor, on dormancy in a polymorphic population must carry the corollary that some members of the population may behave quite differently.

The duration of dormancy in mature seeds is enhanced by low temperatures during seed development and diminished by high temperatures. Exposure of the parent plant to low temperature even before anthesis can induce a gemination suppression in the mature seed (Fig. 3.1). The dormancy is apparent when seeds are germinated at 4°C but not at 20 or 30°C (Sawhney, Quick & Hsiao, 1985). When three distinct strains were grown at either constant 15 or 20°C the progeny of plants grown at 15°C were equally dormant but those from plants grown at 20°C showed varying amounts of dormancy and even non-dormancy (Peters, 1982a). Each line showed different proportions of dormant to non-dormant grains. Similar responses indicating that high temperatures suppress and low temperatures promote dormancy have been reported (Sexsmith, 1969; Naylor, 1978; Somody, Nalewaja & Miller, 1984a; Larondelle *et al.*, 1987; Adkins & Simpson, 1988). Simply reducing the night temperature by 5°C (20°C day,

15°C night), compared to a constant 20°C day and night during panicle development increased dormancy in the mature seeds. This suggests that high respiratory activity at night may have helped reduce dormancy. Alternatively, it may indicate that dormancy is enhanced by the presence of high levels of photosynthetic activity during the day. Reference will be made to this point later in this section under photoperiodism.

The arbitrary division of genetically distinct lines into phenotypically

Fig. 3.1. The effect of different temperatures prior to anthesis on the subsequent germination of seeds of *Avena fatua* (Line SH 319) incubated at three temperatures; *(a)* 30°C; *(b)* 20°C; *(c)* 4°C (After Sawhney, Quick & Hsiao, 1985).

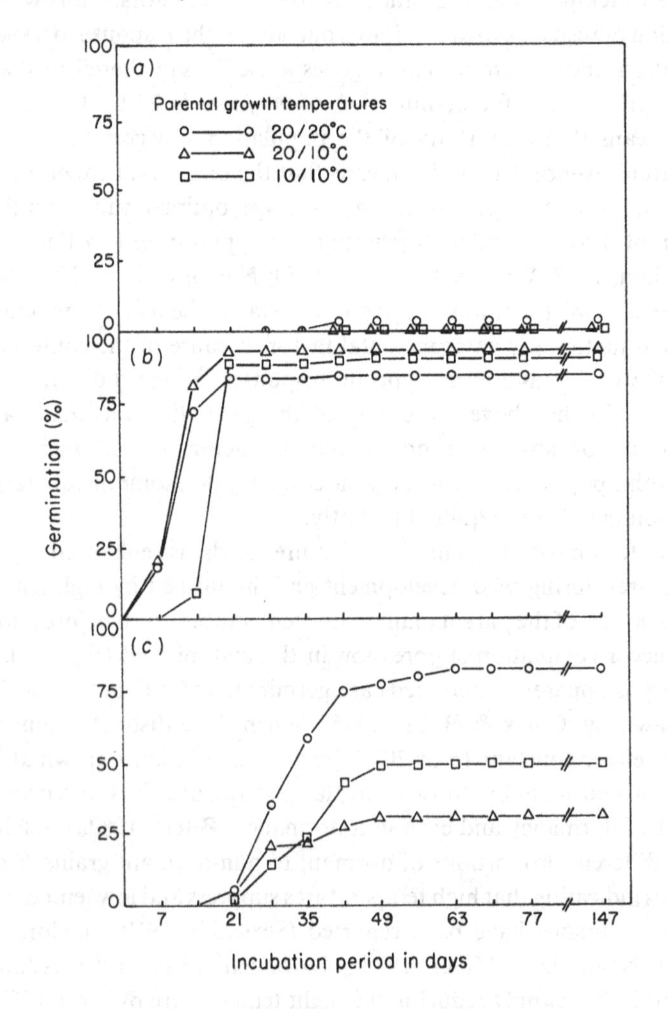

distinct 'dormant' and 'non-dormant' families becomes blurred when the parent plants are exposed to different temperatures during seed development. For example, a number of genetically distinct lines were divided into the above two classes on the basis of the ability of the seeds to germinate at 20°C (Naylor & Fedec, 1978). When the progeny of the dormant group of lines was germinated over a wide range of temperatures, the lines could be separated into sub-groups according to their varied germination responses within the range 4–32°C. In some of the lines germination occurred at all temperatures. These were classified as non-dormant. In others there were varying degrees of suppression of germination over the mid-range of temperatures (Fig. 3.2). However, when the plants from the apparently non-dormant lines were grown at three different day/night temperature regimes (28/22, 20/20, 15/10°C) the progeny showed quite different germination responses at either 4, 20 or 28°C (Sawhney & Naylor, 1980). An example of this response in one of the lines is shown in Fig. 3.3. With increasing reduction in temperature during seed development there was a progressive increase in the dormancy of grains, evident at the highest temperature of germination (28°C). Low temperatures during seed maturation, therefore, induce a sensitivity (dormancy) to high temperatures at the time of germination. In this way the exposure of the parent plant to a particular seasonal regime of day/night temperatures can induce, via the maternal tissues, a particular germination response of the progeny in the subsequent season.

The plasticity of response to temperature within each genetically distinct line undoubtedly has great adaptive value for survival and effectively multiplies the options for germinability in each season for this annual plant.

Within the species *A. fatua* there is clearly considerable genetic diversity in the expression of dormancy in response to temperature. Some lines have very little response to temperature changes while others are very sensitive (Sawhney & Naylor, 1979). The lack of expression of dormancy in some lines exposed to high temperatures during development, as opposed to imposition of dormancy in the same lines at low temperatures, can be thought of as a form of thermal dormancy. That is to say, deeper dormancy is expressed as a broadening of the range of temperatures, either side of some optimal value, in which germination cannot be expressed. The decrease and ultimate removal of dormancy involves a narrowing of the range of temperatures that inhibit germination. High temperatures during seed maturation narrow this range of inhibition temperatures for the mature seed. Low temperatures during seed maturation appear to broaden the range of inhibitory temperatures for germination by increasing the expression of dormancy at progressively lower temperatures. It is for this reason that only very low temperatures (below about 5°C) actively support, rather than inhibit, germination in lines described as dormant on the basis of their germination at 16°C (Fig. 3.4). This type of response where only very low temperatures support germination and higher temperatures are inhibitory, can be found in other grasses; it will be discussed later in this section. Because the genetically distinct lines that would reflect an intermediate response between complete germination, and non-germination, at 20°C have not yet been tested for germination over a wider range of temperatures, it is not possible to say exactly how they would respond. It seems very likely that there would be a continuum of response to temperature during development within each genetically distinct line, and

Fig. 3.2. Germination response to temperature among genetically pure lines of *Avena fatua* and *A. sativa* cultivars Harmon and Torch. The data represent germination during an incubation period of 8 months (After Naylor & Fedec, 1978).

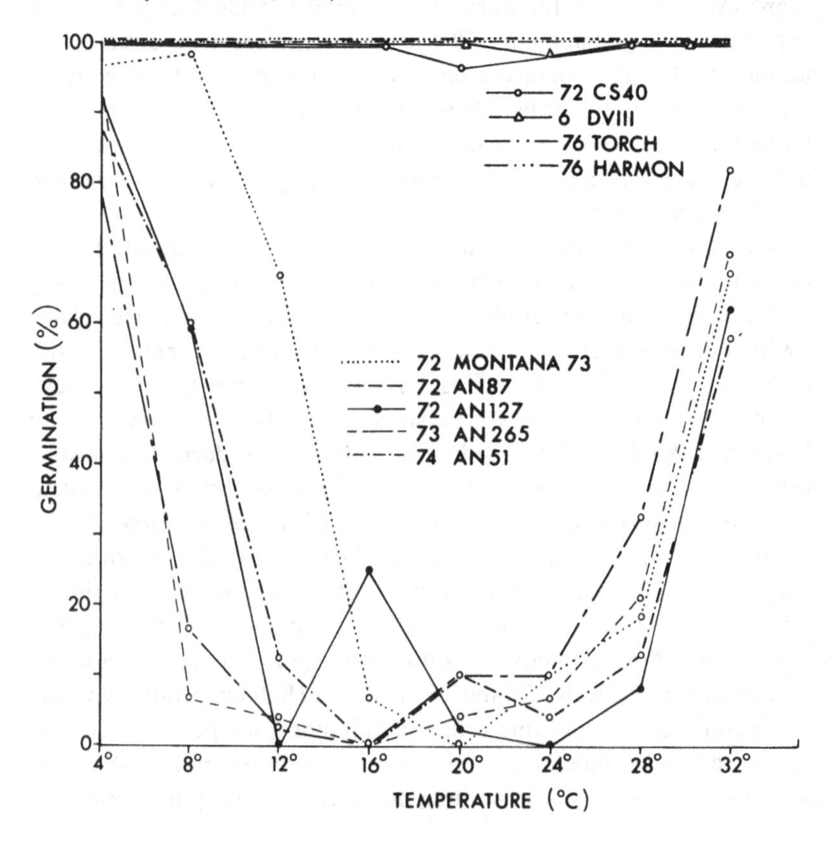

across lines, that would affect germinability of the freshly harvested seed in a pattern similar to that depicted in Fig. 3.5.

The second major effect of temperature is related to the persistence of dormancy in the mature seed. Different strains of *A. fatua* and *A. ludoviciana* after-ripen at different rates at any single temperature (Quail, 1968; Adkins *et al.*, 1986). This genetically based diversity of response underlies some of the apparently conflicting reports about the influence of temperature on after-ripening and loss of dormancy. In addition, the

Fig. 3.3. Influence of incubation temperature on the germination of *Avena fatua*, line SH 430, matured at day/night temperatures of: *(a)* 28/22°C; *(b)* 20/20°C; *(c)* 15/10°C (After Sawhney & Naylor, 1980).

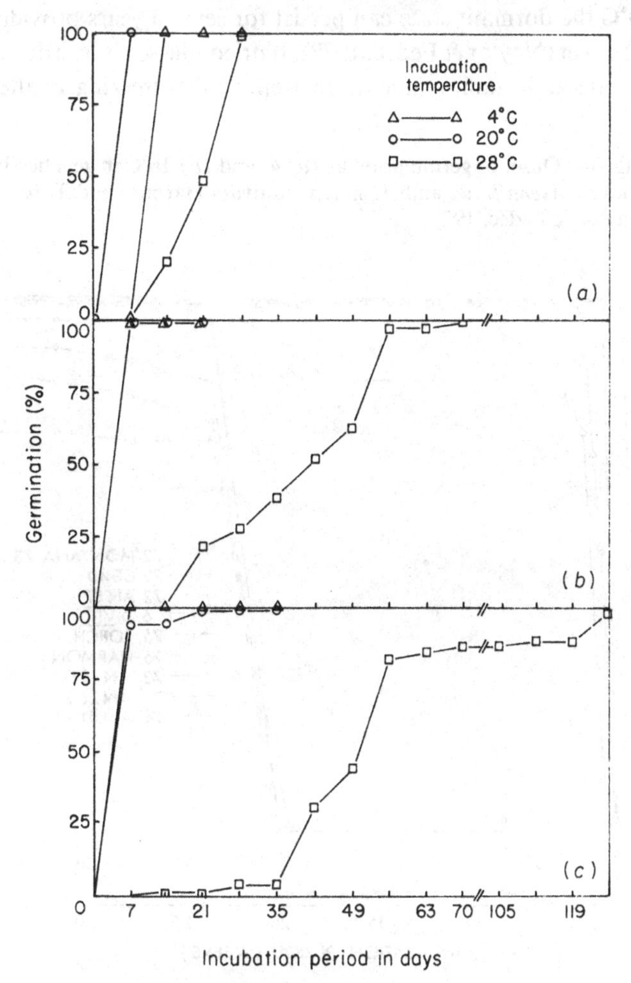

interpretation of temperature effects on loss of dormancy can be confused if a distinction is not drawn between seeds stored dry and seeds that are soaked on water. When seeds are simply air-dried and stored dry, low temperatures prolong dormancy and high temperatures accelerate the loss of dormancy (Johnson, 1935b; Adkins & Ross, 1981a,b; Adkins & Simpson, 1988). Much of this temperature effect can be interpreted as an influence on the state of hydration of the grain. Low temperatures favour high relative humidities and *vice versa*. Low temperatures would thus slow down and high temperatures would accelerate grain dehydration. It has been well demonstrated that prolonged desiccation induces loss of dormancy by influencing hull and seed coat structure. If dormant seeds are soaked in water immediately on removal from the parent plant and are then kept at 20°C the dormant state can persist for several years provided the seeds remain wet (Naylor & Fedec, 1978). If dried, the seeds lose dormancy in a few months at the same temperature. Reports that freezing, or alternate

Fig. 3.4. Onset of germination at *(a)* 4° and *(b)* 16°C in genetically pure lines of *Avena fatua* and *A. sativa*, cultivars Harmon and Torch (After Naylor & Fedec, 1978).

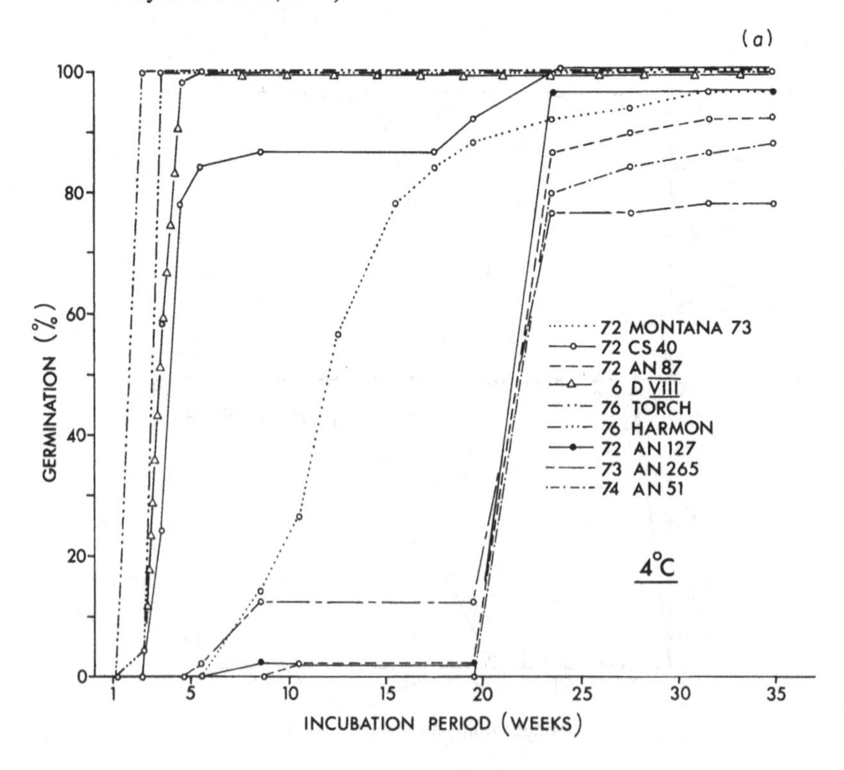

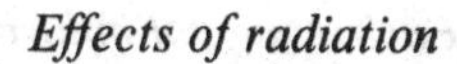

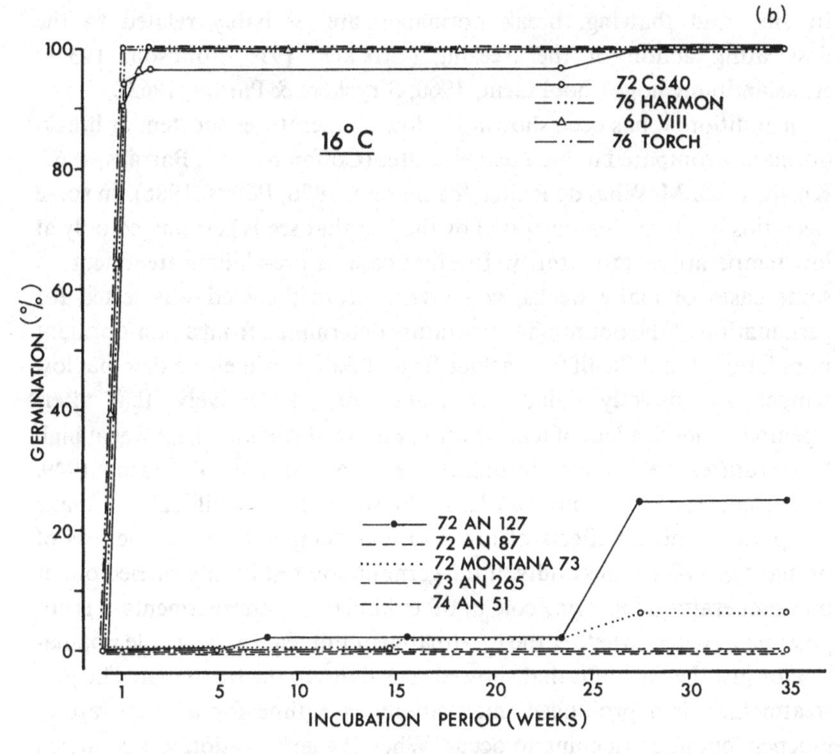

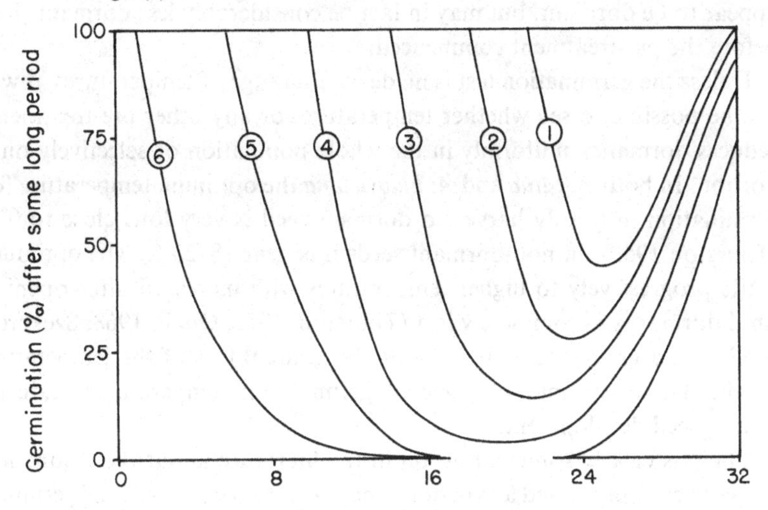

Fig. 3.5. A hypothetical response to incubation temperature among six lines of *Avena fatua* expressing degrees of dormancy ranging from low (1) to high (6).

freezing and thawing, break dormancy are probably related to the desiccating action of the freezing (Crocker, 1916; Johnson, 1935b; Rijkslandbouwhogeschool Gent, 1960; Stryckers & Pattou, 1963).

In addition, it has been shown that low temperatures shorten, or break, dormancy compared to high temperatures (Coffman, 1961; Barralis, 1965; Kurth, 1965; McWha, de Ruiter & Jameson, 1976; Peters, 1986). In some cases this has been demonstrated by the fact that seeds germinated only at low temperatures at maturity. In other cases a pre-chilling treatment, in some cases of many weeks, was given before the seed was tested for germination at the optimum temperature determined from a non-dormant population. It is difficult to conclude from these experiments either that low temperature directly reduces dormancy or, alternatively, that after-ripening is independent of temperature. Firstly, if seeds are kept wet at high temperatures secondary dormancy can be induced (Petersen, 1949; Andersen, 1965; Symons *et al.*, 1986) so that it is difficult to make comparisons of the effects of low and high temperatures on the loss of primary dormancy. Secondly, if the germination test is only carried out at one temperature, following cold or other kind of pre-tretreatments, it is not possible to detect whether the pre-treatment influenced the whole population or just the part of it that showed sensitivity to the treatment. The pre-treatment, when prolonged, may simply give time for a temperature-independent after-ripening to occur. When the germination test is carried out, only that portion of the population that has lost the thermal sensitivity to higher temperatures will germinate. The rest of the population will appear to be dormant, but may in fact be considerably less dormant than before the pre-treatment commenced.

Unless the germination test is made over a range of temperatures, it will not be possible to see whether temperature, or any other pre-treatment, reduces dormancy uniformly in the whole population or selectively on a portion. In both *A. fatua* and *A. ludoviciana* the optimum temperature for germination in freshly harvested dormant seed is very low, close to 0°C (Thurston 1952). In non-dormant seeds it is near 18–20°C. The optimum shifts progressively to higher temperatures with increased after-ripening until dormancy is completely lost (Thurston, 1952; Quail, 1968; Szekeres, 1982). To a great extent this reflects the gradual loss of the temperature sensitivity, in the middle range of germination temperatures, induced during seed development.

There is clearly some confusion in the literature about the distinction between cold-promoted loss of dormancy, and 'apparent' optimal germination at a low temperature simply because high temperatures inhibit germination. The issue will be clarified, by reference to other grass species, later in this section.

A third important effect of temperature on dormancy is the ability of high temperatures (above 23°C) to introduce secondary dormancy in mature grains when they are fully soaked with water, or have a high moisture content (Hyde, 1935; Thornton, 1945; Kiseleva, 1956; Cobb & Jones, 1962; Hay, 1962; Andersen, 1965; Barralis, 1965; Thurston, 1965; Watkins, 1969; Sawhney, Hsiao & Quick, 1984; Tilsner & Upadhyaya, 1985; Symons *et al.*, 1986). There are also many reports indicating that wet seeds exposed to anaerobic conditions (anoxia, nitrogen atmosphere, soaking in unaerated water, etc.) can be brought into a state of secondary dormancy at high temperatures (Christie, 1956; Black, 1959; Mullverstedt, 1963; Kiewnick, 1964; Barralis, 1965; Voderberg, 1965; Hart & Berrie, 1967; Hay, 1967; Fykse, 1970a; Andrews & Burrows, 1972; Schonfeld & Chancellor, 1983; Parasher & Singh, 1985; Symons *et al.*, 1986). The persistence of buried seeds in soil for periods as long as seven years has been attributed to this state of secondary dormancy (Banting, 1966). The depth of the secondary dormancy is inversely proportional to the length of after-ripening (Thurston, 1965; Tilsner & Upadhyaya, 1985; Symons *et al.*, 1987) and seeds that have been after-ripened for a long time lose the capacity to gain secondary dormancy (Thurston, 1957a).

Only seeds that have the genetic capacity for primary dormancy can be induced into secondary dormancy, and the range in persistence of the secondary dormancy is proportional to the persistence of primary dormancy (Tilsner & Upadhyaya, 1985; Symons *et al.*, 1986, 1987). The relationship between genetic origin, persistence of primary dormancy and sensitivity to induction of secondary dormancy is illustrated in Fig. 3.6. In this comparison one line (SH430) had no primary dormancy and could not be induced into secondary dormancy at any temperature, with or without anoxia. Other lines, selected for their sensitivity to induction of dormancy when wetted under anoxia, were also subjected to a range of temperatures at a very high relative humidity but in the presence of oxygen. As temperatures increased, a secondary dormancy was induced that became fully expressed at the highest temperatures. The temperature at which secondary dormancy began to be induced was lower for the line with deeper primary dormancy (Fig. 3.7): the line with deeper dormancy (AN51) was also more sensitive to the induction of secondary dormancy under anoxic conditions. While anoxia induced secondary dormancy at all temperatures, the thermo-induced dormancy was only evident at high temperatures. There is thus a distinction between thermally induced secondary dormancy and that dormancy arising from anaerobic conditions.

It has been argued that the effect of high temperature on promoting secondary dormancy is simply related to the reduced solubility of oxygen at high, compared to low, temperatures (Mullverstedt, 1963). However, it seems more probable that high temperatures directly influence the metabolic system so that oxygen requirement is changed. Both anaerobic conditions and high temperatures directly affect the caryopsis and hulls are not involved (Kommedahl *et al.*, 1958; Hay, 1962). Some of the reports

Fig. 3.6. Germination at 20°C of seed from pure lines of *Avena fatua* after the induction of dormancy by anoxia at two stages of after-ripening. Stage 1 (AR-I), when primary dormancy was almost lost; stage 2 (AR-II), 100 days after primary dormancy was lost. Open bars, air control; shaded bars, anoxia. Induction temperatures were 20° and 26°C (After Symons, Simpson & Adkins, 1987).

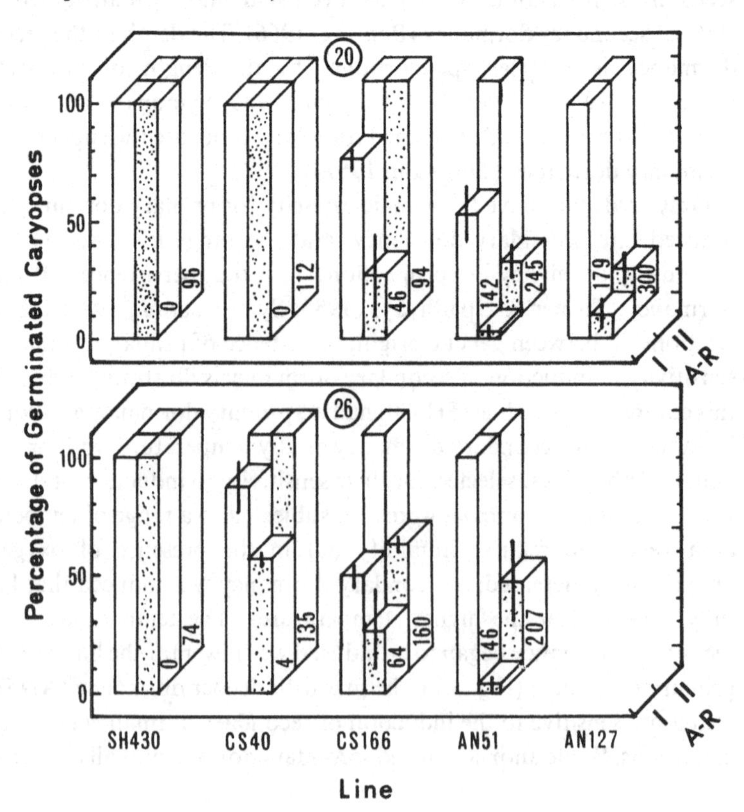

suggest that induction of secondary dormancy may only be a prolongation of primary dormancy under certain conditions such as absence of oxygen. For example, after-ripening involves some oxidative changes in embryos because dry seeds placed in an atmosphere of nitrogen, after-ripen more slowly than in air and, in turn, in an atmosphere high in oxygen (Simmonds, 1971). Thus, keeping germinating seeds in an atmosphere of nitrogen to achieve anoxia, slows the after-ripening so that the seeds appear to germinate later than those in air. The longer the pre-treatment of anoxia the larger the gap will be between treated and untreated germination percentages.

A distinction between thermally-induced and anaerobically-induced secondary dormancy has become apparent since genetically distinct lines have become used as experimental material. Secondary thermo-dormancy seems to be similar to the thermo-dormancy induced during the development of the seed on the parent plant. It is not expressed unless the germinating seed is kept at high temperatures (23–30°C) and the seeds will germinate normally if transferred to lower temperatures. In this sense it is not as persistent as the secondary dormancy induced by anoxia. Thermo-dormancy may be induced after primary dormancy is completely lost and

Fig. 3.7. Germination at 20°C in two genetically distinct lines of *Avena fatua* (CS 166, AN 51) showing the induction of secondary dormancy by anoxia and thermodormancy from high temperatures in air (After Symons, Simpson & Adkins, 1987).

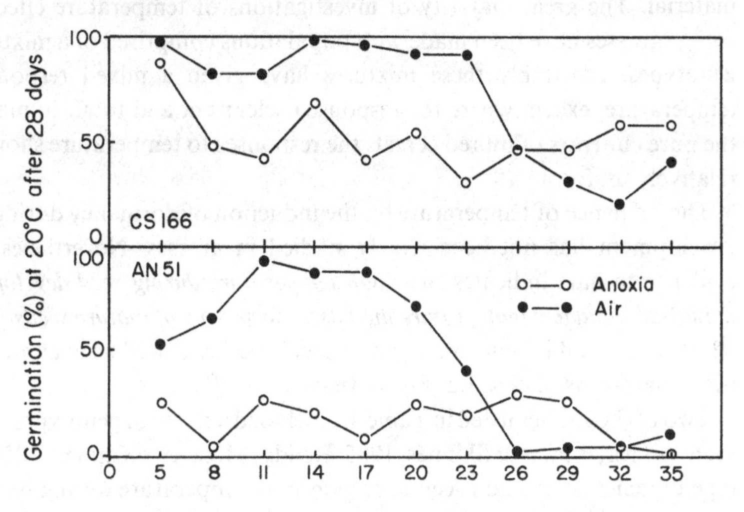

the secondary dormancy of anoxia can no longer be induced. For example, *A. ludoviciana* seeds after-ripened for seven years could not be reintroduced into dormancy at 23 or 25°C, but they would not germinate at 27°C. When transferred back to 7°C they germinated normally (Thurston, 1965). Thermodormancy can be expressed in both dormant and non-dormant lines that are genetically distinct on the basis of their primary dormancy: transfer to a low temperature relieves the dormancy. In some genetic lines the genetic thermo-dormancy is more persistent than in other lines (Sawhney *et al.*, 1984). In others the interaction between thermo-dormancy and anoxia can produce a very persistent secondary dormancy.

In a dry state wild oats are extremely resistant to high temperatures. The upper lethal limit is 105°C (Hopkins, 1936) and there is one report of some seeds surviving 135°C (Williams & Thurston, 1964). All reports of induction of secondary dormancy indicate that the tissue must be hydrated which suggests that anoxia- and high temperature-induced secondary dormancy are related to either membrane integrity or metabolism of active embryonic tissues.

Other grasses:

The same four distinctive effects of temperature seen in *A. fatua* are found in other grasses. Where comparisons have been made between genetically distinct lines or cultivars the degree of post-maturation dormancy varies among genotypes. Thus, the interpretation of temperature effects among grasses must always take account of the genetic nature of the material. The great majority of investigations of temperature effects on forage grasses have been made with populations comprised of a mixture of genotypes. Inevitably these mixtures have given a mixed response to temperature, except where the response is clear-cut and total. In many of the pure cultivars of inbred cereals the responses to temperature should be relatively uniform.

The influence of temperature on the induction of dormancy during seed development has not been widely studied in grasses. Nevertheless, the evidence to date indicates that *high temperature during seed development diminishes, and low temperature increases, dormancy of mature seeds.* These effects are found in temperate and tropical species as well as in annual and perennial forms (Table 3.2, Fig. 3.8).

Two of the studies listed in Table 3.2 included a range of genotypes within each species (Buraas & Skinnes, 1985; Reddy, Metzger & Ching, 1985). The type of genetic variation seen in response to temperature during development, previously found in *A. fatua*, was also noted within these species

Table 3.2. *The influence of temperature during seed development on the induction of dormancy found in mature seeds of 15 grass species. + = enhancement; − = diminishment. Reference source in brackets.*

Species	Temp regime	Effect on dormancy
Aegilops kotschyii	22/17 °C, 17/12 °C	+
(Wurzburger & Koller, 1976	(day/night)	
	22/22 °C, 27/22 °C	−
Agropyron elongatum	Freezing plants	+
(Thornton, 1966a)		
Dactylis glomerata	Constant 15 or 20 °C	+
(Pannangpetch & Bean, 1984)	Constant 25 °C	−
(Probert *et al.*, 1985b)	Constant 6, 11, 16, 26 °C	+, −
Echinochloa frumentacea		
and *E.utilis*	5/10 °C, 10/15 °C,	
(Hughes, 1979)	15/20 °C	+
	(day/night)	
	20/25 °C, 25/30 °C	−
Festuca arundinacea	Constant 15 °C	+
(Boyce, 1973)	Constant 27 °C	−
Hordeum vulgare	Winter greenhouse	+
(Norstog & Klein, 1972)	Warm-day chamber	−
	Mid-summer field conditions	−
(Buraas & Skinnes, 1985)	Constant 9 °C	+
	Constant 15 or 21 °C	−
(Rauber, 1984a)	17/12 °C	+
	(day/night)	
	27/22 °C	−
Lolium spp.	Constant 15 °C	+
(Wiesner & Grabe, 1972)	Constant 30 °C	−
Oryza sativa	Constant 15 °C	+
(Hayashi & Hidaka, 1979)	Constant 30 °C	−
Pennisetum spp.	5/10 °C, 10/15 °C,	
(Hughes, 1979)	15/20 °C, 25/30 °C	+
	(day/night)	
	20/25 °C	−
Phleum pratense	Ecotypes from a cool area	+
(Yumoto *et al.*, 1980)	Ecotypes from a warm area	−
Setaria spp.	Constant 22, 25, 28 or 31°C	−
(Mohamed *et al.*, 1985)	Constant 19 °C	+
Triticale	Constant 9 °C	+
(Buraas & Skinnes, 1985)	Constant 15 or 21 °C	−
Triticum aestivum	Constant 15 °C	+
(Reddy *et al.*, 1985)	Constant 26 °C	−

Table 3.2. *(cont.)*

Species	Temp regime	Effect on dormancy
(Buraas & Skinnes, 1985)	Constant 9 °C	+
	Constant 15 or 21 °C	−
Zea mays (Hughes, 1979)	5/10 °C, 10/15 °C (day/night)	+
	15/20 °C, 20/25 °C, 25/30 °C	−

(*Triticum aestivum*, *Hordeum vulgare* and Triticale). Each species showed a different sensitivity to the same temperatures. *H. vulgare* and Triticale both developed deeper dormancy than wheat from low temperatures during development. In the cultivars with marginal levels of dormancy at maturity, dormancy was only seen at the high germination temperatures. Where deeper levels of dormancy were induced during seed development the dormancy was seen over the full range of temperatures, including the lowest used for germination. One of the most interesting studies, using hulless barley, indicated that temperature differences (27/22°C and 17/22°C day/night) applied during grain development had no influence on the degree of dormancy or germination pattern measured at 6, 18 or 30°C (Rauber, 1985). Nevertheless, some thermo-dormancy was still showing at 30°C germination temperature in both treatments after one year of after-ripening but was absent at lower temperatures. This could indicate that thermo-dormancy is primarily an attribute of the caryopsis and that hulls are the site of the low temperature promotion of dormancy during seed development.

The range of low temperatures that favours imposition of dormancy during seed development is linked to the temperature at which seeds are germinated through the expression of thermal inhibition at high germination temperatures. Mature seeds may either germinate directly at all temperatures (absence of dormancy), partially at low but not at higher temperatures (thermal dormancy), or not at any temperature (full dormancy). Transfer of seeds from the high temperature, in which thermal dormancy is expressed, to a low temperature may lead to normal germination in seeds with shallow dormancy. Repeated cycles of alternation from a high to a low temperature during a 24 h period often lead to accelerated loss of dormancy, compared to allowing the seed to after-ripen only at a single temperature. The physiological nature of the thermal sensitivity induced by low temperatures during seed development is not yet understood.

While high temperatures during seed development, compared to low, favour the absence of dormancy in mature seeds, there is some indication that excessively high temperatures can again lead to the induction of dormancy. In wheat the response to accumulation of heat units during the final stages of seed maturation appears to be bimodal (Grahl, 1975). Dormancy is expressed after both high and low levels of accumulated heat units and is absent at intermediate values. A similar effect of heat units has been noted in *Pennisetum* (Hughes, 1979).

The second major effect of temperature, on the rate at which mature seeds lose primary dormancy (an aspect of after-ripening), has been widely investigated in grasses. Reports of the main seed testing associations

Fig. 3.8. Effect of temperature during seed maturation on the germination response to light and alternating temperatures in seeds from hybrid plants of *Dactylis glomerata* (After Probert, Smith & Birch, 1985b).

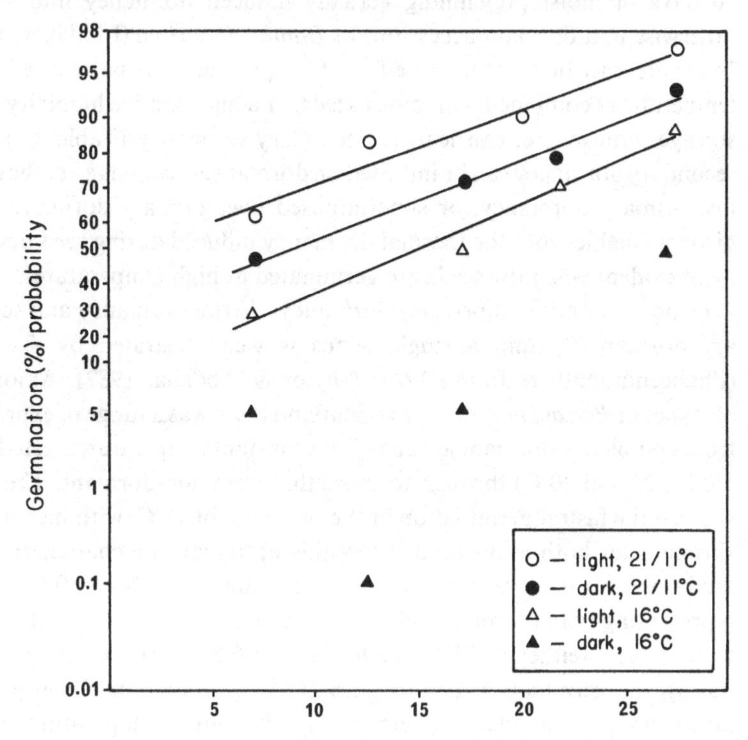

contain many references to the search for particular temperatures, or combinations of alternating temperatures, that promote germination of freshly harvested dormant grass seed. The details of the specific temperatures, or combinations, that effectively overcome dormancy have been well summarized in the *Handbook of Seed Technology for Genebanks* (Ellis, Hong & Roberts, 1985). It is difficult to see patterns in much of this information because the temperature regimes are generally applied to intact florets so that the interactions between hull, seed coat, endosperm and embryo become intertwined. There are surprisingly few references, for example, to studies of temperature on the after-ripening of excised embryos devoid of surrounding structures. Despite these limitations it is possible to discern the same pattern of response to temperature of seeds in many grasses that is also found in *A. fatua*.

The loss of dormancy during storage is accelerated at high temperatures, particularly if the seeds and the environment are dry (Table 3.3). On the other hand, provided seeds are kept dry, primary dormancy is sustained by low temperatures, particularly below 0°C. There is at least one case where 30 days of moist pre-chilling actually induced dormancy into several otherwise non-dormant accessions of *Danthonia sericea* (Lindauer, 1972). In the presence of moisture the effect of temperature may be reversed. High temperatures combined with moist seeds, or a high relative humidity in the storage atmosphere, can lead to secondary dormancy (Table 3.3). This secondary dormancy can be imposed on dormant accessions after they have lost primary dormancy, or superimposed over primary dormancy: it is distinguishable from the thermal dormancy induced during seed development evident when the seeds are germinated at high temperatures.

Genetic variation in primary dormancy, thermodormancy and secondary dormancy within a single genus is well illustrated by *Poa* spp. (Phaneendranath & Funk, 1981; Naylor & Abdalla, 1982). Among 50 ecotypes of *Poa annua* collected in Scotland there was a range of expression from completely dormant at each of the constant temperatures tested (2, 5, 15, 20, 25 and 30°C) through to those that were non-dormant. The latter showed the fastest germination at the optimum of 15°C, with increasingly slowed rates both above and below this optimum, the characteristic bimodal temperature response for non-dormant seeds. Seeds from plants representing both dormant and non-dormant types exposed to alternating temperatures (either 15/25°C or 20/30°C for 16/8 h during each 24 h cycle) during the germination period gave 100% germination. Thus primary dormancy was completely overcome by alternating temperatures within several days of after-ripening. When freshly harvested seed of an essentially

Table 3.3. *The influence of high or low temperature on the persistence of primary dormancy, and induction of secondary dormancy, during the after-ripening of mature grass seeds.*

Species	References
(a) Examples of high temperature promoting the loss of dormancy during after-ripening.	
Agropyron elongatum	Thornton, 1966a
Bromus sterilis	Hilton, 1984a
Bromus tectorum	Thill *et al.*, 1980
Chloris orthonothon	Solange *et al.*, 1983
Chrysopogon fallax	Mott, 1978
Chrysopogon latifolius	Mott, 1978
Digitaria milanjiana	Hacker *et al.*, 1984
Echinochloa crus-galli	Harper, 1970
	Shimizu *et al.*, 1974
	Sung *et al.*, 1983
Eleusine coracana	Shimizu *et al.*, 1974
Eragrostis spp.	Nakamura, 1962
Hordeum vulgare	Hewett, 1958
Oryza sativa	Nakamura, 1963
	Cohn & Hughes, 1981
Oryzopsis hymenoides	Toole, 1940b
Panicum maximum	Smith, 1979
Sorghum bicolor	Stanway, 1959
Sorghum plumosum	Mott, 1978
Sorghum stipoidium	Mott, 1978
Sporobolus cryptandria	Toole, 1941
Triticum aestivum	Hagemann & Ciha, 1987
Themeda australis	Mott, 1978
12 sub-tropical grass spp.	Taylorson & Brown, 1977
(b) Examples of low temperatures accompanied by dryness sustaining the primary dormancy of freshly harvested mature grass seeds.	
Andropogon furcatus	Coukos, 1944
Andropogon scoparius	Coukos, 1944
Bouteloua curtipendula	Coukos, 1944
Bromus inermis	Coukos, 1944
Dactylis glomerata	Probert *et al.*, 1985d
Danthonia sericea	Lindauer, 1972
Echinochloa crus-galli	Shimizu *et al.*, 1974
Eragrostis ferruginea	Fujii & Yokohama, 1965
Festuca spp.	Kearns & Toole, 1938
Oryza sativa	Cohn & Hughes, 1981
Poa annua	Phaneendranath & Funk, 1981

Table 3.3. *(cont.)*

Species	References
Sorghastrum nutans	Coukos, 1944
Triticum aestivum	Hagemann & Ciha, 1987
(c) Examples of high temperatures, together with moisture, inducing secondary dormancy in mature after-ripening grass seeds.	
Agropyron pauciflorum	Nakamura, 1962
Avena sativa	Hyde, 1935
Bouteloua curtipendula	Sumner & Cobb, 1962
Hordeum vulgare	Grahl, 1965
	Andersen, 1965
	Rauber, 1987
Poa annua	Naylor & Abdalla, 1982
Poa pratensis	Phaneendranath & Funk, 1981
Setaria faberii	Taylorson, 1982
Sorghum bicolor	Wright & Kinch, 1959
Triticum aestivum	Hewett, 1959
	Grahl, 1965

non-dormant line was heated to 40°C before the germination period, secondary dormancy (20% germination) was introduced in a population that was otherwise 60% germinable.

Freshly mature seed of two *Poa pratensis* cultivars, illustrating one type with very little dormancy (Fig. 3.9) and another with some thermal sensitivity to germination in high alternating temperatures (20/35°C) but not lower (10/20°C, 15/25°C) (Fig. 3.10) were exposed to a variety of after-ripening conditions (Phaneendranath & Funk, 1981). Both types showed a pronounced induction of secondary dormancy that increased progressively with time at temperatures between 21 and 38°C., provided the relative humidity of the storage was high. Thermal sensitivity, that is the inability to germinate at the high alternating temperatures, was removed in the first 6–8 days of after-ripening at both high and low storage temperatures. Conversely, the progressive increases in temperature and humidity, when prolonged beyond a few days, led to a secondary dormancy that could not be broken by any alternating temperature regime.

The above two examples indicate the problem of distinguishing between direct effects of temperature, moisture and their interactive effects on after-ripening, and any separate effects on the germination process.

Alternating temperatures have been widely used for germinating seeds that show dormancy when freshly harvested. Most grasses respond more rapidly to alternating temperatures than to often lengthy, moist 'pre-chill'

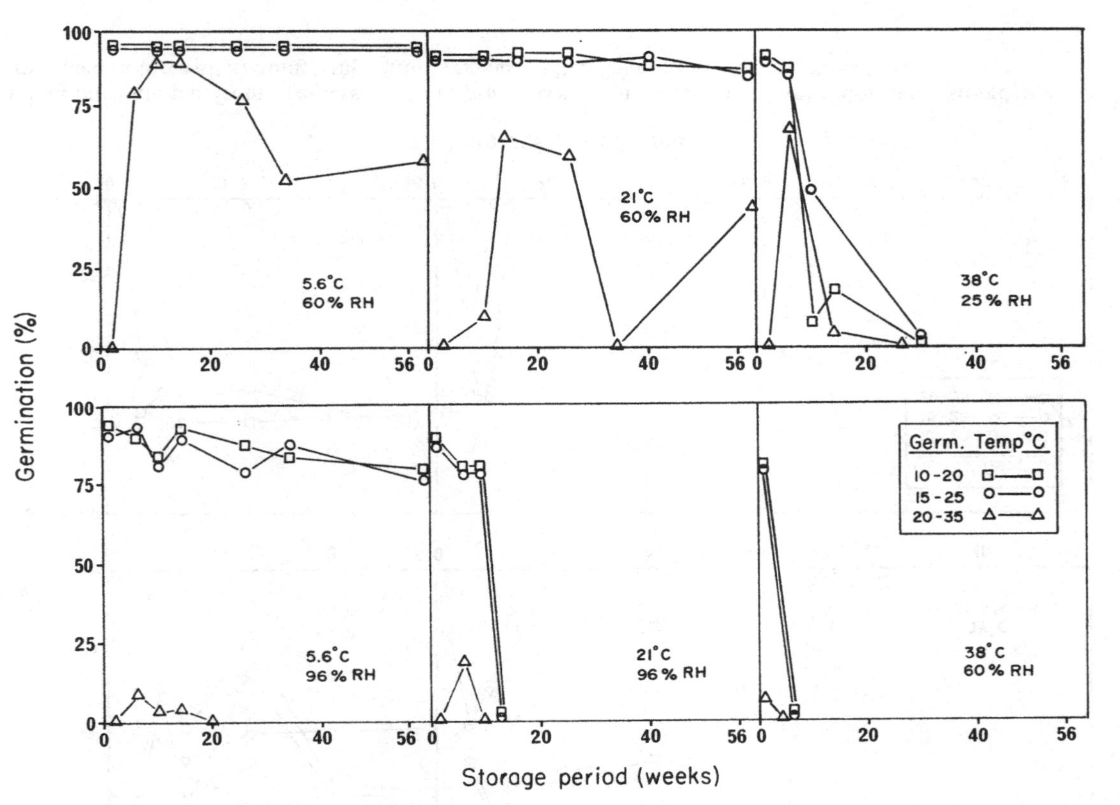

Fig. 3.9. Germination of samples of an accession of *Poa pratensis* with shallow dormancy stored in six conditions for 56 weeks then germinated in three alternating temperature regimes (After Phaneendranath & Funk, 1981).

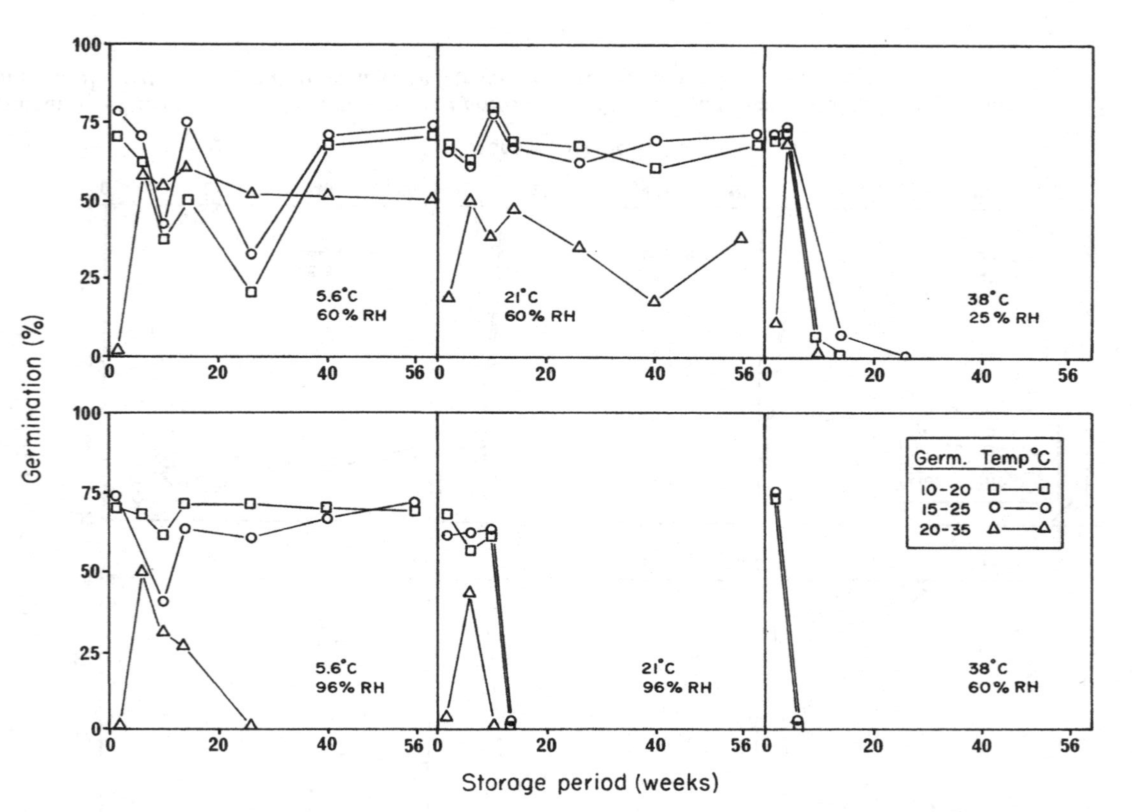

Fig. 3.10. Germination of samples from an accession of *Poa pratensis* with an intermediate level of dormancy stored in six conditions for 56 weeks then germinated in three alternating temperature regimes (After Phaneendranath & Funk, 1981).

that is a form of after-ripening at low temperature (Nakamura, 1962). With alternating temperatures the maximum germination occurs when the warm part of the cycle is of minimum length (Toole & Koch, 1977; Goedert & Roberts, 1986). A comparison, among 40 species of grasses, of both of these treatments that accelerate after-ripening showed that the treatments probably bring about quite separate changes, possibly in different tissues. The six species that could not be made to germinate with alternating temperatures (wheat, barley, oat, African millet, foxtail millet and meadow bromegrass) could be germinated with pre-chilling at low temperature. Of the six different species that could not be germinated with pre-chilling, all could be germinated by alternating temperatures.

Many of these temperature effects are undoubtedly related to structural changes in the hulls, grain coat and endosperm. In several cases (*Hordeum vulgare*, Dunwell, 1981a; *Oryza sativa*, Cohn & Hughes, 1981) it has been demonstrated that excised embryos can germinate over a wide range of temperatures but when in the intact caryopsis, or with hulls present, they can only germinate in a narrow range of temperatures. Hull removal, pericarp damage and scarification often achieve the same acceleration of germination as manipulation of temperature (Ellis, Hong & Roberts, 1985a,b). It is possible that high temperature promotes the breakdown of these barriers to water permeability, both within and outside the caryopsis. Atterberg noted as early as 1907 that a single exposure to ethyl ether vapour was sufficient to overcome much of the dormancy in *Hordeum vulgare* indicating an effect on fat-related structures. More recently, attention has been drawn to the influence of other anaesthetic compounds that influence membranous structures that can also be influenced by temperature changes (Hendricks & Taylorson, 1976; Taylorson & Hendricks, 1981; Taylorson, 1982, 1988).

Example of Avena fatua:

(b) Light Exposure to light during grain development can influence dormancy and directly determine the germinability of mature grain.

Most *Avena* species originated in a Mediterranean environment as winter annuals that germinate in the autumn under short days. They often have a requirement for vernalization in the vegetative phase to be able to flower under long days in the following spring (Sampson & Burrows, 1972; Darmency & Aujas, 1986). *A. fatua* is now dispersed in a great range of latitudes and micro-climates and thus shows considerable variation in response to daylength for flowering habit, particularly where it is a weed in spring-sown crops. Because of this polymorphic behaviour, accessions or ecotypes grown in a single natural, or artificial, environment invariably show great diversity in flowering time, length of grain development and thus depth of grain dormancy (Griffiths, 1961; Thurston, 1963; Odgaard, 1972; Paterson, Boyd & Goodchild, 1976; Sharma, McBeath & Vanden Born, 1977; Somody, Nalewaja & Miller, 1984a). Accessions from low latitudes flower earlier and are more dormant than those of high latitudes (Thurston, 1963; Odgaard, 1972; Paterson, Goodchild & Boyd, 1976). Shortening of the photoperiod increased dormancy of the mature grains (Richardson, 1979). Photoperiod interacts with temperature during grain development to alter the expression of grain dormancy in each genotype (Somody, Nalewaja & Miller, 1984a; Adkins, Loewen & Symons, 1987). For example, when pure lines, chosen to represent a range of expressions of dormancy in the mature seeds, were grown in the same environment there was a positive correlation between the length of period of seed development (a function of photoperiod) and the depth of seed dormancy (Adkins, *et al.*, 1987).

Some of the confusion about light effects on dormancy in the mature seed of *A. fatua* are undoubtedly related to this genetic diversity in response to daylength and temperature during seed development. Variation is found among different accessions, between different years within one genotype or among individual seeds within a single sample taken at random from a field. In addition it is known that the effects of light related to dormancy are only expressed, in many genotypes, following a particular period of after-ripening when dormancy is already partially lost from the sequential removal of some of the blocks to germination (Johnson, 1935b; Cumming & Hay, 1958; Hart & Berrie, 1966; Quail, 1968; Hilton & Bitterli, 1983). Fully after-ripened, completely non-dormant seeds of *A. fatua* are not promoted by light.

Artificial white light, or sunlight, may inhibit the germination of some types of seeds imbibed with water at an optimum temperature for germination (Cumming, 1957; Cumming & Hay, 1958; Hay & Cumming, 1959; Hart & Berrie, 1966; Murdoch & Roberts, 1982; Cairns, 1984). There are genetic differences for this response to light. For example, accessions from Europe were promoted by white light but Canadian and southern African accessions were inhibited (Cairns, 1984). Germination of three Canadian lines with no dormancy in darkness could, nevertheless, be inhibited by white light when freshly harvested (Sawhney, Hsiao & Quick, 1986). Seeds in which secondary dormancy was induced were inhibited by white light (Hay & Cumming, 1959).

The hulls and seed coat interact to attenuate the effects of light known to directly inhibit an excised embryo (Black & Naylor, 1957). In the intact floret white light can penetrate through hulls and influence the embryo: light, compared to darkness, inhibits the intact floret; removal of hulls enhances the inhibition (Szekeres, 1982). When hulls are present, continuous white light, or a short photoperiod alternating with darkness (within 24 hours) are more inhibitory than a long photoperiod. When hulls are removed there is no difference between the light treatments. It is possible that the hulls modify these light effects by altering the water status of the caryopsis. Florets imbibed with restricted amounts of water were inhibited by white, blue, red, and far-red light compared to darkness, but in seeds with a high amount of water the same light treatments promoted germination (Hsiao & Simpson, 1971). Pricked caryopses failed to respond to light in the above fashion but those with seed coats intact remained inhibited indicating an interaction with water status (Cumming & Hay, 1958). Further indication of a connection between light and moisture status of the seed is seen in a report of partially after-ripened florets where the germination rate was 58% in darkness and 40% in white light (Holm & Miller, 1972). Germination was reduced to 10% when seeds were stressed in mannitol for 24 h. The water stress was negated by red light that induced a germination of 45%.

Dormancy imposed by long periods of exposure to low intensities of white light persists after hulls are removed and the coat of the caryopsis is broken. This indicates that the ultimate effect of any interaction between light and water is on the embryo (Thurston, 1963).

Following the discovery of the significance of phytochrome as the receptor pigment for light responses in seeds it has been shown that some of the effects of light in controlling dormancy in *A. fatua* are related to the response of this pigment to red and far-red light. The promotion of germination by white light seen in certain genotypes of *A. fatua* noted in the UK (Adkins, 1981; Hilton & Bitterli, 1983) shows the classical red/far-red (R/FR) light control indicative of the presence of phytochrome. Light that produces either high or low levels of P_{FR} promoted, or inhibited germination, respectively. The germination in one seed stock exposed to a single 10 min irradiation of R, following soaking on water in darkness, could be immediately reversed by FR (Hilton & Owen, 1985). The promotion of germination was increased by successive daily irradiations with R light. This particular genotype may reflect an adaptation to a specific local microclimate where seeds on the surface, or placed beneath the surface by cultivation, are not exposed to high temperatures. In western Canada, where seeds are exposed to high soil-surface temperatures in late summer and sub-zero temperatures in winter, seeds from genetically non-dormant strains can develop a state of secondary dormancy by exposure to diffuse white light and high temperatures.

While some of the differences found in behaviour of seeds responding to white light may be due to genetic differences expressed in different climatic and agronomic conditions, it is more likely that phytochrome is the key receptor for light of low intensity when the R/FR ratio favours promotion, or inhibition. Frankland (1981) has shown that even when a seed is exposed to white light for a prolonged period, the promotory effect of P_{FR} can still be demonstrated but it is dominated by the more powerful antagonistic effect of a high-irradiance photo reaction probably associated with cyclic oscillation and continuous interconversion of the P_R and P_{FR} forms of phytochrome. Phytochrome is definitely present in *A. sativa* seeds indicating that *A. fatua* can be expected to be similar in constitution (Hilton & Thomas, 1987; Tokuhisa & Quail, 1987).

In summary, the effects of light, although attenuated by the presence of hulls, are on the caryopsis and possibly directly on the embryo. There is a link between light effects and respiratory activity in the caryopsis. When partially after-ripened seeds are inhibited by white light, the inhibition can be offset by increase in the partial pressure of CO_2 to 3%, from that of normal air (0.03%) (Hart & Berrie, 1966). The inhibition of white light may be due to a decrease in fixation of CO_2 into malic acid. Malic acid increases when dormancy is lost and decreases when secondary dormancy is induced (Hart & Berrie, 1967). Respiratory activity is increased by R light, prior to the occurrence of germination, in partially after-ripened seed in which R light significantly promoted germination compared to darkness (Hilton & Owen, 1985). Also the stimulatory effects of KNO_3, known to affect the respiratory system directly (Adkins, Simpson & Naylor, 1984b) are much more pronounced in light than in darkness (Hilton, 1984b). It is clear that the early reports suggesting that light has no effects on dormancy and germination of *A. fatua* are incorrect (Atwood, 1914; Banting, 1962). This species shows the classical phytochrome response, although at present it is only demonstrated in a limited number of cultivars, and the inhibitory effects of white and blue light are probably due to the complicated temperature and light interactions that lead to secondary dormancy. The polymorphic nature of wild oat populations undoubtedly adds to the difficulties in interpreting light responses in this species.

Other grasses:

To interpret the role of light in inducing, or terminating, seed dormancy in grasses it must be recognized that there are several factors that modify light effects. Firstly, the dark-inhibition (dormancy) of seeds that can be overcome by light, declines during after-ripening and eventually disappears (Bass, 1951; Delouche, 1958, 1961; Koller & Roth, 1963; Fujii & Yokohama, 1965; Hagon, 1976). This probably accounts for a number of reports in the literature where no response to light could be found. Fully after-ripened seeds germinate equally well in light and dark over a wide temperature range.

Secondly, there are both genetically and environmentally induced differences, among ecotypes and accessions within a species, in the degree of response to light (Delouche & Bass, 1954; Kinch, 1963; Nakamura, 1963; Toole & Borthwick, 1971; Taylorson & McWhorter, 1969; Lindauer, 1972; Juntilla, 1977; Naylor & Abdalla, 1982). For example, in a comparison of *Poa pratensis* accessions it was noted that no single light intensity could stimulate germination in all the accessions because they were of different genetic background and age (Bass, 1951). The sensitivity to light decreased with age of the seed. A comparison of accessions of *Dactylis glomerata* from different geographical regions of north and south Europe showed that seed from the northern types was dependent on light for germination (Probert, Smith & Birch, 1985b). The southern types could germinate in darkness but if exposed to long days became dormant. It was concluded from the study that dormancy in the northern types was due to low levels of P_{FR} in mature seeds that allowed a sensitivity to low levels of light energy. Alternatively, the southern types were induced to become dormant because of high levels of P_{FR}, combined with a sensitivity to high irradiance energy (HER). In another comparison among a wide range of accessions of *Eleusine coracana* tested at maturity, some types could be induced to germinate by light at a high temperature (30°C) and others only at 15°C (i.e. they were dormant to light at 30°C) (Shimizu & Mochizuki, 1978). Still others were inhibited at 15 and 30°C in the presence of light. In this way temperature interacted with light, among different genotypes within the species, to give a range of different responses to any single environment.

Thirdly, light can influence seeds with secondary dormancy as well as with primary dormancy (Fukuyama, Takahashi & Hayashi, 1973; Lush & Groves, 1981; Taylorson, 1982). Thus at high temperatures, or other conditions that induce secondary dormancy, light sensitivity can be re-introduced in seed that was previously non-dormant.

It is quite probable that all grasses with seed dormancy show sensitivity to light based on the plant pigment phytochrome. A survey of the action of light on seed dormancy among grass species (Table 3.4) indicates that 67 of the 80 species respond positively to light by losing dormancy. In 28 of these light-promoted species the phytochrome responses to R/FR light were found to be present. Of the 8 species in which light inhibited germination,

Table 3.4. *The action of light on seed dormancy, among and within grass species, estimated from documented evidence of light promotion (+) or inhibition (−) of germination. Proof of phytochrome (P) or probable high energy irradiance (HER) involvement in the response is included. Reports indicating no response to light (N), genetic, climatic and geographical diversity in the response (G) or absence of a light response in fully after-ripened seed (A) indicate that other environmental conditions during maturation, storage and germination can interact with genotype to modify the influence of light on dormancy. Reference source in brackets.*

Species	Germination	Mechanism	Condition
Agropyron			
– *cristatum*	+		
(Nakamura, 1962)			
– *cristatum*	–		
(Nakamura *et al.*, 1960)	–		
– *lehmanniana*	+		
(Wright, 1973)			
– *pauciflorum*	+		
(Nakamura, 1962)			
– *repens*	+		
(Andersen & Drake, 1944)			
– *repens*	+		
(Williams, 1968)			
– *repens*	+	P	
(Holm & Miller, 1972)			
– *smithii*	+		
(Bass, 1959)			
– *smithii*	+		G, N
(Kinch, 1963)			
– *smithii*	+	P	
(Schultz & Kinch, 1976)			
– *smithii*	+		G
(Bass, 1955)			
– *smithii*	–		
(Bass, 1959)			
– *smithii*	–		
(Kinch, 1963)			
– *smithii*	–		
(Nakamura, 1962)			
– *smithii*	–		
(Delouche & Bass, 1954			
– *smithii*	+	P	
(Toole, 1976)			

Table 3.4. *(cont.)*

Species	Germination	Mechanism	Condition
Agrostis			
– *capillaris* (Williams, 1983a)	+	P	
– *gigantea* (Williams, 1973)	+		
– *palustris* (Toole & Koch, 1977)	+	P	G
– *tenuis* (Nakamura, 1962)	+		
– *tenuis* (Andersen, 1944)	+		
– *tenuis* (Pierpoint & Jensen, 1958)	+	P	
– *tenuis* (Williams, 1968)	+		
– *tenuis* (Toole & Koch, 1977)	+	P	G
Alopecurus			
– *myosuroides* (Froud-Williams *et al.*, 1984)	+	P	
– *pratensia* (Nakamura, 1962)	+		
Arrhenatherum			
– *elatius* (Nakamura, 1962)	+		
– *elatius* (Froud-Williams *et al.*, 1984)	+	P	
Avena sativa (Nakamura, 1962)	+, –		
Axonopus			
– *compressus* (Nakamura, 1962)	+		
– *compressus* (Andersen, 1941)			A
Bothriochloa			
– *macra* (Hagon, 1976)	+		
Bouteloua			
– *eriopoda* (Wright & Baltensperger, 1964)	–		

Table 3.4. *(cont.)*

Species	Germination	Mechanism	Condition
Bromus			
– *catharticus*	–	P	
(Froud-Williams & Chancellor, 1986)			
– *commutatus*	+		
(Nakamura, 1962)			
– *commutatus*	–	P	
(Froud-Williams & Chancellor, 1986)			
– *inermis*			N
(Grabe, 1955)			
– *mollis*		+	
(Nakamura, 1962)			
– *mollis*	–		
(Ellis, Hong & Roberts, 1986)			
– *ramosus*	+		
(Drake, 1949)			
– *ramosus*	+		
(Nakamura, 1962)			
– *sterilis*	–	P	
(Hilton & Owen, 1985a)			
– *sterilis*	–		
(Hilton, 1984a)			
– *sterilis*	–		
(Froud-Williams *et al.*, 1984)			
– *sterilis*	–	HER ?	
(Ellis, Hong & Roberts, 1986)			
– *sterilis*	=,+	P	
(Hilton, 1987)			
– *sterilis*	–,+	P	
(Hilton, 1982)			
Chloris			
– *gayana*	+		
(Sharir, 1971)			
– *orthonothon*	+	P	
(Solange *et al.*, 1983)			
Cynodon dactylon	+		
(Nakamura, 1962)			
Cynosurus cristatus	+	P	
(Williams, 1983a)			
Dactylis			
– *glomerata*	+		G
(Juntilla, 1977)			
– *glomerata*	+		G

Table 3.4. *(cont.)*

Species	Germination	Mechanism	Condition
(Pollard, 1982)			
Danthonia sericea (Lindauer, 1972)	–		G
Danthonia spp. (Hagon, 1976)			N
Digitaria ciliaris (McKeon, 1985)	+		
Echinochloa			
– *crus-galli* (Holm & Miller, 1972a)	+	P	
– *crus-galli* (Nakamura, 1962)			N
– *turnerana* (Conover & Geiger, 1984a)	+		
– *turnerana* (Conover & Geiger, 1984b)	+		
Eleusine			
– *coracana* (Shimizu & Tajima, 1979)	–		G
– *coracana* (v. dormant strains) (Shimizu & Tajima, 1979)		P	
– *corocana* (v.dormant strains) (Shimizu & Tajima, 1979)			N
– *indica* (Fulwider & Engel, 1959)	+		
Elymus triticoides (Gutormson & Wiesner, 1987)	+		
Eragrostis			
– *curvula* (Toole & Borthwick, 1971)	+	P	G
– *ferruginea* (Isikawa *et al.*, 1961)	+	P	
– *ferruginea* (Fujii & Yokohama, 1965)			A
Eremochloa ophiuroides (Delouche, 1961)	+		N, G
Festuca			
– *arundinacea* (Nakamura, 1962)	+		
– *arundinacea*	+	P	

Table 3.4. *(cont.)*

Species	Germination	Mechanism	Condition
(Delouche, 1961)			
– *octoflora*	+		
(Hylton & Bass, 1961)			
– *octoflora*	–		
(Huang & Hsiao, 1987)			
– *ovina*	+		
(Nakamura, 1962)			
– *pratensis*	+		
(Nakamura, 1962)			
– *rubra*	+		
(Nakamura, 1962)			
– *rubra*	+	P	
(Williams, 1983a)			
Holcus			
– *lanatus*	+		
(Nakamura, 1962)			
– *lanatus*	+		
(Williams, 1983a)			
Hordeum			
– *vulgare*	+, –	P	
(Norstog & Klein, 1972)			
– *vulgare*			N
(Nakamura, 1962)			
Lolium			
– *multiflorum*	+		
(Nakamura, 1962)			
– *rigidum*	+		
(Gramshaw, 1972)			
– *perenne*	+		
(Nakamura, 1962)			
– spp.	+		
(Andersen, 1947b)			
Oryza sativa	–	HER	G
(Nakamura, 1963)			
Oryzopsis			
– *hymenoides*	–		
(Clark & Bass, 1970)			
– *miliacea*	+, –	P, HER	
(Negbi & Koller, 1964)			
Panicum			
– *dichotomiflorum*	+	P	

Table 3.4. *(cont.)*

Species	Germination	Mechanism	Condition
(Taylorson & Hendricks, 1979a)			
– *dichotomifolium*	+	P	
(Taylorson, 1980)			
– *dichotomifolium*	+		
(Baskin & Baskin, 1983			
– *fasciculatum*	+		
(Andersen, 1961)			
– *maximum*	+		
(Harty *et al.*, 1983)			
– *ramosum*	+		
(Andersen, 1962)			
– *turgidum*	+		A
(Koller & Roth, 1963)			
– *virgatum*	+	P	
(Holm & Miller, 1972)			
Pennisetum glaucum	+		
(Andersen, 1958)			
Phalaris			
– *arundinacea*	+	P	
(Juntilla *et al.*, 1978)			
– *arundinacea*	+		
(Landgraff & Juntilla, 1979)			
– *arundinacea*	+		
(Colbry, 1953)			
– *arundinacea*	–		
(Nakamura, 1962)			
– *tuberosa*	+, –		
(Nakamura, 1962)			
Phleum pratense	+		
(Gordon, 1951)			
Poa			
– *annua*	+	P	
(Froud-Williams *et al.*, 1984)			
– *annua*	+		
(Naylor & Abdalla, 1982)			
– *annua*	+	P	
(Toole & Borthwick, 1971)			
– *capillata*	+		
(Kearns & Toole, 1938)			
– *compressa*	+		
(Fryer, 1922)			
– *compressa*	+		

Table 3.4. *(cont.)*

Species	Germination	Mechanism	Condition
(Andersen, 1938)			
– *pratensis*	+		G, A
(Bass, 1951)			
– *pratensis*	+		
Toole & Borthwick, 1971)			
– *pratensis*	+		
(Bass, 1950b)			
– *pratensis*	+		G
(Naylor & Abdalla, 1982)			
– *pratensis*	+		
(Andersen, 1955)			
– *pratensis*	+		A
(Delouche, 1958)			
– *trivialis*	+	P	
(Froud-Williams *et al.*, 1984)			
– *trivialis*	+		
(Froud-Williams *et al.*, 1986)			
– *trivialis*	+	P	
(Williams, 1983a)			
– *trivialis*	–		
(Hilton *et al.*, 1984)			
Setaria			
– *viridis*	+	P	
(Holm & Miller, 1972)			
– *viridis*			N
(Nakamura, 1962)			
– *faberii*	+	P	
(Taylorson, 1982)			
– *faberii*	+	P	
(Hendricks & Taylorson, 1978)			
Sorghastrum nutans			
(Emal & Conard, 1973)	+	P	
Sorghum			
– *halapense*	+		
(Taylorson & McWhorter, 1969)			
– *halapense*	+, –		
(Huang & Hsiao, 1987)			
– *halapense*	+	P	
(Taylorson, 1975)			
– *halapense*	+		G
(Taylorson & McWhorter, 1969)			
Sporobolus			

Table 3.4. *(cont.)*

Species	Germination	Mechanism	Condition
– *contractus* (v.dormant) (Toole, 1941)			N
– *flexuosus* (v. dormant (Toole, 1941)			N
– *giganteus* (v.dormant) (Toole, 1941)			N
– *wrightii* (Toole, 1941)	+		
Stipa			
– *viridula* (Nifenegger & Schneiter, 1963)	+,–		
– *viridula* (Fulbright *et al.*, 1983)	–		
– *bigeniculata* (Hagon, 1976)			N
Themeda australis (Hagon, 1976)			N
Triticum			
– *aestivum* (Khan *et al.*, 1964)	+	P	
– *aestivum* (Nakamura, 1962)			N
Zoysia japonica (Nakamura, 1962)	+		

four were shown to have the P system. One of these species (*Bromus sterilis*) may be unique in that the P_{FR} can be inhibitory to germination (Hilton, 1982). In most known cases of P influence in plants P_{FR} is promotory to germination (Toole, 1961). *Oryza sativa* has been considered to be inhibited by light by some other mechanism than P, possibly the HER effect (Nakamura, 1963).

Thus, in 75 of the 80 species, light has either a promoting or inhibiting effect in breaking or inducing dormancy, respectively. The phytochrome system has been demonstrated in one half of that number and it seems highly probable that the rest will be shown to be similar.

The identification of the precise physiological action of light in either stimulating or inhibiting germination of the embryo of grasses is complicated by the interactions with the surrounding covering structures.

The hulls and coat of the caryopsis modify permeability to moisture and gases governing the metabolism of the embryo. Accelerated after-ripening at high temperatures (Taylorson & Brown, 1977), alternating temperatures (Probert *et al.*, 1985a), nitrate (Bass, 1951), chemical scarification (Huang & Hsiao, 1987), desiccation (Fujii & Yokohama, 1965) and ethanol (Taylorson & Hendricks, 1979a) each interact strongly with light. The response to these factors disappears, along with the light dependency, as after-ripening proceeds. Simply removing the hulls and seed coats may remove dark-inhibition of germination. For example, in *Echinochloa turnerana* after-ripened seed could not be germinated in darkness but germinated in light (Conover & Geiger, 1984a). De-hulling increased germination from 2 to 100% showing that light enables the embryo to overcome the repressive influence of hulls. In *Agropyron smithii* light inhibits the germination of intact florets compared to darkness (Bass, 1959). When the lemma and pericarp are pierced to expose the embryo, germination is virtually complete. In this case hulls enhance the light inhibition of the embryo. In *Sorghastrum nutans* failure of intact florets to germinate in darkness can be partially overcome by light (34%) and considerably enhanced by chemical scarification (62%); the scarification had no effect on dark germination. When the seeds were after-ripened for a longer period germination was complete in light, whether florets were intact or scarified. Embryos were still repressed in darkness (50%) with no difference between the intact and scarified treatments. Thus the hulls repressed the light promotion of germination.

The covering structures (hulls and pericarp) were considered to be the main determinants of dormancy in *Sorghum halapense* (Huang & Hsiao, 1987). They regulated moisture and gas exchange that in turn modified the response to both temperature and light. The effects of light were inhibitory at 22°C, promotory at 35°C and without effect at the intermediate temperature of 28°C. Breaking the seed coverings by scarification or accelerated after-ripening at 50°C removed these interactions and promoted germination in both light and darkness.

In many cases the effects of light during the early stages of after-ripening are additive to the stimulatory effects of either alternating temperatures (Hylton & Bass, 1961; Koller & Roth, 1963; Williams, 1973; Wright, 1973; Toole & Koch, 1977; Roberts & Benjamin, 1979; Probert *et al.*, 1985d) or nitrate (Colbry, 1953; Delouche & Bass, 1954., Hylton & Bass, 1961; Cohn, Butera & Hughes, 1983). In some cases all three factors are additive (Roberts & Benjamin, 1979). The timing of the separate actions is different (Taylorson, 1975), with the light effect following the changes induced by alternating temperatures. The alternating temperatures seem to be necessary for removing the thermo-dormancy expressed at high temperatures. There is a distinct separation of part of the aspect of dormancy imposed on the grain during maturation, and part that is sensitive to light and temperature at maturation, or during the early stages of after-ripening. A study with *Dactylis glomerata* showed that the effects of light and temperature are additive at the time of germination and quite independent of the temperature of grain development (Probert *et al.*,1985b). The temperature- and light-promoted loss of dormancy are linked in some way because cultivars after-ripen at the same rate at a low temperature (15°C, 15% RH) quite independently of the level of dormancy expressed at maturity. This dormancy, discussed previously in the sub-section on temperature, was shown to be dependent on the temperature during seed development.

Further evidence that the structures surrounding the embryo and caryopsis cause the limitations on embryo germination can be seen in cases where the embryo is excised from these structures. Excised embryos of *Oryza sativa* show no dormancy and the combination of light and nitrate necessary to achieve germination in the intact floret is without effect (Gutormson & Wiesner, 1987). On the other hand, the embryo is clearly the site of primary action of phytochrome because dormancy can be induced in the non-dormant excised embryos of *Hordeum vulgare* (Norstog & Klein, 1972) and *Agropyron smithii* (Delouche & Bass, 1954) by exposure to red light.

There is evidence for more than one kind of light reaction influencing the state of dormancy in grass seeds (Hylton & Bass, 1961; Toole & Borthwick, 1971; Toole, 1976; Pollard, 1982; Ellis, Hong & Roberts, 1986) and the following example of *Poa pratensis* illustrates these separate actions (Toole & Borthwick, 1971). Seeds germinate well in darkness provided they are given a regime of alternating temperatures. They do not germinate at constant temperature. In a 15–25°C regime the promotive effects of temperature are achieved in 5 or 6 cycles when the daily period at 25°C is between 4–14 h. Brief, daily high intensity fluorescent light could induce germination at 20°C, an action typical of the P response. On the other hand, continuous lighting, of medium to a high intensity promoted germination that was weak and seeds already potentiated to germination were inhibited, an action typical of HER. One cultivar of *Poa pratensis* was more inhibited by this light treatment than the other.

An alternative explanation of the dual effects of white light on germination has been proposed for *Oryzopsis miliacea* (Negbi & Koller,

1964). Short periods of irradiation promote germination under all temperature conditions, but continuous irradiation is either less promotory or even inhibitory. While P_{FR}, the promotive form for germination, was considered the form of P found at equilibrium in white light, the explanation of the dual effect was thought to be related to the different stages of after-ripening experienced within the population of seeds. Short irradiations with FR produced inhibition in the early hours and promotion in the later hours of the experiment. The FR acted first on the fraction of the population normally able to germinate in the dark (i.e. partially after-ripened) by inhibiting germination. Later, the FR light promoted the remaining light requiring fraction of the population because the P_R form of the pigment was being gradually activated. Thus the blue and FR parts of the continuous white light were inhibitory and the R fraction promotory, both via the P system due to the natural variation within the population of mature seed.

There is very little information about the effects of radiation of wavelengths shorter than visible light on seed dormancy in grasses. Osborne (1980) has shown that senescence occurs in both dry and moistened (dormant) grains when they are stored for long periods. There is a gradual loss of membrane integrity and enzyme activity, as well as deterioration of the DNA template. It is not clear whether DNA stability, which is higher in moistened than dry grains, is due to repair or prevention of damage. The factors that lead to senescence are not understood but it is well known that gamma- and X-rays can alter DNA and cause lesions that produce mutant seedlings. There is little evidence yet to indicate that such treatment is of natural significance in either inducing or breaking dormancy in grass seeds. There is a report that low doses (2–4 kR) of gamma-radiation stimulated germination in *Avena fatua*: high doses (20–50 kR) killed all viable seeds (Suss & Bachthaler, 1968). Dormant seeds from varieties of *Triticum aestivum*, *Hordeum vulgare* and Triticale lost dormancy after exposure to 10 kR of gamma-radiation (Rana & Maherchandani, 1982).

Microwaves killed seeds of *Triticum aestivum* and *Avena fatua* by their heating action when the moisture content was high (Lal & Reed, 1980) but there was no evidence for any ability to break dormancy at low dosages.

Various claims have been made for enhanced germinability and seedling growth leading to increased grain yields from treating cereal seeds with a high intensity magnetic field. These remain to be substantiated.

(c) Photoperiodism In nature, photoperiod changes cyclically during both summer and winter at latitudes north and south of the equator. These changes in photoperiod affect grass species, and variants within a species, in

different ways and also interact with temperature to control the timing of flowering through the action of the P system (Evans, 1964). It was shown in the earlier parts of this section that timing of flowering and seed maturation has an important influence on the level of dormancy found in mature seeds. Photoperiod also influences the release from dormancy, and germination, via the P and HER systems found in seeds. While some of these effects were discussed earlier in separate sections they are brought together for discussion in this section.

Example of Avena fatua:

There is genetic diversity within *A. fatua* for the timing of flowering in response to photoperiod. For example, 29 accessions screened by growing them in four different photoperiods, demonstrated considerable diversity in both length of time to anthesis and depth of dormancy at grain maturity (Somody *et al.*, 1984a). Long days shorten the time to flowering (Sampson & Burrows, 1972), and short days delay flowering and time to grain maturation, thus increasing the level of grain dormancy (Sharma, McBeath & Vanden Born, 1977; Richardson, 1979). When accessions from a range of latitudes were compared at a single location in the UK the plants from low latitudes flowered sooner than those from high latitudes so that there was considerable variation in expression of seed dormancy (Thurston, 1963). When seeds of *A. fatua* and *A. ludoviciana* were incubated in a long day (16 h), 18% of the *fatua* and zero of the *ludoviciana* germinated (Thurston, 1963). Under a short photoperiod (8 h), dormancy was more easily overcome and 53% of the *fatua* and 92% of the *ludoviciana* germinated. The inhibitory effect of the long days persisted even after the seeds were removed from long days, and were dehusked and the caryopses pricked. These seeds, when returned to the long day conditions, had low germination but did germinate fully under the short day regime. A similar effect has been noted with European accessions (Szekeres, 1982).

Other grasses:

Many temperate grass species have both a vernalization and photoperiodic requirement for the induction and timing of flowering (Evans, 1964). The timing of germination of progeny seed, in relation to flowering and grain maturation on the parent plant, is determined by the degree of primary grain dormancy and is also related to the temperature and photoperiodic changes of the environment after grains reach maturity. In some cases the grain may take on secondary dormancy. Taken together these possibilities ensure that germination is not only prolonged through time but also that flushes of germination occur in significant portions of the total population providing adaptive strategies for competition and survival. This can be illustrated with the winter annual weed grass *Bromus japonicus* (Baskin & Baskin, 1981). Seeds of this temperate species normally germinate in the autumn. Plants overwinter in the rosette form and become vernalized under the low temperatures and short days, so that they flower under the long days of late spring and early summer. About one third of the mature grains have primary dormancy and most grains that fall to the ground germinate immediately, or after-ripen quickly and then germinate in the autumn. Nevertheless, a majority of the grains remain on the plant and are not dispersed until winter. The combination of low temperatures and short days at the soil surface induces secondary dormancy during the winter. This dormancy persists until the late summer so that a flush of germination occurs in the autumn. Thus, most seeds germinate a year later than they developed, so that germination is spread across two seasons.

It is generally assumed that cyclic temperature changes in the seasons are very significant in controlling periodic germination from seed banks in the soil. However, recent evidence suggests that photoperiodic responses of grains to light, coupled with light intensity (photon flux), may have a very significant role in controlling the germination of some grass seeds. These effects, generally inhibitory to germination, are in addition to the fact that low energy light promotes germination through the presence of P in most grass seeds. Genetic adaptation to photoperiod can be illustrated with *Poa annua* (Naylor & Abdalla, 1982) (Fig. 3.11). Three populations, germinated in darkness and three different photoperiods, each produced the highest germination rate and total germination in an 8 hour photoperiod. The two populations with complete dormancy in darkness were inhibited by long periods of light. The population with partial dormancy in darkness was not inhibited by long periods of light. This kind of photoperiodic light inhibition of germination is probably related to the inhibitory action of light associated with the HER reaction known to inhibit germination in seeds of *Brachiaria humidicola*, *Echinochloa turnerana*, *Panicum maximum*, *Eragrostis tef*, *Bromus mollis* and *B. sterilis* (Ellis, Hong & Roberts, 1986).

3.3 Latitudinal influences on dormancy – tropical and temperate zones

Evidence from the detailed studies on *Avena fatua* indicates that this annual weed has penetrated deeply into world agriculture in the temperate, Mediterranean and even tropical zones (Thurston & Phillipson, 1976). *A. fatua* probably originated in south-west Asia. In a similar manner most of the common cereals and forage grasses, together with their grassy weeds, have achieved widespread distribution through these zones far beyond their centres of origin. This distribution is related to the mercantile activities of the fifteenth to eighteenth centuries (Simpson, 1981). For example, the majority of cultivated forage grasses originating in the Mediterranean or northern Eurasian zones, were transferred to the Americas, southern Africa and Australasia before the end of the nineteenth century (Grigg, 1974). This process still continues unabated (Hawkes *et al.*, 1976).

It might be expected that a great deal of adaptation and diversification in the expression of seed dormancy would occur within each of these transported grass species. A grouping of grasses under physiological traits that reflect adaptation to tropical and temperate zones has been made, for example, with the C_3 and C_4 types of respiratory metabolism. It is of interest to consider whether the traits that confer seed dormancy in grasses are similar, or different, within the groups adapted to tropical or temperate zones.

Aside from the strictly structural aspects of the control of dormancy, it is clear from the earlier sections of this chapter that there is considerable genetic diversity in the expression of dormancy, partly regulated by photoperiodism and temperature differences related to latitude. The

Fig. 3.11. The influence of photoperiod during germination of seeds from three distinct populations of *Poa annua* (After Naylor & Abdalla, 1982).

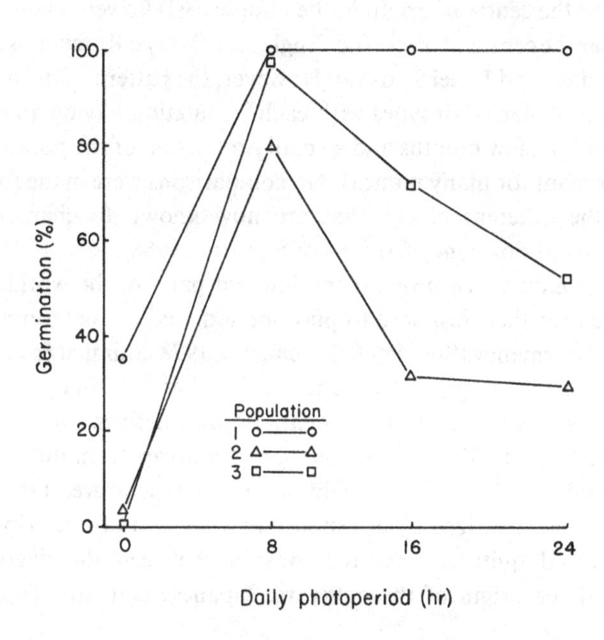

question could then be raised: are there similar mechanisms of grass seed dormancy in the tropical and temperate zones? Alternatively, are there more mechanisms for the expression of dormancy in grasses of the temperate zone than the tropical, or *vice versa*? At the present time the information to answer these questions is meagre, in part because there has been no agreement about classification of sub-types of dormancy, and in part because few comparisons for the trait of seed dormancy have ever been made among populations of different geographical origin within a single species.

Avena fatua and *A. barbata* are found in the Mediterranean climate of California where they were probably introduced from Spain in the late sixteenth century. There is considerable polymorphism for many characters, including seed dormancy, distributed even within small microclimatic zones (Marshall & Jain, 1969). Both of these species are also found in Western Australia where they were probably introduced in the 1880s by goldminers from California (Paterson, Goodchild & Boyd, 1976). In Western Australia, *A. barbata* does not show seed dormancy but *A. fatua* does from the southern areas much more than accessions from the north. In eastern Australia, which has a more temperate climate in the south and tropical in the north, *A. fatua* occurs mainly in the south, whereas *A. sterilis* and *A. ludoviciana* are more widespread in the north. Samples of Australian *A. fatua* were compared with European and Mediterranean types at one latitude in England (Thurston, 1961). Selections from the higher latitudes (with respect to the centre of origin in the middle east) flowered later than those from near the centre of origin (i.e. England 125 days, Russia 124 days, Australia 114 days and Israel 93 days). However, the pattern of dormancy appeared to be similar in all types with each population having an after-ripening period of a few months and a small proportion of the population remaining dormant for many months. No comparisons were made for the presence of the different blocks that are now known to characterize different degrees of dormancy (Adkins & Simpson, 1988).

Selections of *Eleusine coracana* from different parts of the world have been compared for their response to photoperiod, and temperatures that overcome seed dormancy (Shimizu & Mochizuki, 1978; Shimizu & Tajima, 1979). Seeds from Japan generally germinated well at 30°C in light. Seeds from Africa germinated at 15°C in light. Improved forms from India germinated well at both 30 and 15°C, but native cultivars from India could not be induced to germinate in light at any temperature. Depth of dormancy was characterized by a response to a method of after-ripening and this revealed quite distinct relationships between the degree of dormancy and the origin of the cultivars: Japanese cultivars displayed medium dormancy, many African cultivars medium or deep dormancy, Indian improved cultivars shallow dormancy, and Indian landrace cultivars medium or deep dormancy. Some of the differences in dormancy undoubtedly reflect selection against dormancy during crop improvement.

Comparisons among different ecogeographic races of rice, *Oryza sativa* and *O. glaberrima*, suggest that those subjected to the least selection pressure are the most dormant (Ellis *et al.*, 1983). The type *indica* is tropical, *japonica* is sub-tropical and *javanica* intermediate. *Glaberrima* is confined to west Africa. In order of decreasing depth of dormancy they were found to be *glaberrima, indica, javanica* and *japonica*. Nevertheless, the character of the dormancy appeared to be similar in all types on the basis of the dormancy breaking treatments.

Populations of *Dactylis glomerata* from different parts of Europe were found to be similar in having very little dormancy with the exception of the northernmost accessions from Denmark (Juntilla, 1977). Another study with this species also indicated that types from the northern latitudes of Europe were more dormant than those from lower latitudes (Probert *et al.*, 1985a). The northern types needed both light and alternating temperatures whereas those from the south germinated well in complete darkness at a constant temperature.

Selections of *Setaria lutescens* collected from different parts of the USA showed great variation in the levels of dormancy (Norris & Schoner, 1980), and 9 genotypes of *Panicum maximum* selected from Africa, Australia and the USA displayed very different mechanisms and degrees of dormancy based upon the variety of chemicals used for stimulating germination (Smith, 1979).

These limited examples suggest that different mechanisms and levels of expression of dormancy can occur within the latitudinal boundaries of dispersion of each species. Where heavy selection pressure has been applied by plant breeding, for example among some of the important cereal crops, dormancy has been reduced significantly. Now that there is a range of tests available for identifying different types of dormancy (Ellis, Hong & Roberts, 1985a,b) it would be of interest to investigate further the relationship between latitudinal dispersion, mechanisms and degree of dormancy within grass species.

3.4 Gases – oxygen and anoxia, carbon dioxide, ethylene, volatile organics; pressure.

The conclusion has often been drawn from studies of the role of hulls and seed coats in grass seed dormancy that they are significant barriers to gas exchange (Chapter 2). The loss of dormancy during after-ripening has been interpreted as an increase in permeability of these structures that eliminates the restrictions on availability of atmospheric O_2 for, as well as the egress of CO_2 from, the embryo. This section discusses this general concept. In addition the effects of gases, and some volatile organic compounds, on metabolic activity of the caryopsis and embryo are examined in relation to the release of primary dormancy and induction of secondary dormancy.

Example of Avena fatua:

There have been at least 26 reports demonstrating that an increase in partial pressure of oxygen, above that of the normal atmosphere, increases the percentage of germination of dormant seeds (Simpson, 1978). There have also been six reports that anoxia can lead to either the deepening of primary dormancy or the induction of secondary dormancy (Simpson, 1978). With the exception of one of these reports (Black, 1959), all the interpretations of the effects of oxygen, and anoxia, are based on the premise that metabolism is affected and this in turn affects germinability.

The results in the first study of O_2 effects on dormancy in *A. fatua* were interpreted as proof that the caryopsis coat, and surrounding hulls, were barriers to the entry of O_2 (Atwood, 1914). Pricking the grain coat promoted germination in freshly mature dormant grain at every level of O_2 between 4 and 20%. Below 8%, O_2 germination was reduced in intact florets. In pricked florets the reduction began at 12%. Above these values, in each case, increasing the O_2 had little effect on promotion of germination in freshly harvested grain. It was not until the grain had after-ripened for some time that higher O_2% than air stimulated germination. Pricking improved germination considerably but this response could well have been due to improved water uptake. In freshly harvested grains even the highest O_2% combined with pricking the grains could not induce germination of the whole population. This indicated a component of dormancy independent of O_2 in freshly harvested seed. With after-ripening the differences between pricked and unpricked seeds disappeared.

Later observations showed that in very dormant strains, even when the hulls were removed and the caryopsis coat was broken, there was still residual dormancy at any partial pressure of O_2 (PPO) above the normal atmosphere (Baker & Leighty, 1958; Hart & Berrie, 1966). Proof that improved germination in increased PPO is linked to O_2 uptake required for metabolism was presented by changing both temperature and PPO for germinating intact florets (Mullverstedt, 1961). At temperatures close to 20°C germination increased proportionately with increase in PPO over the range 6–15%. The highest germination reached was 75%. However, at very low O_2 (1.8%) all the seeds died at a constant 20°C, but if the temperature was alternated between 10 and 24°C, 65% germinated. Alternating temperatures between 6 and 15°C maintained 100% viability of the grains (indicated by a tetrazolium test) but only permitted 2% of them to germinate.

Anoxia, achieved with hydrogen or nitrogen substituted for a normal atmospheric gas mixture, causes secondary dormancy (Mullverstedt, 1963b; Fykse, 1970a; Ellis, Hong & Roberts, 1985a; Symons *et al.*, 1986), but only in populations that have primary dormancy at seed maturity. Beyond a certain point in the after-ripening, anoxia is not able to induce a persistent secondary dormancy (Fig. 3.12).

Examination of the behaviour of excised embryos in different gaseous environments suggests that dormancy is not just a reflection of hull and seed coat impermeability to O_2. There are differences in O_2 uptake in excised embryos and endosperm tissues between pure lines differing in degrees of dormancy (Simpson, 1978). However, these are not necessarily related to the degree of dormancy expressed within each line at any time. When the O_2 uptake is monitored in embryos of a very dormant line over several years of

Fig. 3.12. Genetic differences, among six pure lines of *Avena fatua*, in the expression of dormancy in seeds germinated at 20°C. Solid bars: Persistence of primary dormancy following maturity and dry after-ripening at 26°C and 25% relative humidity. Hatched bars: Period when anoxia can induce secondary dormancy. Open bars: Period when anoxia cannot induce secondary dormancy (After Symons, Simpson & Adkins, 1987).

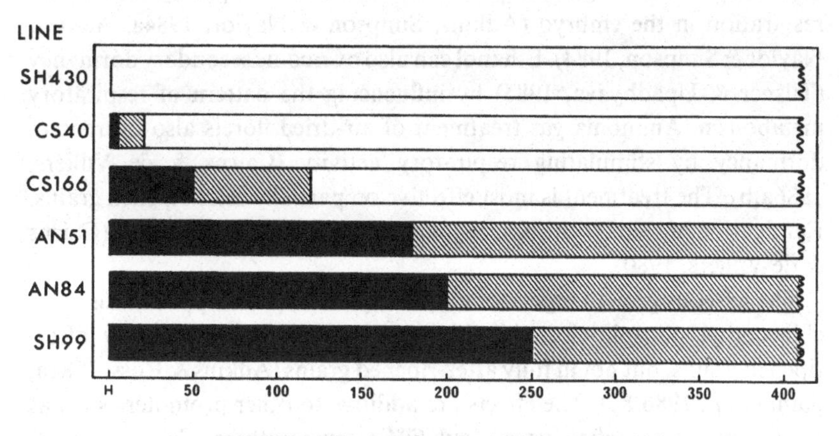

after-ripening the small changes in uptake do not reflect the marked changes that occur in germinability.

Excised embryos from very dormant grains can be induced to germinate slowly in either air or O_2 if placed on a nutrient medium solidified with agar (Simmonds & Simpson, 1972). These embryos cannot germinate under anoxia (N_2 gas) in the same conditions, but if GA is added to the medium they can. Thus stimulation of anaerobic metabolism by a hormone, combined with a specific water potential of the environment, can by-pass the inhibitory effects of anoxia.

Carbon dioxide produced from respiration may interact with PPO to influence germination. Altering the concentration of CO_2, within several levels of O_2 above and below normal PPO in the atmosphere, had no influence on the germination of intact florets in darkness (Koch, 1968b). However, in partially dormant seeds an absence of CO_2, or very low levels, increased the light-promoted inhibition of germination (Hart & Berrie, 1966). At zero CO_2, over a wide range of O_2 concentrations, white light inhibited intact florets and also dehulled caryopses. At 3% CO_2 the florets were inhibited but not the dehulled caryopses. At 20% CO_2 there was no effect of light on either intact florets or dehulled caryopses. This light reaction interacting with CO_2 disappeared with full after-ripening. In all combinations of CO_2, germination was always proportional to O_2 concentration. The inhibition in light was attributed to a decrease in fixation of CO_2 into malic acid: malic acid increases as dormancy is lost and decreases when secondary dormancy is induced (Hart & Berrie, 1967).

Ethanol can prevent the induction of secondary dormancy produced by anoxia (Hay, 1967). More recently ethanol has been shown to overcome the dormancy blocks to germination in a range of genotypes by stimulating respiration in the embryo (Adkins, Simpson & Naylor, 1984a; Adkins, Naylor & Simpson, 1984). Ethanol can also overcome secondary dormancy (Tilsner & Upadhyaya, 1985) by influencing the pattern of respiratory metabolism. Ammonia gas treatment of air-dried florets also overcomes dormancy by stimulating respiratory activity (Cairns & de Villiers, 1986a,b). The treatment is most effective on partially after-ripened grains. Fumigation of seeds with aluminium phosphide has a similar effect (Cairns & de Villiers, 1980).

The gaseous growth regulator ethylene (derived from applications of 2–chloroethyl phosphonic acid) stimulates germination in partly after-ripened grains, but not in fully after-ripened grains (Adkins & Ross, 1981a; Saini *et al.*, 1986/87). The effects are additive to other promoters such as KNO_3 and most effective at low (7°C) temperatures (Saini, Bassi &

Spencer, 1986). Thus ethylene alone is not able to completely break dormancy. When either ethylene or hydrogen gases were used as atmospheres for germination, at pressures of 10000 psi germination was markedly increased and equivalent to dehulling and cutting the caryopsis coat (Hoffman, 1961). Increasing the pressure above that of the normal atmosphere reversed the germination promotion of anaesthetics such as ethanol, ethyl ether and chloroform in *Panicum* spp. (Hendricks & Taylorson, 1980). Because hydrogen gas is considered to be inert, it must be assumed that the effects of ethylene and hydrogen were related to the pressure and not to a direct physiological effect of these gases. No comparisons were made with ethylene and hydrogen at atmospheric pressure (Hoffman, 1961).

In summary, the evidence to date indicates that loss of dormancy is associated particularly with significant changes in respiratory activity in both embryos and endosperm tissues. The induction of secondary dormancy is also associated with changes in respiration. Respiratory activity can be influenced by changing the gaseous composition of the atmosphere in which grains are germinated and also by removing the barriers to exchange of gases between the embryo and environment.

Other grasses:

In general, among grasses, increasing PPO above the normal atmosphere breaks dormancy and promotes germination of both newly mature and partially after-ripened seeds. However, there is increasing evidence that in the particular grasses adapted for germination and early seedling growth in an aquatic environment, anaerobic conditions are needed for dormancy to be broken (Table 3.5). In the case of rice, traditionally germinated in the water of paddy fields, there has been conflicting evidence in the literature about the role of O_2 in relation to dormancy and germination. It has been reported that seed dormancy at maturity in rice cultivars increases in order of the sub-types *japonica, javanica, indica, perennis* and *glaberrima* (Ellis, Hong & Roberts, 1985a). In *O. glaberrima*, dormancy is removed when the hulls are removed indicating that germination is improved by the increased availability of O_2 to the embryo (Sugawara, 1973). A similar response to increased O_2 is found in *O. indica* (Roberts, 1962; Hayashi & Morifuji, 1972). On the other hand, in a very dormant cultivar ('Hadsaduri') of *O. japonica*, the freshly harvested dormant seeds can only be germinated in anaerobic conditions (Kono *et al.*, 1975). If the dormancy is broken by accelerated after-ripening for 3 weeks at 40°C, seeds can germinate in both aerobic and anaerobic conditions. In

Table 3.5. *Grass species in which dormancy is overcome, and germination promoted, by imbibing seeds under anaerobic conditions. Reference source in brackets.*

Species	Ancillary information
Agrostis nebulosa	
(Morinaga, 1926)	Germinates equally well under anaerobic and aerobic conditions.
Axonopus compressus	
(Morinaga, 1926)	Germinates equally well under anaerobic and aerobic conditions.
Cynodon dactylon	
(Morinaga, 1926)	Germinates best under anaerobic conditions.
Echinochloa crus-galli	
(Kennedy *et al.*, 1983)	Dormancy released faster at 30 °C than at 50 °C.
(Arai & Miyahara, 1960)	Dormancy overcome in soil submerged in water at 5 °C; at higher temp. secondary dormancy induced.
(Kennedy, Fox & Siedow, 1987)	
(Rumpho & Kennedy, 1983)	
Echinochloa turnerana	
(Conover & Geiger, 1948b)	Germination inhibited in dark, promoted in light in anoxia. Secondary dormancy induced in darkness.
Eragrostis ferruginea	
(Fujii, 1963)	Germination promoted in dark, inhibited in light under anoxia.
Oryza sativa	
(Kono *et al.*, 1975)	Very dormant cultivar. Germination under anaerobic conditions. Germination after thermal loss of dormancy in both anaerobic and aerobic conditions.
(Takahashi, 1985) (Takahashi & Miyoshi, 1985)	High 0_2 inhibits germination of *japonica* cvs. but not *indica* or *perennis*. Dehulling inhibits *japonica*.
Phleum pratense	
(Morinaga, 1926)	
Poa compressa	
(Morinaga, 1926)	Germination promoted by flooding.
Scholochloa festucacea	
(Smith, 1972)	Application of ethanol breaks dormancy. Production of ethanol under anoxia.
Zizania aquatica	
(Svare, 1960)	Oxygen suppresses germination.
(Simpson, 1966a)	

the dormant seeds the metabolism is locked into a pattern of aerobic respiration. After dormancy is broken both aerobic and anaerobic respiration can occur.

This diversity in response to O_2 has been clarified by Takahashi (Takahashi, 1984, 1985; Takahashi & Miyoshi, 1985). When *indica*, *japonica*, *O. perennis* and *O. perennis* f. spontanea were compared for the level of dormancy at seed maturity, the *japonica* types were non-dormant and the rest dormant. However, when all the types were dehulled and germinated at the normal O_2 concentration of air the *japonica* types were found to be dormant. The other types were all less dormant than the intact floret but still showed varying degrees of dormancy that disappeared with after-ripening. The dormancy induced by dehulling in *japonica* types was shown to be due to the inhibitory effect of increased availability of O_2 (Fig. 3.13). A similar inhibitory effect of high O_2 tension could also be seen in

Fig. 3.13. The inhibitory effect of oxygen on germination of *Oryza sativa*, cultivar Sasanishiki. *(a)* germination on filter paper or immersed in water; *(b)* germination in sealed vessels containing mixtures of oxygen and nitrogen (After Takahashi, 1985).

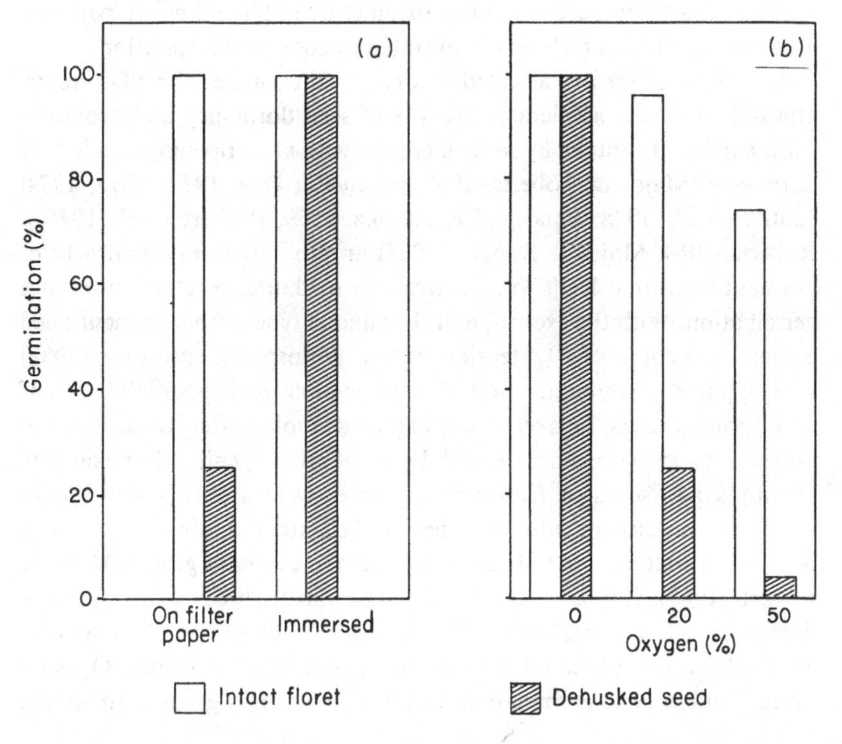

intact florets at the highest concentration (50%). The *japonica* types also had a much higher moisture content in the mature floret (30%) than other types of rice (*circa* 25%). This may have aided the retention of the respiratory system dominant during the last stages of seed development. After hulls were removed, the water content of the caryopses of *japonica* types became similar to that of the other dehulled types (*circa* 12%). Thus for a certain period immediately after maturity, anaerobic conditions favoured the germination of intact florets because of their hypersensitivity to O_2. There is thus a form of adaptation to both aquatic (anaerobic) and terrestrial (aerobic) habitats in *japonica* types of rice.

Echinochloa crus-galli, a common grass weed of rice paddies is also adapted to germinate under anaerobic conditions through the ability to carry on respiration, primarily via glycolysis, with partial operation of the tricarboxylic acid cycle and no oxidative phosphorylation (Kennedy, Rumpho & Vanderzee, 1983). As with *O. japonica* carbon dioxide and ethanol production are lower under anoxia than in air. Isolated mitochondria of *E. crus-galli* appear to be different from mitochondria of terrestrial plants in that they have a form of respiration that is neither cyanide nor SHAM sensitive (Kennedy, Fox & Siedow, 1987). When both these oxidative pathways are inhibited together the still unknown pathway accounts for at least 66% of the overall mitochondrial respiration.

Among most grasses adapted to dry habitats, increasing PPO above atmospheric levels accelerates the loss of seed dormancy and promotes germination. O_2 uptake by seeds increases with after-ripening and loss of dormancy (Major & Roberts, 1968; Heichel & Day, 1972; Mott, 1974; Juntilla *et al.*, 1978; Landgraff & Juntilla, 1979; Probert *et al.*, 1985c). Roberts (1964; Major & Roberts, 1968) proposed that any treatment of dormant seeds that leads to an increase in oxidation reactions promotes germination. With the exception of the aquatic types of plants mentioned earlier, reduction of O_2 tension below atmospheric pressure (20%) generally leads to induction or prolongation of dormancy, or inhibition of non-dormant seeds. This sensitivity of the metabolic system to variation in O_2 levels varies with genotype and degree of after-ripening. For example, O_2 applied to seeds of *Hordeum vulgare* early in after-ripening breaks dormancy of dormant cultivars; when applied later it inhibits. However, it does not inhibit later in fully after-ripened non-dormant grains (Major & Roberts, 1968). Thus in some seeds O_2 is necessary to break some aspects of dormancy but is simultaneously inhibitory to other specific reactions essential for germination. Once dormancy is completely broken O_2 is no longer inhibitory to germination. In this type of grass grain dormancy is broken by metabolic poisons that specifically inhibit some aspect of respiration (e.g. cyanide, azide, carbon monoxide) yet are highly inhibitory to non-dormant grains (Roberts, 1964).

The oxidative activity of microorganisms (Gaber & Roberts, 1969) and peroxidases found in hulls (Renard & Capelle, 1976) can reduce the availability of O_2 to the embryo and induce a response characteristic of anoxia, depending on the degree of after-ripening.

Carbon dioxide can influence dormancy in several ways. It can enhance the inhibitory effect of anoxia through exclusion of O_2 in the environment of the grain (Roberts, 1962). Increasing the CO_2 tension to 450 mm at 30°C promoted loss of dormancy in 22 cultivars of rice, regardless of whether hulls were present or absent (Tseng, 1964). As with *Avena fatua*, light can interact with CO_2 to overcome dark dormancy in *Panicum dichotomiflorum* (Taylorson, 1980). More grains germinated in the presence of light in stoppered than in open bottles. The stoppered bottles accumulated CO_2. Germination was reduced from 95 to 56% by removing the CO_2 from the bottles with KOH. Light promotion required CO_2 and enhancement of CO_2 concentration promoted germination. It is probable that the light induced loss of dormancy noted in 14 species of grasses treated in stoppered bottles at 50°C was in part due to this CO_2 enhancement of germination (Taylorson & Brown, 1977) because CO_2 of non-metabolic origin is produced by light under these conditions (Taylorson, 1979b).

Ethanol produced by anaerobic respiration has a striking ability to overcome the dormancy of some grass seeds. There appears to be a common link, located in the pathways of the respiratory system, between the actions of ethanol, CO_2, O_2 and various metabolic inhibitors in breaking seed dormancy. A number of volatile organic compounds, including ethanol, ethyl ether and chloroform can break dormancy in caryopses of *Panicum capillare* (Hendricks & Taylorson, 1980). Because the germination promotion of these compounds can be reversed by increasing the gaseous pressure around the seeds it has been argued that their anaesthetic properties can, in some way, influence the structure of cell membranes when they are in a dry state (Taylorson, 1982). On the other hand, there are indications in several other species such as *Hordeum vulgare* (Deunff, 1983), *Oryza sativa* (Cohn, Boullion & Chiles, 1987) and *Avena fatua* (see earlier in this section) that ethanol can have a specific effect in the respiratory system that is distinct from a general membrane effect. In addition ethanol may have a desiccant effect similar to that described for methanol where light sensitivity is induced in *Eragrostis ferruginea* by increasing seed coat permeability to O_2 (Fujii & Yokohama, 1965).

Ethylene can partially break dormancy in a few grass species (Ellis, Hong & Roberts, 1985b) but in others it has no effect (Roberts, 1963b) or even inhibits germination (Leshem, 1977). Ethylene chlorhydrin seems to be more effective than ethylene in breaking dormancy of grass seeds (Gianfagna & Pridham, 1951; Delouche & Nguyen, 1964; Daletskaia & Nikolaeva, 1987). The basis of the effects of ethylene in grass seeds is unclear. It is also unclear whether the promoting effects of such compounds as ethylene bromide, 1,2–dibromo-3–chloropropane (Miller, Ahrens & Stoddard, 1965) and nitrogen dioxide (Cohn & Castle, 1983) are related to their physiological action or to their ability to exclude, or dilute, O_2 and CO_2.

3.5 Growth regulators – promoters, inhibitors, limiting and allelopathic factors

Many different substances have been investigated for the practical purpose of terminating seed dormancy, also to produce more uniform germination at planting time. Since the discovery of naturally-occurring growth regulators in plants much effort has been expended on testing the hypothesis that seed dormancy is initiated, maintained and terminated through changes in the balance of promoters and inhibitors. These various substances will be discussed under the two classes, promoters and inhibitors, respectively. This is an arbitrary division because in general compounds that promote at low concentrations (e.g. hormones), with perhaps the exception of the gibberellins, can also be inhibitory at high concentrations. For a compendium of practical knowledge about substances that can break dormancy in grasses the reader is referred to the substances and conditions recommended for breaking seed dormancy in managing gene banks (Ellis, Hong & Roberts, 1985a,b). Inorganic compounds such as nitrates will be considered later.

Example of Avena fatua:

The first observation that a plant hormone could significantly promote the loss of seed dormancy was made with a gibberellin, GA_3 (Green & Helgeson, 1957; Helgeson & Green, 1958). Since then there have been more than 50 reports focusing on the actions of exogenously applied GA. There have been indirect estimates of naturally occurring GAs in the grains of *A. fatua* and the several reports using chromatography have identified GA_4/GA_7 (Taylor & Simpson, 1980), GA_1 (Metzger, 1983) and GA_3 (Kaufman *et al.*, 1976). This confusion needs to be resolved.

Application of GA_3 to maturing inflorescences prevents the induction of dormancy in mature seed (Black & Naylor, 1959; Bowman, 1975; Peters, Chancellor & Drennan, 1975). The sensitivity to GA_3 during the development of the embryo and endosperm can be eliminated by maturing excised panicles in either sucrose or maltose, but not glucose or fructose. This implies that some of the action of GA in preventing dormancy is related to carbohydrate metabolism (Cairns, 1982; Cairns & de Villiers, 1982).

Dormancy can be overcome by soaking freshly harvested intact florets in a solution of GA_3 (Renard, 1960; Cobb & Jones, 1962; Koch, 1968a; McWha *et al.*, 1976; Sharma *et al.*, 1976). The response of grains varies with age and among strains (Corns, 1960; Chancellor, Catizone & Peters, 1976; Hsiao & Quick, 1985; Upadhyaya, Hsiao & Bonsor, 1986). GA_3 breaks both primary and secondary dormancy of *A. fatua* and 'dormoats' produced from crossing *A. fatua* and *A. sativa* (Andrews & Burrows, 1972; Tilsner & Upadhyaya, 1987). Secondary dormancy produced in *A. ludoviciana* by anaerobic conditions at 21°C could not be broken by GA_3. The dose-response curve for applied GA_3 is characteristically bimodal in partially after-ripened grains (Naylor & Simpson, 1961a). In freshly mature seeds of a highly dormant strain there is only a single optimum at a high concentration of GA_3 (Naylor & Simpson, 1961b). If these very dormant seeds are given sucrose along with the GA the optimum requirement for GA shifts to a low concentration. Either GA_3 alone at this concentration, or sugar alone, cannot break dormancy but together they produce a synergistic effect that breaks dormancy quickly. At high concentration GA_3 produces sugar from the endosperm reserves and at low concentration GA_3 is required for the utilization of this sugar by the embryo. This was confirmed later by observations that starch granules in the endosperm of dormant grains cannot be broken down by the hydrolases α- and β-amylase in the absence of a third enzyme α-glucosidase (maltase). This latter enzyme, scarcely detectable in dormant seeds, is increased markedly by a high concentration of GA_3 (Simpson & Naylor, 1962).

Further observations on applied GA_3 showed that it overcomes several blocks to germination within the caryopsis. Dormancy in excised embryos can be broken by GA_3 (Andrews & Simpson, 1969; McWha & Jackson, 1976). The hypothesis that naturally occurring GAs are absent in dormant and present in non-dormant grains has been examined. Dormant and non-dormant strains have been compared for their ability to synthesize an endogenous 'GA-like' substance identified on the basis of ability to trigger hydrolysis of endosperm reserves in non-dormant caryopses (Simpson, 1965). At maturity, dormant embryos cannot synthesize a GA-like substance. The ability to synthesize such a substance is fully developed after about 24 months of after-ripening (Fig. 3.14). This development of activity can be inhibited by (2–chloroethyl) trimethylammonium chloride (CCC), an inhibitor of GA synthesis (Simpson, 1966b). When CCC is applied to the developing inflorescence it prevents the synthesis of endogenous GA in the embryo and deepens the dormancy of the mature caryopsis (Metzger, 1983). Plants grown from these deeply dormant grains are dwarfed.

In genetically pure lines with very deep grain dormancy GA applied alone cannot break dormancy without a previous period of after-ripening. In these very dormant seeds a combination of sodium azide and GA_3 can break dormancy; either compound alone is without effect (Upadhyaya *et al.*, 1982a). In less dormant lines azide alone can break dormancy and the addition of CCC prevents the azide action implying that the naturally-produced endogenous GA is not capable alone of breaking dormancy without further after-ripening. These kinds of observations have led to the conclusion that several blocks exist in mature, highly dormant caryopses and that GA can influence these blocks.

Attempts to identify the way GA influences metabolism, particularly the respiratory pathways that oxidize glucose to CO_2, have shown (1) that GA-promoted respiration occurs but not via a stimulation of the pentose-

Fig. 3.14. The synthesis of a gibberellin-like substance at different stages of after-ripening in a genetically dormant line of *Avena fatua*. Gibberellin-like activity measured by its stimulation of amylolytic activity and sugar production (After Simpson, 1965).

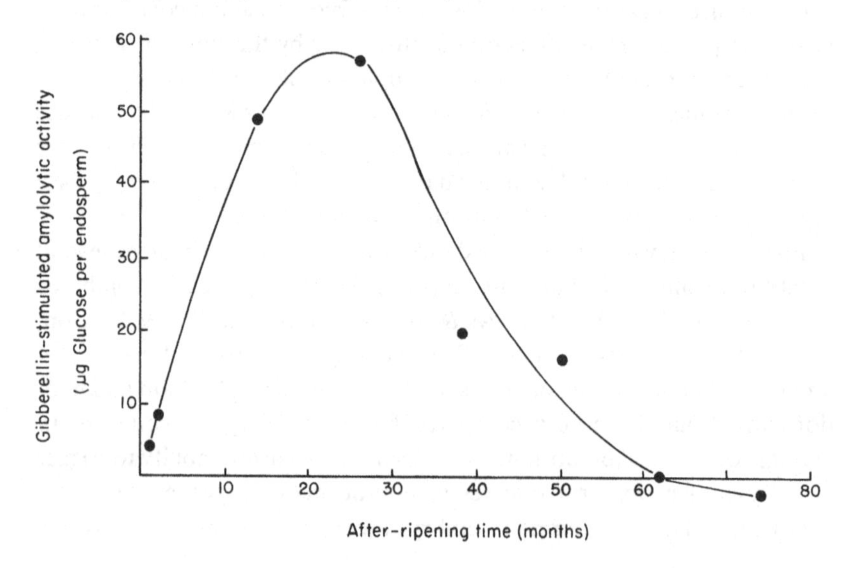

phosphate pathway (PPP) (Adkins, Gosling & Ross, 1980; Fuerst *et al.*, 1983). (2) That adenosine-tri-phosphate (ATP) levels are not increased. (3) That the normal cytochrome oxidase and alternate pathways of electron transport can both be influenced (Simmonds & Simpson, 1971, 1972). (4) That inhibition of the alternate pathway in endosperm tissue by salicylhydroxamic acid (SHAM) prevents GA-promoted α-amylase synthesis. The promotive effect of GA_3 is not directly on the alternate pathway but elsewhere in the steps leading to synthesis of α-amylase (Upadhyaya, 1987). (5) That GA promotes RNA and protein synthesis, both in the embryo (Simpson, 1965; Chen & Park, 1973; Keng & Foley, 1987) and endosperm tissues (Chen & Park, 1973; Hooley, 1984a,b; Keng & Foley, 1987). (6) That increase of α-amylase and α-glucosidase activities are caused by synthesis (Simpson & Naylor, 1962; Chen & Park, 1973; Anonymous, 1983; Jong, 1984; Hooley & Zwar, 1986; Keng & Foley, 1987). Collectively these findings indicate that an important action of GA is the stimulation of production of hexose sugars from starch reserves and their stepwise degradation via the mechanisms of glycolysis, Krebs cycle and electron transfer. It is not clear whether GAs have multiple effects or whether each GA has a single effect at a central part of the system, such as transcription or translation of nucleic acids.

Exogenous application of GA does not overcome all the expressions of grain dormancy found among different genetic strains or at different times after grain maturation of the same strain. In cases of prolonged dormancy the action of GA in overcoming dormancy is always slow, and some cases may take weeks. This suggests that other limiting processes are involved, especially during the early stages of after-ripening.

The single tentative identification of cytokinins in *A. fatua* grains (Taylor & Simpson, 1980) indicated zeatin and its two conjugated derivatives, zeatin-9–glucoside and zeatin riboside. The amounts in mature seeds were highest in the least dormant lines. There were no changes during after-ripening and the amounts increased quickly during the period of soaking prior to germination. The few reports of exogenous application of cytokinins (kinetin and benzyladenine) indicated, variously, no effect on dormancy (Holm & Miller, 1972; Hurtt & Taylorson, 1978), a weak effect in overcoming dormancy (Chancellor & Parker, 1972; Sharma *et al.*, 1976) and a quite strong effect in overcoming secondary dormancy (Tilsner & Upadhyaya, 1985). In this latter case isopentenyl adenine and kinetin induced 80% germination within 2 days, compared to 18% with KNO_3 and zero with ethanol. Thus cytokinins appear to have powerful effects in overcoming secondary dormancy.

These differences in response to cytokinins probably indicate genetic differences in the material under investigation; in some cases difficulties in penetration of cytokinins into the grains because of poor solubility and penetrability. There have been no reports of the effects of applying zeatin, or its derivatives, to this species. Benzyladenine and zeatin applied to maturing inflorescences reduced the level of dormancy in mature seeds by a small amount (Sharma *et al.*, 1976).

Thiourea has varied effects on grain dormancy. In several cases it had no effect (Holm & Miller, 1972; Bowman, 1975) and in another it caused a marked stimulation of germination (Sharma *et al.*, 1976).

The limited information on indole-3–acetic acid (IAA) does not suggest an important role in dormancy or germination of *A. fatua*. IAA applied to grains inhibits germination of *A. fatua*. IAA has been detected in *A. sativa* (Percival & Bandurski, 1976; Zimmerman, 1978) but the amounts do not change during germination (Zimmerman, Siegert & Karl, 1976).

The most significant inhibitors that have been identified with certainty in *A. fatua* and some other species belong to the family of abscisins. *Cis,trans*-abscisic acid (ABA) has been identified in both primary and secondary florets of *A. fatua* (Quail, 1968; Berrie *et al.*, 1979), *A. sterilis* (Tal, 1977) and the hulls of *A. sativa* (Ruediger, 1982). The amounts found do not have an obvious relationship to the presence of dormancy. A study of the quantities present during the development of the seed from anthesis showed high levels immediately after anthesis followed by a rapid diminution to low levels that remained constant until seed maturity (Berrie *et al.*, 1979). ABA has a powerful inhibitory effect on the germination of non-dormant embryos (Andrews & Burrows, 1972; Cuming & Osborne, 1978b) as well as intact florets (Holm & Miller, 1972). ABA can reverse the GA_3-stimulated germination of dormant *A. ludoviciana* (Quail, 1968) as well as the GA_3-stimulated formation of α-amylase in barley aleurone: this same inhibitory effect has been demonstrated in aleurone protoplasts derived from *A. fatua* where GA_4 stimulated mRNA formation specific for α-amylase indicating ABA control over gene expression (Zwar & Hooley, 1986). Other isomers of ABA, and derivatives such as phaseic and dihydrophaseic acid have not been investigated for their potential inhibitory or promotory properties in relation to dormancy in *Avena* species.

The concept that a balance of promoters and inhibitors control dormancy in *A. fatua* has not been seriously tested on the basis of the fragmentary approach used to date. Accumulated evidence indicates that the formation, or release, of gibberellins is limited in dormant seeds. While applied GA can break dormancy and by-pass the natural after-ripening requirement, it is not clear whether endogenous GA(s) are produced before, or after, dormancy is lost. The indication that ABA can modify GA-stimulated enzyme formation in endosperm tissue, which is probably a post-germination event, may have no relation to the role of these same compounds in determining the state of grain dormancy.

Other grasses:

Among the promoters, several GAs have been identified in the seeds of a number of grass species (Pharis & King, 1985). The composition of the bioactive and relatively inactive GAs appears to be very different in some species of grasses. In mature seeds the proportion of bioactive GAs appears to be very low. The role of this family of GAs is not at all clear and mainly inferred from observations following applications of a single GA. After the discovery of GA there was a large number of reports showing germination promotion of dormant seeds by GA_3. A cross-section of some of these reports is summarized in Table 3.6 to illustrate several important points. In about two thirds of the species GA breaks dormancy to some degree. In most of these cases the response was slow, often taking several weeks, and the population was still not completely germinated when the experiment terminated. In seeds that respond positively to light, the effect of GA is often obscured. Some species, according to these experiments, showed no response to GA although in other reports they did respond positively. This discrepancy is probably due to genetic and ecological differences between the samples which were disregarded.

In some of the species that failed to respond to GA_3 it is possible that other GAs such as A_1, A_4, A_7 and A_9 (Dathe, Schneider & Sembdner, 1978) have a role. On the other hand in many of the examples the failure of GA to break dormancy was simply due to the fact that embryos were not dormant, were insensitive, or the hulls and/or seed coats were the physical cause of dormancy so that unless these were first removed the effects of GA on the caryopsis could not be seen (Cardwell, Oelke & Elliott, 1978; Sung, *et al.*, 1983). It is clear from those studies where both a range of genotypes and a range of periods of after-ripening were examined that GA is most effective at an intermediate stage of after-ripening (Groves, Hagon & Ramakrishnan, 1982). In some cases where a genotype is particularly dormant there is no response at all to GA unless there is first a period of after-ripening (Evans & Young, 1975).

The general role of GA in overcoming dormancy in grasses has been largely inferred from the many observations on GA action on endosperm reserves in cereals (Briggs, 1973; Jacobsen, 1983). It is clear from studies on

Table 3.6. *The influence of gibberellin on seed dormancy in different grass species. + = improvement in germination; − = no influence on germination; V = variation in the response among genotypes or accessions. Reference source in brackets.*

Species	Response to GA	Observations
Andropogon gerardii (Kucera, 1966)	+	Maximum effect on new seed.
Aristida contorta (Mott, 1974)	+	80% germination when dehulled.
Arrhenatherum elatius (Nakamura, 1962)	+	
Agropyron cristatum (Nakamura, 1962)	+	
(Kucera, 1966)	−	GA inhibits germination.
Agropyron pauciflorum (Nakamura, 1962)	−	
Agropyron smithii (Schultz & Kinch, 1976)	+	Incomplete germination.
(Nakamura, 1962)	−	
Agrostis spp. (Ludwig, 1971)	+	28 days to get 100%
Agrostis gigantea (Nakamura, 1962)	−	
Agrostis tenuis (Nakamura, 1962)	−	
Alopecurus pratensis (Nakamura, 1962)	−	
Avena sativa (Nakamura, 1962)	+	
(Kahre *et al.*, 1965)	−	
Axonopus compressus (Nakamura, 1962)	−	
Bothriochloa macra (Hagon, 1976)	+	Maximum germination 81% in dark, 92% in light after 14 days.
Bromus inermis (Nakamura, 1962)	+	
Bromus commutatus (Nakamura, 1962)	−	
Bromus mollis (Nakamura, 1962)	+	A marked response.

Table 3.6. *(cont.)*

Species	Response to GA	Observations
Bromus ramosus (Nakamura, 1962)	+	A marked response.
Bromus tectorum (Evans & Young, 1975)	–	Germination depressed.
Chrysopogon fallax (Mott, 1978)	+	Maximum germination 86%.
Cynodon dactylon (Nakamura 1962; (Young *et al.*, 1977)	+	Partial germination.
Dactylis glomerata (Ludwig, 1971)	+	
(Haight & Grabe, 1972)	–	After alternating temp.
(Nakamura, 1962)	–	
(Juntilla, 1977)	V	Cultivar differences.
Danthonia spp. (Hagon, 1976)	+	Maximum germination 81% dark, 75% light after 14 days.
Digitaria spp. (Gray, 1958)	+	
Digitaria milanjiana (Baskin *et al.*, 1969)	+	Intact floret 0%. Dehulled, increases from 45 to 96%.
Digitaria pentzii (Baskin *et al.*, 1969)	+	Intact floret 0%. Dehulled increases from 42 to 93%.
Distichlus spicata (Amen, 1970)	–	Negative response.
Echinochloa crus-galli (Sung *et al.*, 1987) (Nakamura, 1962)	–	
Elymus canadensis (Nakamura, 1962)	–	
Eragrostis curvula (Nakamura, 1962)	–	
Eremechloa ophiuroides (Delouche, 1961)	+	14 days to 50% germination.
Eleusine coracana (Nakamura, 1962)	–	
Festuca ovina (Nakamura, 1962)	+	
Festuca rubra (Button, 1959; Nakamura, 1962)	+	Shallow dormancy.

Table 3.6. *(cont.)*

Species	Response to GA	Observations
Festuca spp.	+	28 days to germinate 100%.
(Ludwig, 1971)		
Festuca arundinacea	–	
(Nakamura, 1962)		
Festuca pratensis	–	
(Nakamura, 1962)		
Holcus lanatus	–	
(Nakamura, 1962)		
Hordeum glaucum	+	Incomplete germination.
(Popay, 1981)		
Hordeum murinum	+	Incomplete germination.
(Popay, 1981)		
Hordeum spontaneum	+	
(Renard, 1960)		
Hordeum vulgare	+	Marginal effect.
(Nakamura, 1962; Kahre *et al.*, 1965)		
	–	2- and 6-row types.
(Kahre *et al.*, 1965)		
	–	Inhibits in excess H_2O.
(Anonymous, 1956)		
	+	Excised embryos.
(Bulard, 1960)		
Lolium multiflorum	–	
(Nakamura, 1962)		
Lolium perenne	–	
(Nakamura, 1962)		
Oryza sativa		
(Agrawal, 1981; Deore & Solomon, 1981)	+	
	V	Differential response for *javanica*, *indica*, *japonica*, & *glaberrima*.
(Ellis, Hong & Roberts, 1983)		
	+	21 days to maximum germination.
(Roberts, 1963b)		
Oryzopsis hymenoides	+	Incomplete germination.
(Clark & Bass, 1970)		
	+	Excised embryos.
(Barton *et al.*, 1971)		
	+	
(McDonald & Khan, 1977)		
	V	2 accessions germinate, others very dormant.
(Young & Evans, 1984)		
	V	

Table 3.6. *(cont.)*

Species	Response to GA	Observations
(Zemetra *et al.*, 1983)		
Panicum ramosum (Andersen, 1962)	+	Maximum germination 85%.
Panicum maximum (Smith, 1979)	+, −, V	9 genotypes tested.
Panicum virgatum (Nakamura, 1962)	−	
Paspalum dilatatum (Nakamura, 1962)	−	
Paspalum notatum (Nakamura, 1962)	−	
Pennisetum typhoides (Sandhu & Husain, 1961; Burton, 1969)	+	Equivalent to dehulling.
Phalaris spp. (Ludwig, 1971)	+	28 days to maximum germination.
Phalaris arundinacea (Nakamura, 1962)	−	
(Juntilla *et al.*, 1978)	V	Maximum germination 25% in darkness, 65% in light for florets.
Phalaris tuberosa (Nakamura, 1962)	−	
Phleum pratense (Kahre *et al.*, 1962)	+	Incomplete germination.
(Nakamura, 1962)	−	
Poa nemaralis (Nakamura, 1962)	−	
Poa pratensis (Nakamura, 1962)	+	
(Wiberg & Kolk, 1960)	−, +, V	
Poa trivialis (Nakamura, 1962)	−	
Secale cereale (Kahre *et al.*, 1965)	+	Marginal response from both spring & winter forms.
Setaria lutescens (Kollman & Staniforth, 1972)	+	
Setaria viridis (Nakamura, 1962)	−	
Sorghum bicolor (Aisien & Palmer, 1983)	−	No effect on embryo or endosperm.
	−, V	No effect on intact or scarified

Table 3.6. *(cont.)*

Species	Response to GA	Observations
(Gritton & Atkins, 1963a)		floret.
(Wright & Kinch, 1959)	+, −, V	Variation in types.
Sorghum halapense (Huang & Hsiao, 1987)	+	Maximum germination 76%.
Sorghum plumosum (Mott, 1978)	+	Maximum germination 94%.
Sorghum stipoideum (Mott, 1978)	+	Maximum germination 89%.
Sorghum nutans (Emal & Conard, 1973)	+	Maximum germination 65% in darkness, 95% light.
Stipa bigeniculata (Hagon, 1976)	+	Maximum germination 45% dark, 71% light for the intact seed.
Stipa viridula (Fulbright *et al.*, 1983)	+	Light improves germination.
Themeda australis (Mott, 1978)	+	Maximum germination 68% dark, 71% light after 14 days.
(Hagon, 1976)	+	
(Groves *et al.*, 1982)	V	Maximum germination with intermediate stage of after-ripening.
Triticum aestivum (Nakamura, 1962)	+	
(Kahre *et al.*, 1965)	+, −	
Triticum vulgare (Norstog & Klein, 1972)	+	
Zea mays		
(Prasad *et al.*, 1983)	+	Overcomes secondary dormancy but incomplete germination.
Zea perennis (Mondrus-Engle, 1981)	+	White coloured seed less responsive than dark.
Zizania aquatica (Simpson, 1966a)	−	Very dormant seed.
(Cardwell *et al.*, 1978)	+	When dehulled increased from 31 to 51%.
Zoysia japonica (Nakamura, 1962)	−	

barley and wheat that GAs have an important role in promoting the synthesis of proteins, principally hydrolytic enzymes such as α-amylase. At seed maturity bioactive GAs appear to be in short supply so that exogenously supplied GA can increase α-amylase activity in both the scutellum and aleurone tissues. Sensitivity of these tissues to GA also increases as seeds become desiccated (Nicholls, 1979, 1986) or after-ripened (Crabb, 1971). While embryos are considered to be the source of GA, externally supplied GA overcomes the dormancy of excised embryos (Bulard, 1960; Barton, Roe & Khan, 1971) indicating a deficiency of active endogenous GAs. It is unclear whether the natural role of GA is to overcome dormancy, or after dormancy is terminated by some other agent, to mobilize endosperm reserves. There is no convincing evidence yet that a balance of GAs and inhibitors controls dormancy. However, in one study an unusual interaction of exogenous GA has been found in a number of wheat cultivars (McCrate *et al.*, 1982). An embryo assay demonstrated the presence of a water-soluble inhibitor in the caryopses at maturity. This inhibitor did not diminish with after-ripening but the response of the embryo to the inhibitor declined with after-ripening; this sensitivity to the inhibitor could be counteracted with GA. The sensitivity was unrelated to α-amylase activity. The response of the excised embryos to the inhibitor paralleled cultivar differences in resistance to pre-harvest sprouting.

An important action of GA in breaking dormancy may be achieved through the modification of osmotic potential achieved through promotion of formation of low molecular weight mono- and di-saccharides, within the embryo and in the surrounding endosperm. It has been demonstrated in *Themeda triandra* that the hulls create a mechanical restraint on the emergence of the radicle (Martin, 1975). A similar constraint can be achieved by decreasing the water potential of the germination medium. Both of these constraints can be overcome by the application of GA (Fig. 3.15). When germination occurred in non-dormant grains, GA increased three-fold over a period of 48 h. The content of endogenous GA was low in dormant seeds and remained constant over the same period. It is possible therefore that GA assists the intracellular generation of negative water potentials that aid radicle emergence. The generally slow response to GA treatment is probably related to the time lag for enzyme synthesis, then hydrolysis of reserves and the slow diffusion, or active transport, of the metabolites.

Among other growth promoters, the naturally-occurring cytokinins, zeatin and the conjugates zeatin riboside and glucoside were first isolated from a grass (*Zea mays*) (Letham & Palni, 1983). There are reports of the isolation of cytokinins from other grasses but there are few studies that relate these endogenous levels to the state of seed dormancy. When synthetic cytokinins are applied to dormant seeds there is some evidence for promotion of germination but in most cases the response is not complete and there are just as many species that fail to respond (Table 3.7).

The organic compound thiourea has significant effects in breaking dormancy in some dicotyledenous species (Bewley & Black, 1982) but appears to be of little significance in grasses. Of 40 species of grasses examined for a response to thiourea only seven gave a significant positive response, two gave a weak response and the remainder were unaffected (Nakamura, 1962). Other reports with rice suggest that under some circumstances, with particular genotypes, thiourea has some influence in overcoming dormancy (Roberts, 1963b; Tseng, 1964).

It is a curious fact that the hormone family of naturally occurring auxins, that stimulate both root formation and elongation in many plant species, do not appear to be of any importance in overcoming dormancy in grasses.

The hormonal growth inhibitor ABA (Table 3.8) and one of its metabolites, 4–dihydrophaseic acid (Dathe *et al.*, 1978) has been identified in several grass species. Nevertheless, a clear relationship between the presence of these compounds and seed dormancy has not been established. Detailed studies on the presence of ABA during seed development of *Triticum aestivum* (King, 1976) demonstrated a 40–fold increase in ABA during the late stages of development of the floret over the period when

Fig. 3.15. Effect of gibberellin (GA), glume removal and osmotic potential on dormant spikelets of *Themeda triandra* (After Martin, 1975).

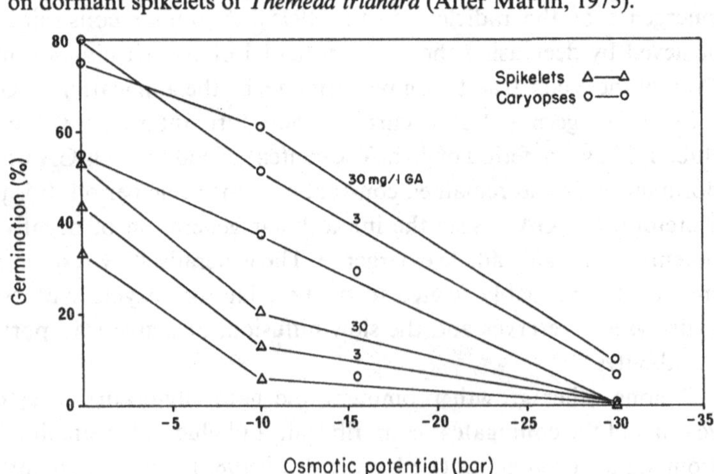

Table 3.7. *The influence of cytokinins in overcoming seed dormancy in grasses. + = promotion of germination. − = no effect on germination. Reference source in brackets.*

Species	Effect	Compound & influence
Agropyron smithii (Toole, 1976)	+	Kinetin; additive with GA_3.
(Schultz & Kinch, 1976)	+	Kinetin; germination incomplete.
Bothriochloa macra (Hagon, 1976)	−	Kinetin.
Danthonia spp. (Hagon, 1976)	−	Kinetin.
Distichlus spicata (Amen, 1965)	−	Kinetin.
Festuca arundinacea (Boyce, 1973)	−	Kinetin.
Oryza sativa (Paul & Mukherji, 1977)	−	Kinetin; waterlogged & secondary dormancy.
(Cohn & Butera, 1982)	+	Kinetin, benzyladenine, isopentenyladenine, zeatin; weak response.
(Roberts, 1963b)	+	Kinetin; weak response.
Oryzopsis hymenoides (Clark & Bass, 1970)	−	Kinetin.
Sorghum bicolor (Aisien & Palmer, 1983)	−	Kinetin; on embryo and endosperm.
Stipa bigeniculata (Hagon, 1976)	+	Kinetin.
Themeda australis (Hagon, 1976)	−	Kinetin.
Triticale (Weidner, 1984)	+	Kinetin; on embryos.
(Weidner *et al.*, 1984)	+	Kinetin; equal to GA_3.
(Weidner, 1984)	−	Benzyladenine.
Zea mays (Prasad *et al.*, 1983)	+	Kinetin; waterlogged with secondary dormancy.
Zizania aquatica (Cardwell, 1978)	−	Kinetin; pricked seed.

Table 3.8. *Grass species with* cis, trans-*abscisic acid in developing or mature seeds.*

Species	Reference
Avena sativa	Berrie *et al.*, 1979.
Hordeum vulgare	Dashek *et al.*, 1977.
	Dunwell, 1981a.
Oryza sativa	Hayashi, 1979
	Hayashi & Tanaka, 1979.
	Hayashi, 1987.
Oryzopsis hymenoides	McDonald & Khan, 1977.
Secale cereale	Dathe *et al.*, 1978
Triticum aestivum	Leshem, 1978.
	King, 1976.
	Walker-Simmons, 1987.
Zea mays	Neill, Horgan & Rees, 1987.
Zizania aquatica	Albrecht, Oelke & Brenner, 1979.

germinability of the embryo declined. The prevention of precocious germination was attributed to the presence of ABA. On the other hand ABA declined with seed maturation in both *Avena fatua* and *A. sativa* (Berrie *et al.*, 1979). Both free and conjugated forms of ABA declined during after-ripening in *Zizania aquatica* as dormancy was lost (Albrecht *et al.*, 1979). The embryo and pericarp contained fifty times the amount of ABA found in the hulls and endosperm, and amounts were much less in non-dormant than dormant seeds. In a dormant cultivar of *Hordeum vulgare* free ABA could not be identified (Dunwell, 1981a). The inhibitory effect of exogenously applied ABA on excised embryos of this cultivar could be overcome by GA_3 but not in the intact caryopsis. In *Festuca arundinacea* exogenous ABA can inhibit germination and the inhibition is overcome by GA_3 (Boyce, 1973). While ABA appears to play some role in the normal development of the caryopsis in *Zea mays* there are no differences in endogenous ABA between normal and viviparous types and the conclusion has been made that embryo dormancy is unrelated to the presence of ABA (Neill,*et al.*, 1987). A natural inhibitor, later identified as ABA (Hayashi, 1979), decreased with after-ripening in both the hull and caryopsis of dormant *Oryza sativa* (Hayashi & Himeno, 1973).

ABA can prevent the formation of α-amylase in isolated wheat aleurone (Varty *et al.*, 1983), a property that has also been demonstrated in *Hordeum vulgare* (Khan, 1969; Leshem, 1978), *Avena fatua* (Zwar & Hooley, 1986) and extensively reviewed (Jacobsen, 1983).

In summary the above evidence does not yet allow the conclusion that ABA is a significant naturally-occurring inhibitor responsible for seed dormancy in grasses. Nevertheless, because there are several isomeric forms of ABA and a number of metabolites, such as phaseic and dihydrophaseic acids, that could potentially play a role in the control of germination, it is premature to rule out the possibility that abscisins have an important role in seed dormancy.

The accumulated information about promoters and inhibitors found in grass seeds is neither qualitatively nor quantitatively sufficient to support the argument (Amen, 1968) that a balance of promoters and inhibitors controls dormancy. There is no doubt that they have an influence on seed development and early seedling growth but their exact role in the induction, maintenance and release of dormancy is not well understood in grasses at present.

There are certainly other limiting factors for seed germination in grasses that have important influences on the expression of dormancy. Aside from the naturally occurring hormones considered above there are many reports of inhibitory compounds, or their inferred presence, in grass seeds. Some of these substances may play a role in the suppression, or promotion, of germination. In some cases these substances may have an inhibitory effect on the germination of other species. It is also likely that there are compounds present in the structures of grass seeds that have allelopathic properties protecting the grain from destruction from microorganisms during long periods of dormancy. Some of these factors are considered in the remainder of this section.

Example of Avena fatua:

Several oxidized forms of nitrogen that occur commonly in soils have promoting effects on germination. Sodium (Adkins, Naylor & Simpson, 1984), potassium (Hay & Cumming, 1959) and ammonium (Rademacher & Kiewnick, 1964) nitrates can overcome some aspects of dormancy. Sodium nitrite, sodium azide (Adkins, Naylor & Simpson, 1984), and urea (Chancellor, Catizone & Peters, 1976) break dormancy in some, but not all lines, at certain stages of dormancy. Organic compounds commonly found associated with disintegrating plant tissues such as malic (Hart & Berrie, 1967; Adkins, *et al.*, 1985), citric, pyruvic, lactic, fumaric (Adkins, Simpson & Naylor, 1985) succinic and malonic acids (Simmonds & Simpson, 1972; Adkins, *et al.*, 1985) all promote the loss of dormancy. Ethanol, acetaldehyde (Hay, 1967; Adkins, Naylor & Simpson, 1984), thiourea, hydroxylamine chloride (Adkins, Naylor & Simpson, 1984), turkey manure (Somody *et al.*, 1984b) and hydrogen cyanide (Kamalavalli *et al.*, 1978) can also overcome some expressions of dormancy.

There is evidence for inhibitory substances found within seed tissues that promote dormancy. An inhibitor from the hulls of *A. fatua* prevented germination of caryopses (Stryckers & Pattou, 1963). Nonanoic acid (Berrie *et al.*, 1976; Metzger & Sebesta, 1982), triterpenoid glycoside (Haggquist *et al.*, 1984), maltose, and glucose (Cairns, 1982) all inhibit germination and are found in caryopses. There are reports of an inhibitor in the hulls of *A. sativa* that prevents the germination of other seeds but not *A. fatua* (Elliott & Leopold, 1953; Koves, 1957; Karl & Rudiger, 1982; Lohaus *et al.*, 1982). Coumarin, found in a number of grass species and legumes, has a powerful inhibiting effect on the germination of *A. fatua* seeds (Naylor & Simpson, 1961b). Dormant seeds of *A. fatua* can persist for long periods in soil while they are dormant without being destroyed by microorganisms. There are indications that hulls may contain substances with allelopathic properties. For example, seeds can survive for long periods in compost (Wiberg, 1959; Metz, 1970) and the hulls contain soluble substances with phytotoxic properties (Davis, 1962; Helgeson & Davis, 1963). Wild oat plants have an allelopathic influence on the distribution of adjacent plants (Tinnin & Muller, 1971, 1972). In laboratory investigations intact florets, after long periods of incubation, have a low incidence of fungal infection that is probably related to the presence of water-soluble substances that can strongly inhibit elongation of seminal roots of wheat (personal observations of the author).

Other grasses:

Non-specific inhibitory substances have been partially identified in a number of grass species (Table 3.9). In addition there are reports invoking inhibitors, to explain dormancy, that did not provide proof of their presence. Evidence linking allelopathic substances to dormancy has been demonstrated with *Hordeum vulgare* (Gaber & Roberts, 1969; Rauber & Isselstein, 1985) where microorganisms, present on the surface of the grains, compete for available oxygen and induce secondary dormancy by depriving the embryo of available oxygen. Treatment with an antibiotic prevented this induction of secondary ('water') dormancy (Gaber & Roberts, 1969). Removal of the hulls of *Agropyron intermedium* reduces seed germination due to the infection of the caryopsis by fungi, suggesting an anti-fungal role for the hulls (Kilcher & Lawrence, 1960). The outer glumes of *Sorghum vulgare* protect the caryopsis through antibiosis (Wright & Kinch, 1959). A relationship between seed dormancy and

Table 3.9. *Reports of non-specific inhibitors of germination in some grass species.*

Species	Reference
Aegilops kotschyii	Wurzburger & Leshem, 1967, 1969.
	Wurzburger, Leshem & Koller, 1974.
Bouteloua curtipendula	Sumner & Cobb, 1962.
Dactylis glomerata	Fendall & Canode, 1971.
Digitaria sanguinalis	Gianfagna & Pridham, 1951.
Elymus caput-medusae	Nelson & Wilson, 1969.
Eragrostis lehmanniana	Wright, 1973.
Oryza sativa	Mai-Tran-Ngoc-Tieng & Nguyen-Thi-Ngoc-Lang, 1971.
	Uotani *et al.*, 1972.
	Mitra *et al.*, 1975.
	Hayashi, 1976, 1979, 1980, 1987.
	Hayashi & Tanaka, 1979.
	Takahashi, Kato & Tsunagawa, 1976.
	McDonald & Khan, 1983.
Setaria glauca	Yokum, Jutra & Peters, 1961.
Stipa viridula	Frank & Larson, 1970.
Triticum aestivum	Miyamoto, Tolbert & Everson, 1961.
	Khan *et al.*, 1964.
	Hubac, 1966
	Leshem, 1978.
	McCrate *et al.*, 1982.
Uniola paniculata	Westra & Loomis, 1966.
Zea mays	Ortega-Delgada *et al.*, 1983.

allelopathic action has been demonstrated (Ruediger & Lohaus, 1985) in several species.

3.6 Soil and agricultural practice

There are several soil-related factors that have important, and in some cases subtle, long-term effects on dormancy in grass seeds. At maturity the dispersal units of grasses fall or are blown to the surface of the ground. In the case of hydrophytes the seeds may fall directly into water and sink quickly into the anaerobic environment of the muck soil. In either case, the physical characteristics of the soil, for example texture, will vary in different locations. Availability of O_2, temperature regime, chemical composition, pH, and level of activity of microorganisms can all influence the persistence of dormancy by affecting rate of after-ripening and induction of secondary dormancy. Many seeds become buried by such actions as ploughing, tillage, wind and water erosion, deep cracking of clay soils and the activity of small and large animals. Seeds of some forage and weed species are ingested by grazing animals and scarified by stomach acids; the non-dormant seeds are then spread to other environments in the faeces. Other factors, such as the global use of herbicides in modern agriculture, and particular patterns of crop management and rotation, have created new selection pressures on grasses favouring survival of plant populations with seed dormancy.

Each of the above factors has a modicum of influence on the expression of seed dormancy and plays a role in the survival of the species. Because of the great diversity of environments occupied by grasses, and the great number of species, it is not possible in a limited text to deal extensively with each of these factors. Emphasis is placed on presenting information on how these factors influence dormancy in *A. fatua* with some limited references to other species.

Mortality of seeds lying on the soil surface is generally high (Banting, 1962; Ministry AFFDAS, 1974; Kohout, 1977; Hsiao & Quick, 1983). Exposure to light may induce some species to germinate quickly and others to stay, or become dormant (Roberts, 1986). Seeds buried in the soil generally have a maximum germination at some optimum depth. Below this depth secondary dormancy can be induced by specific moisture content and gas mixtures (Johnson, 1935b; Cumming & Hay, 1958; Anghel & Raianu, 1959; Barralis, 1965; Banting, 1966). Temperature fluctuations at depth in the soil are attenuated compared to the soil surface. With soil crusting and increase in depth, oxygen availability decreases and carbon dioxide increases (Koch, 1968b). The deeper that seeds are buried the longer they will survive (Tingey, 1961; Kropac, Havranek & Dobry, 1986). However, emergence from great depth in the soil is limited by seed size which governs the energy reserves for growth of the shoot, or mesocotyl, and surrounding coleoptile. Thus shoots of the large cereal grains can emerge from considerable depth. *Avena fatua* can emerge from depths between 15 and 25 cm (Tingey, 1961; Kurth, 1967; Kohout & Pulkrabek, 1977) and *Zizania aquatica* can extend the first aerial leaf one metre to the surface of water (Simpson, 1966a). The probability of water-logging, or burial beneath a water table, with consequent risk of induction of secondary dormancy, increases with soil depth (Allan, Vogel & Craddock, 1961; Lewis, 1961; Schafer & Chilcote, 1970; Williams, 1973). Seeds with primary dormancy do not necessarily persist longer than non-dormant seeds because the pattern of loss of viability is similar in both types (Banting, 1966). Factors such as freezing (Kuhnel, 1965; Voderberg, 1965), alternate wetting and drying, and pathogens (Voderberg, 1965) are important for the loss of viability. A specific temperature can interact with excessive soil moisture and cause secondary dormancy. For example, *Echinochloa crus-galli* seeds germinate well in moist soil at 5°C but are induced into secondary dormancy at 25°C (Arai & Miyahara, 1960).

Differences in soil type and texture can have broad effects on the geographical distribution of grass species that influence the expression of seed dormancy (Quail & Carter, 1968; Fykse, 1970a; Von Prante, 1971; Paterson, 1976; Kropac, 1980). On the other hand the germination of non-dormant seeds may be relatively independent of these factors (Pejka, 1971; Odgaard, 1972). Differences in germination of *A. fatua* populations between sandy and loam soils were not as significant as the effects of water availability and depth of the water table (Manson, 1932; Lewis, 1961).

The natural chemical composition of the soil, the added chemicals from agricultural fertilization, and the accelerated release of ions such as nitrate with summer-fallowing have very significant effects on the termination of seed dormancy in many grass species. There is considerable literature indicating the promotory effects of various inorganic salts, particularly nitrates, used as soil additives (Ellis, Hong & Roberts, 1985a,b). Table 3.10 lists common soil additives used as fertilizers, that are known to break dormancy in the single species *Avena fatua.* Salts containing a nitrate ion seem to be the most effective substances under field conditions, although cyanamide is equally effective (Fykse, 1970a). Different ecotypes within the species are influenced differently by nitrate ions (Fykse, 1970a; Adkins *et al.*, 1984a). Therefore, prolonged application of nitrogen fertiliser can apply selection pressure favouring particular ecotypes. The practice of summer-fallowing in the drier climates of North America greatly stimulates release of nitrate ions from the soil. This is probably a significant factor in the promotion of weeds such as *A. fatua.*

While the nitrate ion seems to be very effective in breaking grass seed dormancy, a study comparing the different forms of nitrogen fertilizer found that germination was stimulated up to 0.03% N in each form. This suggests that a common nitrogen derivative is formed by the metabolism of the seed and this triggers the promotion of germination. The intermediate, hydroxylamine, found in nitrogen transformations by soil microorganisms as well as plant cells, does not appear to be active (Adkins *et al.*,1984a). There are more than 100 reports of nitrogen compounds promoting the germination of seeds of various grass species (Simpson, 1988). The conclusion can therefore be drawn that there is a common reaction among

Table 3.10. *Substances used as soil additives in agriculture that stimulate the germination of dormant seeds of* Avena fatua.

Common name	Formula	Reference
Ammonia gas	NH_3	Cairns & de Villiers, 1986a.
Ammonium nitrate	NH_4NO_3	Sexsmith & Pittman, 1963.
		Rademacher & Kiewnick, 1964.
		Watkins, 1966.
		Fykse, 1970a.
		Chancellor, Catizone & Peters, 1976.
		Sharma, McBeath & Vanden Born, 1976.
Ammonium sulphate	$(NH_4)_2SO_4$	Sinyagin & Teper, 1967.
		Fykse, 1970a.
Calcium ammonium nitrate	$CaNH_4NO_3$	Watkins, 1966.
Calcium nitrate	$CaNO_3$	Fykse, 1970a,b.
Cyanamide	NH_2Cn	Fykse, 1970a,b.
Potassium nitrate	KNO_3	Schwendiman & Shands, 1943.
		Hay & Cumming, 1959.
		Sinyagin & Teper, 1967.
		Fykse, 1970a,b.
		Morgan & Berrie 1970.
		Sharma, McBeath & Vanden Born, 1976.
		Saini, Bassi & Spencer, 1986.
Potassium sulphate	K_2SO_4	Sinyagin & Teper, 1967.
Sodium nitrate	$NaNO_3$	Hay & Cumming, 1959.
		Sharma, McBeath & Vanden Born, 1976.
		Adkins, Simpson & Naylor, 1984a.
Sodium nitrite	$NaNO_2$	Adkins, Simpson & Naylor, 1984a,b.
Thiourea	$CS(NH_2)_2$	Sharma, McBeath & Vanden Born, 1976.
Urea	$CO(NH_2)_2$	Watkins, 1966.
		Fykse, 1970a,b.
		Chancellor, Parker & Teferedegn, 1971.
		Adkins, Simpson & Naylor, 1984a.

many grass species to nitrogenous fertilizers that indicates a common mechanism for the control of at least part of the expression of seed dormancy.

Deficiencies of specific chemicals may influence the level of dormancy. For example, manganese deficiency, common in oat species, reduces the number of viable seeds and the level of dormancy (Thurston, 1951).

The pH of soil has significant effects on the expression of seed dormancy in grasses. Organic acids produced by decaying vegetation, such as lactic, citric, fumaric, malic and pyruvic acids can induce germination of dormant seeds of *A. fatua* (Adkins *et al.*, 1985). The promotion is from the stimulation of the respiratory system within the caryopsis in a manner similar to the influence of nitrate. Monocarboxylic acids also stimulate the germination of *Oryza sativa* (Wagenvoort & Opstal, 1979; Cohn, Boullion & Chiles, 1987; Cohn *et al.*, 1987) and *Hordeum vulgare* (Jansson, 1960). Species adapted to grow in either acid, neutral or basic soils are also adapted to germinate in these conditions (Stubbendiek, 1974). *Eragrostis curvula*, adapted to acid soils, can germinate over the pH range 4–11.5. *Andropogon hallii* found only on neutral soils can germinate in a pH as low as 2.5 but the optimum is around 6.0. *Panicum antidotale*, found on basic soils has three pH optima: the largest proportion of seeds germinates between 8.5 and 11.5; the two other sub-populations have optima at 8.0–8.5 and 4.0–6.0, indicating genetic variation within the general population.

The ingestion of grass seeds by ruminants is a natural form of acid scarification that can break hull and coat dormancy as well as transport seed to new environments. For example, 12% of the seeds of *Avena fatua* fed to bullocks was excreted in a viable state (Kirk & Courtney, 1972). To avoid the spread of this species a recommendation has been made to avoid grazing on it, as a practical measure of control (Mather & Greaney, 1949). The viability of grass seeds in animal or liquid manure is considerably shortened (Thurston, 1952, Metz, 1970).

Crop rotations involve the inversion of soil by ploughing and frequent tillage, so that weed seeds become buried. Fallowing for periods of up to one year often forms part of a rotational system to permit weed control by tillage and/or herbicides. These actions lead to the selection of weed populations with enhanced seed dormancy that secures survival, despite the control measures. In 1856 Lagreze-Fossat predicted, on the basis of his observations of seed dormancy, that the practice of a two-year cereal rotation with a fallow one year followed by cultivation in the second year would not eliminate the two grassy weeds *A. fatua* and *A. ludoviciana*, even if the soil was cultivated to a depth of 20 cm. The genetic basis of this survival mechanism has been demonstrated, 130 years later (Jana & Thai, 1987). Two heterogeneous populations of *A. fatua* were synthesized, each composed of a different set of true-breeding lines characterized either by dormant or non-dormant seeds in equal initial frequencies. Each of the populations was grown under two cultivation regimes: (a) continuous cropping in each growing season, (b) a two-year rotation comprised of one year of propagation, followed by one year of summer-fallow. The populations responded differently to the two practices. In the summer-fallow regime the relative frequency of lines with dormant seeds increased substantially in both the mixed populations, compared to the continuous-cropping (Fig. 3.16). Non-dormant seeds were eliminated during the fallow year leaving the dormant seeds to dominate the population.

Herbicides also cause genetic changes in the constitution of grass weed populations, particularly the soil-incorporated forms, because of a shift to resistant forms (Jana & Naylor, 1982). While it is not yet fully confirmed, it is highly probable that the use of specific herbicides such as carbamates, has led to an increase in frequency of plants with seed dormancy. Aside from their lethal properties toward the growing plant, carbamates have unusual and striking effects on seed germination of grasses. Diallate (S-(2,3–dichloroallyl)di*iso*propylthiolcarbamate) and butylate (S-ethyl di*iso*butyl-thiolcarbamate) break dormancy in *Setaria faberii* in a manner similar to nitrate, or cyanide, through a stimulation of respiratory activity (Fawcett & Slife, 1975). The carbamate, *iso*-propyl-N-phenylcarbamate (IPC) has opposite effects on the germination of species within the sub-families Panicoideae and Festucoideae (Al-Aish & Brown, 1958). IPC inhibits germination, without exception, in the genera and species within the sub-family Festucoideae. Alternatively, the majority of the species within the sub-family Panicoideae are promoted by IPC. Comparison of IPC with other inhibitors that break dormancy by influencing respiration, such as sodium azide, sodium fluoride, and carbon monoxide, indicate that IPC depresses the respiration of festucoid embryos similar to the lethal effect of very low O_2 tensions. Alternatively, in Panicoid species such as *Oryza sativa* and *Zea mays*, respiration is not reduced to the lethal point by either IPC or low O_2 tension. Aside from demonstrating very important differences in the respiratory systems of Panicoid and Festucoid seeds, this study indicates that herbicidal compounds of the carbamate type are likely to apply strong selection pressure for both seed germination and growth characteristics within genetically diverse populations of grass species. The extensive studies with *A. fatua* indicate different forms of respiratory activity associated with genetically different expressions of seed dormancy. Recent studies indicate that within the species *A. fatua*, resistance to the herbicides diallate and triallate (S-2,3,3–trichloroallyl,N,N-di*iso*propylthiocarbamate) has increased significantly (Jacobsohn & Andersen, 1968; Jana & Naylor, 1982). A range of genetic tolerance to IPC among selections of *A. fatua* was four- to six-fold between the least and most resistant types selected at random among a population (Seely, 1976). From the same general population unexposed to IPC there were at least six discrete

Fig. 3.16. The influence of summer-fallowing and continuous-cropping on the relative abundance of non-dormant wild oats in two heterogeneous populations (*(a)* and *(b)*) originally composed of equal frequencies of lines with either dormant or non-dormant seeds. (After Jana & Thai, 1987).

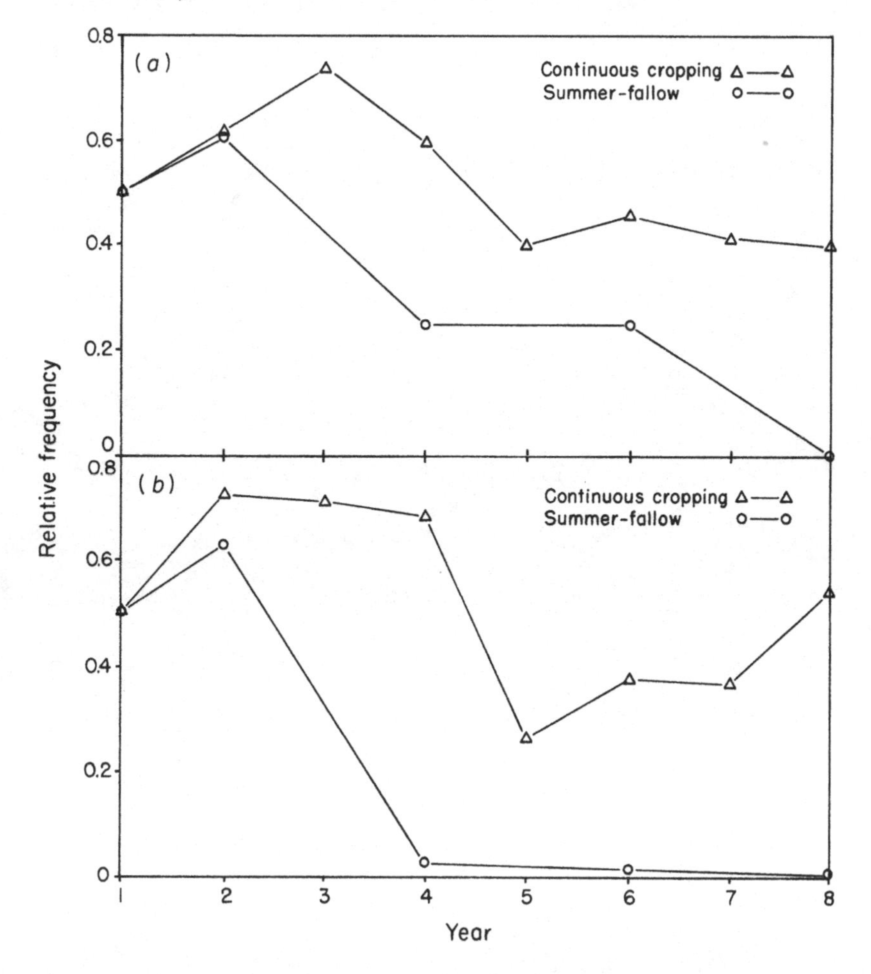

expressions of seed dormancy. Thus the likelihood of linkage of increased seed dormancy with increasing herbicide resistance is high. Studies of population dynamics of *A. fatua* under the influence of triallate were difficult to interpret due to the confusing influence of seed dormancy, possibly increased directly through the selection pressure from triallate (Selman, 1970). Other studies have indicated that the herbicides cfp-methyl and bzp-methyl reduce seed dormancy when sprayed on adult plants shortly before anthesis (Peters *et al.*, 1975).

4

Timing of dormancy

If you can look into the seeds of time,
And say which grain will grow, and which will not,
Speak then to me, who neither beg nor fear
Your favour nor your hate.

(Macbeth 1. 3. 58–61.)

4.1 Semantic considerations

The discussion of seed structure and environment in relation to dormancy (Chapters 2 and 3) has shown that genetic variation in dormancy is expressed as a distribution of germinability over time. Seeds of non-dormant genotypes germinate at maturity, or even before maturity, in a wide range of environmental conditions. Alternatively, at the time of maturity and abscission from the parent plant, seeds of dormant genotypes cannot be germinated within a wide range of temperatures in the presence of water, oxygen and light. With the passage of time the dormant seed becomes sensitive first to a narrow range of environmental conditions that promote germination. Later the range of each environmental factor, within which germination can occur, broadens until germination is limited only by water and extremes of temperature.

The term 'after-ripening' has often been used, loosely, to categorize the collective changes that seeds undergo with time as dormancy is lost. After-ripening has been used as a descriptor for loss of dormancy in a dry, stable, storage environment (Crocker & Barton, 1957), in a variable natural environment such as the soil (Baskin & Baskin, 1981), under conditions suited to optimal germination of non-dormant seeds (Simpson, 1966a), and 'by undefined biochemical changes occurring in seeds' (Baskin & Baskin, 1985). Eckerson (1913) suggested that the term after-ripening should be confined to changes within the embryo. The suggestion arose from the discovery that the structures external to the embryo restrict germination in a way that is distinguishable from dormancy shown by an excised embryo.

In spite of Eckerson's suggestion, the term after-ripening has become increasingly used as a catch-all 'black-box' to denote changes known and unknown, internal and external, environmental, and structural that contribute to loss of seed dormancy. In itself the term after-ripening does nothing to clarify the timing of dormancy. The term 'after-ripen' formed from the combination of the adverb 'after' and the intransitive verb 'ripen' taken literally means 'the ripening process that takes place after some specified event'. The event, never specified, could be, for the sake of argument, the moment of seed maturation marked by abscission of the seed from the parent plant. The word ripen is synonymous with the process of development to maturity. An interpretation of the term 'after-ripen' might therefore be: 'the second development that leads to maturity following the first development that leads to maturity !' Perhaps it is the ambiguous nature of the term that made it appropriate to use as a descriptor for a set of unknown processes. Because of its conceptual fuzziness the term should probably be abandoned. The term has no value for denoting timing of dormancy, whether time is judged by an instrument such as a clock or by sequential developmental changes within a seed.

There is a similar weakness in the use of a sub-term like 'stratification', described as 'after-ripening at specified low temperatures for some period of time' (Mayer & Poljakoff-Mayber, 1963). The simple practice of chilling a seed becomes transformed into a new descriptor that subsumes time, temperature and the seed but without clarifying the process.

The terms primary and secondary dormancy convey an element of timing in the sense that the second must follow the first. Alternatively, secondary might be construed to be of lesser importance than primary. However, if secondary dormancy is qualitatively, or quantitatively, identical to primary dormancy the connotation of numbering is strictly about timing. If the two forms are in reality quite different in quality and quantity, describing them as primary and secondary fails to convey anything about these differences and confounds sequence (or apparent level of significance) with differences in the quality of each state.

These semantic discussions about the appropriatenesss of words used for distinguishing different states of dormancy, and when they begin and end, serve to indicate that fuzziness has existed in this area since the end of the nineteenth century. Some of the fuzziness relates to the origin of the use of the word dormant in association with a perceived failure of a seed to germinate in conditions that common sense suggests should give rise to normal germination. The single term 'dormant' can only describe the observed fact that a seed has failed to germinate in a particular situation at some point in time. It is not possible to convey insight about the cause, site, complexity or timing of dormancy without pointing to these attributes through the addition of an adjective, or adjectives, to the noun dormancy.

From the perspective of our present knowledge of the complex nature of dormancy found in grass seeds the old expressions of after-ripening, primary, secondary, innate and enforced dormancies appear vague and non-specific. The debates, about which term is most appropriate or most acceptable to seed physiologists, or which term transcends all other terms, lose their significance once it is understood that we are dealing with the description of a control mechanism for a complicated, developing, biological system. The adoption of a 'systems attitude' on the part of an observing scientist is important for avoiding the pitfalls of linear and single-factor explanations of complicated systems (Trewavas, 1987).

If non-germinability can be determined by a variety of configurations of the seed-environment system then each of the separate configurations could be described as the state of dormancy. If the rates of change of some, or all, configurations are independent of each other the final act of germination will be determined by the last configuration to change to a form that permits germination. Continuity of the dormancy state from initiation to the point of germination depends on the maintenance of at least one configuration of dormancy. If the test of the presence of dormancy consists of germination under the optimal conditions of a non-dormant genotype, a genotype will be described as dormant if there is no germination. However, if the non-germinating seed is transferred immediately to a new environment it may germinate normally. The question then arises: is this a state of dormancy in the seed or just a poor environment for germination? In systems thinking this is not a valid question. The statement could be made that the seed-environment system interaction does not permit germination. It is the nature of the seed-environment system *in toto* that allows germinability, or non-germinability, not the isolated character of the seed. In this sense an impermeable hull, or seed coat, that prevents water uptake by the embryo is of equal importance in expressing dormancy as an embryo that has several metabolic blocks such that the primary root cannot expand and germinate. Dormancy can be achieved, or lost, by either changing the seed, changing the environment, or both. The problem of describing the timing of dormancy thus becomes a problem in describing how either, or both, the environment and the seed change over time in ways that prohibit or permit germination.

4.2 Timing in the induction, maintenance and release of dormancy

Trewavas (1987) has drawn an interesting parallel between the nature of the introduction, maintenance, reinforcement and loss of dormancy in a typical grass seed (*Avena fatua*) and the process of human memory. His essay, based on a systems perspective, draws analogies between the timing in each of these complex biological systems. The perspective used by him to describe seed dormancy, after some further refinement, can be used to integrate much of the background factual information about seed dormancy in grasses outlined in chapters 2 and 3. This unified concept of timing in dormancy is also applicable to other forms of dormancy found in higher plants.

To understand timing in seed dormancy it is helpful to avoid the human notion of real time measured by a clock. The measure of time for a seed comes from a combination of environmental experience and developmental change. The stage of development reached by a seed, or adult plant, at any one moment is the summation of past environmental experiences integrated through the previous developmental stages into the specific form of its structure. There is a high degree of plasticity in arriving at any particular level of development, whether judged at the level of a population, a single plant, a tissue or components of a cell. While Trewavas (1987) illustrated this plasticity by reference to *A. fatua*, other grasses and probably all living organisms show plasticity in development. In the specific case of seed dormancy in grasses the genotype of each species interacts with the environment, particularly through the maternal tissues, to induce a range of expressions of the dormancy trait. Progeny of even an inbred line exhibit variation in expression of dormancy. In some genotypes a warm maternal environment may produce non-dormant seeds but in others dormant seeds. Conversely a cool temperature may produce dormant or non-dormant seeds according to the interaction between genotype, moisture, soil, nutrient conditions and other factors of the environment. Some proportion of the mature seed lying under or on the soil surface will be dormant. Each developing seed is able to integrate its environmental experience according to genetic make-up through a highly plastic process of development. Plasticity in the development of dormancy is simply an expression of the general capacity of plants to accommodate themselves to environmental change. The widest limits of developmental change are defined by the genetic blueprint defining a species. Within these limits there is considerable plasticity in attaining any structural form or physiological condition. As stated by Koestler (1967) development is like a game of chess where the rules of the game are fixed but the strategies to arrive at completion of the game are infinitely varied. Physiological, structural, and molecular adaptations can occur at any level of the systems network we call a seed. In seeds the accumulated character of these adaptations to past experience is expressed as variation in germinability.

Following abscission of the seed from the parent plant, dormancy can be maintained for considerable periods of time in either a dry or moist seed.

The mechanism for sustaining dormancy, probably largely metabolic, permits the active turnover of RNA, protein, and the central components of the respiratory system. Hulls and seed coat play an important interactive role in excluding water for long periods so that the water potential needed for the final act of germination is held below some critical threshold. During this period of 'after-ripening' there is *an inability to achieve integrated development up to the threshold for germination.* The interaction between genotype of each seed, previous developmental achievement, and environmental experience sustains the expression of varied states of dormancy at the population, tissue, metabolic and molecular levels. *Prolongation of the state of dormancy is thus a condition of prolonged asynchronism in the processes underlying the developmental responses of the seed to the changing environment.* Before germination can occur in any particular environment, synchronism of development must be achieved in some critical proportion of the sub-parts of the physiological system. This requirement for synchronism may exist within any one level, several levels, or between levels, of the total physiological system. The extent of asynchronism within each sub-unit, a functional unit described as a **holon** by Koestler (1967), determines the degree of dormancy. Adjustments in development occur continually at every level of the seed system in response to subtle, or gross, changes in the environment. These adjustments may lead to an increase in synchronism or asynchronism among the sub-units at any level in the system. Primary, secondary or other expressions of the states of dormancy can be discerned by breaking dormancy in a sequential manner with various chemicals such as ethanol, azide, nitrate or gibberellin (Adkins & Simpson, 1988).

Trewavas (1987) concluded that there is no obligatory causal molecular sequence involved in the induction or breakage of dormancy. Instead there appears to be a set of metabolic states, any one of which can be occupied by a dormant seed. In this sense entering or leaving the dormant state is a random stochastic event that can be initiated by altering the developmental balance among the sub-units at any level in the physiological system of the seed. The same developmental or physiological endpoint can be attained through a variety of molecular directions.

The triggering action for attaining each new developmental change, such as germination, requires the synchronizing of some critical number of sub-units of the seed into a stable action unit, or holon, that reflects the properties of wholeness and ability to act as a unit. A holon may exist as a structural unit such as a plastid or cell. Synchronization can be achieved by a range of chemical and environmental factors described earlier in chapter

3. The synchronizing ability of any single factor, illustrated by reference to the action of gibberellin in overcoming dormancy of *A. fatua* (Fig. 4.1) becomes more effective, at any one concentration, as the seed progresses through the developmental phases permitted by different environmental conditions.

The general relationship between development of an array of sub-units in a seed system leading to their synchronism or asynchronism over time is illustrated in Fig. 4.2. The gap of asynchronism is wide, or narrow, in proportion to both the number of sub-units required to complete a functional unit (holon), and the relative stages of development achieved by each of the sub-units. Each holon may exist at any number of positions within a level, or at any level, within the hierarchy of functional levels that constitutes a seed. The concept is equally applicable to aggregations of molecules, or plastids, or cells, as to sub-populations of seeds within the general population of a species.

Consideration of the many environmental factors, internal and external, that can modify dormancy (Chapter 3) suggests that a great range of chemicals and environmental conditions break dormancy at different points in the developmental cycle of the seed. Some sequential order in the loss of different states of dormancy (Adkins & Simpson, 1988) reflects the attainment of separate action thresholds by sub-units of the seed system. At

Fig. 4.1. The increase in synchronizing influence of gibberellin (GA) with seed age in *Avena fatua*. Rates of germination for seeds aged 1, 5, 7 or 24 months (M) under dry conditions and then incubated in the presence or absence of 1 or O.5 μm GA (After Trewavas, 1987, based on data from Naylor & Simpson, 1961b).

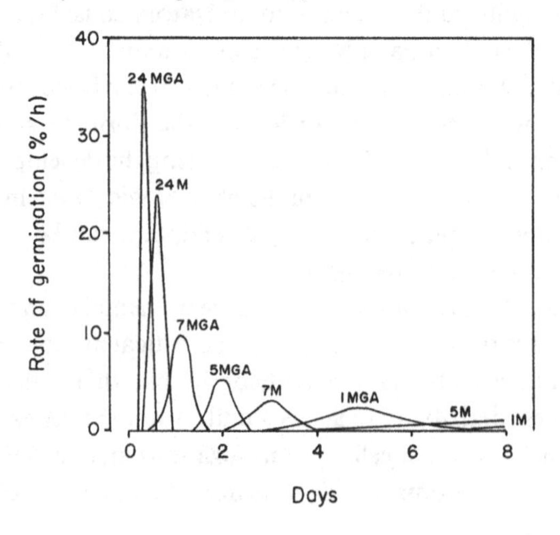

any developmental stage before the attainment of the final threshold for germination, measured as root growth, there is the possibility of environmental, or chemical, retardation of development in one, or several, subparts of the system. This retardation is expressed as asynchronism that leads to a dysfunction in part of the system. In the above sense, states of secondary or tertiary reinforcement of dormancy are quantitatively indistinguishable from primary dormancy but qualitatively different. The same environmental, or chemical, factors that promote synchronism in the late stages of seed maturation may well induce asynchronism at an earlier stage of development and *vice versa.*

An illustration of how environmental factors may retard or enhance synchronism in the development of a holon to a physiologically active stage is depicted in Fig. 4.3. Because temperature is an essential factor in all parts of the system it can be expected to have important positive, or negative, effects on synchronism. Water is important because so many metabolic and exchange functions require water as the milieu. Light has a significant

Fig. 4.2. The relation between stage of development of sub-units and time to reach a threshold for synchronized action as a functional unit (Holon). The threshold marks the boundary between dormancy (asynchronism of sub-units) and germination (synchronism of sub-units).

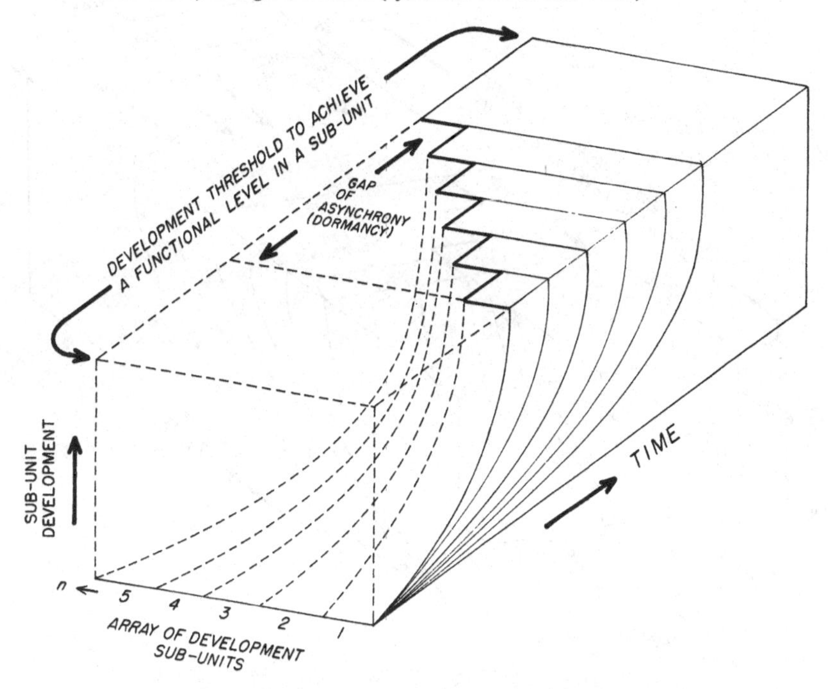

synchronizing influence at the seed population level. Hormones appear to be of particular significance in synchronizing the control of protein synthesis by the genetic template.

Degree, or depth, of dormancy can now be characterized by the degree of disparity in development among the sub-units required for action at a unit level. The extent of the array of sub-units required for physiological action at the unit level could vary at different stages of seed development. As development of the entire system progresses, synchronization becomes simpler because of duplications and cross connections between sub-units with similar functions. The synchronization increases the rate of attainment of the threshold number of sub-units needed to bring about physiological action on the next level.

When viewed from a systems perspective, the notions of primary and secondary dormancy, or induction and release of dormancy, or deep and

Fig. 4.3. The influence of various factors on the rate of attainment of an action threshold for a germination holon comprised of synchronized sub-units. Some units involved in germination may also be part of the action unit in early seedling growth.

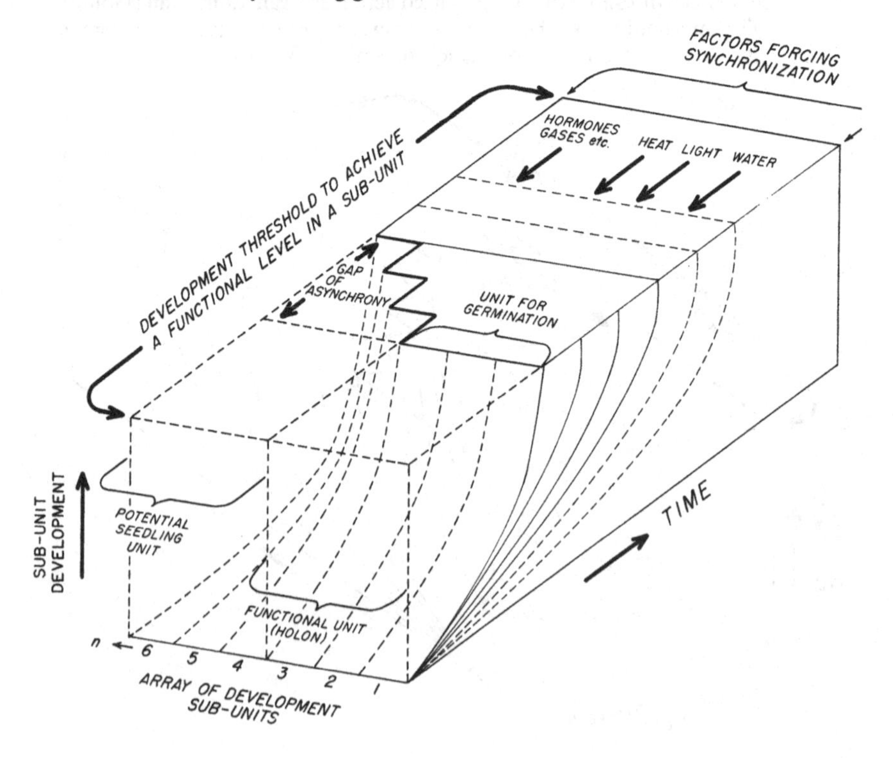

shallow dormancy are seen to be reflections of the homogeneity, or non-homogeneity, in the development of the critical number of sub-units necessary to trigger the next facet of development. The conundrum of why a grass embryo can show dormancy at the earliest stage of development yet continue to grow and develop into a more complex and mature structure is resolved by the realization that growth and development of most sub-parts of the seed continue, albeit very slowly at times, but lack the final act of synchronism for expression of root growth (germination). Artificial manipulation of the environment to attain the synchronism necessary for germination has been demonstrated at all stages of embryo development (Chapter 3). Flexibility of the seed in responding to environmental influence increases with progressive stages of development. A seed can be described as non-dormant once the development threshold for germination has been achieved by synchronism of the sub-units. The only factors limiting germination at this point are absence of water, extremes of temperature, or oxygen deficiency. Once these needs with respect to these factors have been met no limit to germinability exists in the seed.

Control, by the embryo, of sugar production in endosperm tissue is an excellent, much studied, example of synchronism in development. Placed in environmental conditions that favour germination of a whole seed, aleurone tissue will eventually secrete hydrolytic enzymes that break down the starch reserves of endosperm tissue into low molecular weight sugars. The process takes a long time. However the embryo can integrate this process for its own needs by secreting gibberellin that acts both as a hormone ('messenger') and synchronizing agent by speeding up the production of amylases and in turn sugar. Sugar demand and supply are integrated and both accelerate in concert. Two structurally separate sub-units thus become unified in a physiological function called seedling growth. At the cellular level the action of the hormone brings all the individual aleurone cells into a unified delivery of hydrolases. Sugar production is thus greatly accelerated and seedling growth occurs quickly.

This discussion indicates that asynchronism of active physiological processes can be a reasonable explanation for the controlled failure of germination traditionally called dormancy. Nevertheless, the information about the role of seed structures in seed dormancy reviewed in chapter 2 indicates that the hulls and seed coat of grass seeds play a very significant role in preventing germination by controlling the mobility of water and gases. Thus in some grass seeds the embryo dormancy is not very persistent and disappears before the coat and hulls have lost their ability to block germination. The timing of the removal of these structural barriers to germination may be more sporadic and longer lasting than the sequential changes in metabolism underlying the developmental changes in, for example, the embryo. While the changes in these structural barriers cannot be considered as developmental changes, they are nevertheless dependent on sequential environmental changes, such as alternate wetting and drying, and diurnal and seasonal temperature oscillations, that also drive changes in development of the living components of the cell. Soil microorganisms, enhanced by secretions from the living seed tissue, may have a significant role as active agents for the removal of these structural barriers under natural conditions.

If asynchronism of physiological functions underpins the phenomenon of grass seed dormancy, obvious questions arise. How, and when, is asynchronism (or synchronism) programmed and passed to successive generations without creating deleterious disorganization in normal development? One answer could be that dormancy is essentially confined to the meristematic zone of a primary root and thus has no influence on the remainder of the seed. Another answer could be that control of root germination does not require a large number of complex factors. Recent genetic studies of reciprocal crosses between dormant and non-dormant pure lines of *A. fatua* suggest that the dormancy state removed by the action of sodium azide, the first block to disappear as dormancy is lost during after-ripening, has a maternal origin (Jana *et al.*, 1988). As few as three multi-allelic loci may control the trait.

In the early stages of seed dormancy in many grass species cycles of alternating temperatures, under conditions that favour developmental change (sufficient moisture and oxygen), are the only means of breaking dormancy. At later stages as dormancy is progressively lost, germination can occur at a single constant temperature. These facts suggest that alternating temperatures play a particularly significant role in the synchronization of developmental changes, possibly at the enzyme level of the system. Low temperatures during early seed development appear to favour the induction of dormancy in many grasses. On the other hand, high temperatures in the stages of development that occur following abscission from the parent plant can induce secondary or high temperature dormancy. Thus, either extremes of temperature promote dormancy but when combined in alternating cycles they are complementary for promoting synchronism in development.

There do not appear to be any cases where dormancy is broken so rapidly that developmental changes would be ruled out. Generally speaking the removal of dormancy by artificial and natural means requires a long time span with some periods where sufficient moisture and gases provide conditions for metabolism and cell activity. There are similarities in this respect between the synchronizing effect of the environment in induction of vernalization in grass tissues and the removal of dormancy in seeds. Both systems are strongly influenced by gibberellin, low and alternating temperatures, and have a requirement for an energy source and active respiration. Both systems are reversible by high temperatures and anaerobic conditions. There are further analogies between the development of seedling dwarfism, often associated with an extreme expression of seed dormancy, that can be overcome by gibberellin or alternating temperatures.

The desiccation associated with abscission of the apparently mature grain from the parent plant has been associated with the loss of dormancy in many grass species that have non-persistent dormancy, for example common cereals. This dormancy is primarily caused by the hulls, when present, or caryopsis coat in preventing water uptake. The impermeable barriers are broken down by the physical actions of desiccation or alternate wetting and drying. When the grain is subsequently moistened water is freely imbibed by the embryo. In a limited sense the lack of availability of water for the embryo is a limiting factor to coordinated development because metabolic functions appear to continue slowly even under conditions of supposedly dry storage at room temperatures.

For that type of embryonic dormancy state that persists for long periods of time, such as several years, Trewavas (1987) has speculated that some stable molecular mechanism is essential to act as a switch for cells and tissues to progress from one level of development to the next. He has hypothesized a persistent bi-stable compound that cannot switch the system out of the dormancy mode until it changes to the alternative form. Bi-stable compounds of this nature, such as autophosphorylating protein kinases, are known in animals (Lisman, 1985) and dicotyledenous plants (Blowers & Trewavas, 1987). Trewavas believes that in the dormant state of a seed the enzyme is in the unresponsive highly autophosphorylated state, provided ATP is continuously available and the kinase/phosphatase activity ratio is constant. If ATP declines, for example when respiration is blocked, the enzyme switches to the alternative condition that is sensitive to calcium ions. This form promotes formation of membrane proteins and loss of dormancy in a timing action that has some analogies with the timing action of phytochrome when it releases dark dormancy of seeds following a regime of light/dark cycles. Coding of a past environmental experience on the maternal plant could be built into the molecule in its autophosphorylating form because of its ability to be multiply phosphorylated. While these ideas are challenging there is little evidence to date to support the hypothesis, and in any event, systems analysis stresses that the rules for function at one level are generally not the same rules that govern function at other levels. At this point in our understanding of seed dormancy it is not clear exactly at which level, or on how many levels of the seed-environment system, control is exerted.

5
Modelling the induction, maintenance and termination of dormancy in grass seeds

5.1 Modelling theory for biological systems

The analysis of structure and environmental factors affecting expression of grass seed dormancy (Chapters 2–4) indicates considerable complexity and interaction within a system that extends well beyond the boundaries of the unit defined as a seed. In chapter 1 some arguments were presented for using a systems perspective as a first approach to understanding the behaviour of a dormant, or a non-dormant, seed. Because our perception of nature is hierarchical, in order of containment such as tissues, organs, organism, populations, community and ecosystem (Halfron, 1979), the view at each level provides a different perspective of the whole through separations into horizontal and vertical components (Fig. 5.1). Behaviour at each level is explained in terms of the level below but significance is found at the level above. Thus explanation and significance appear as complementary aspects of our perception. At higher levels, changes appear to take place more slowly than at lower levels. Behavioural uncertainty increases when the components of the system become less integrated. Behavioural certainty, or equilibrium within the system, increases with increased integration of the various components. Stability of the system requires that a change in adaptability of one sub-system requires a compensating change in an opposite way in a sub-system elsewhere at the same level. Thus, in Fig.

Fig. 5.1. Perspective in the dimensions of a hierarchically ordered system.

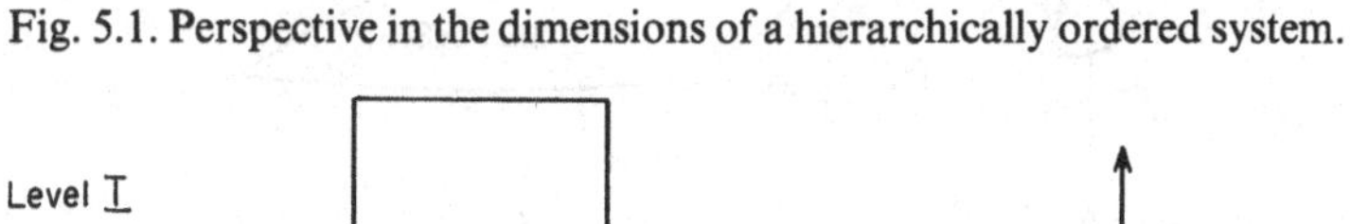
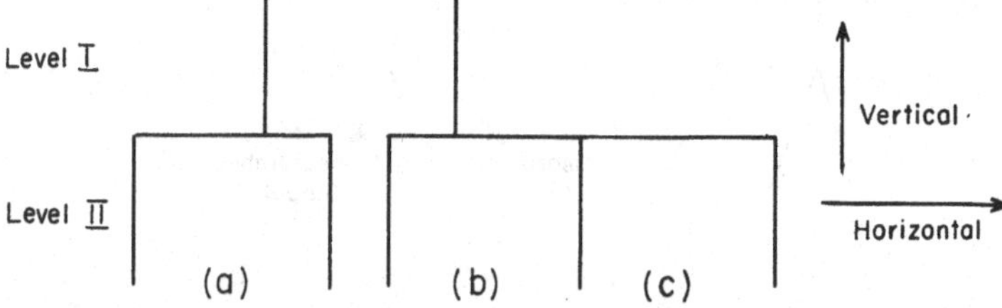

5.1 if (a) changes then (b) or (c) must compensate for stability to exist at level II. *In the same sense the less connected each part is on a given level the less overall change will occur when an external factor impinges on the whole level. Adaptability of the system to environmental uncertainty is maximized when the behaviours of the sub-systems are most independent of each other.*

When a part of the system responds, or is indifferent to some environmental variable it creates a new and specific environment for the sub-parts. In this way water filtered through starchy endosperm, or a membrane, has a different character from water reaching the entire caryopsis from the soil. Similarly, light filtered through a chloroplast, or the pigment phytochrome, will have a different character from the light impinging on the entire floret.

Modelling the behaviour of a seed, or parts of a seed, is a first approach to reducing the confusing complexity of the total system to a figurative or mathematical framework that is simple enough to be comprehensible to the limited capability of the human mind. A 'lumped model' (Halfron, 1979) involves grouping categories, states or compartments, known and unknown, into fewer more general categories to obtain a holistic view. Behavioural changes can then be described quantitatively as transfers of matter, energy or information. A lumped model of a seed system is depicted in Fig. 5.2.

The quantities of energy, mass or information that are stored within the system are called the *state variables* and the system is said to be in

Fig. 5.2. An observer viewing a 'lumped' model of a seed.

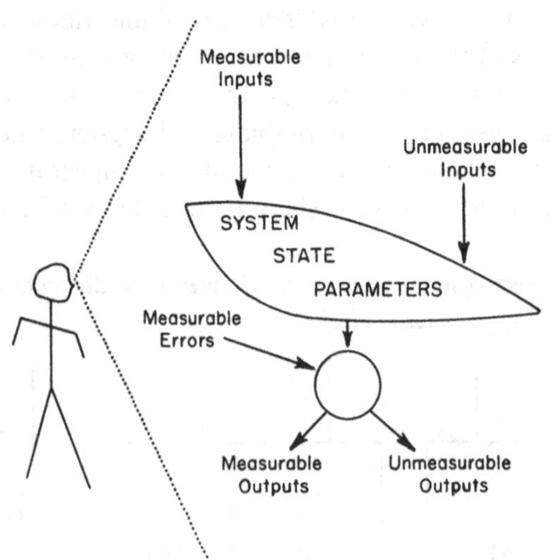

equilibrium when there is a balance between the inputs and outputs. Odum (1983) has indicated how flows of energy in biological systems develop as webs of energy transformation, feedback interaction and recycling. These webs create a hierarchy of transformations of energy in different forms. The quality in each form is measured in terms of the energy in one type that can generate a flow of another type. Systems either need an outside force or a storage depot within the system to generate the rapid flows of energy needed to power change. If the potential energy to power the system is too small, energy degrades without the work transformation that comes when automatic feedback loops guide rapid transformations powered by large sources of energy. Systems with small storage capacity change quickly and *vice versa. In this sense a resting seed with a large store of energy changes slowly whereas a coleoptile or root with little storage capacity can change rapidly. If power is energy flow per unit of time then dormancy is a state of low power and germination and seedling growth are states of high power.*

Modelling a grass seed, with the behavioural states of dormancy and normal rapid germination using energy flows, involves accounting for stored energy, flows of energy, energy transformations and energy embodied in the information that feeds back and repays the 'web' by directing the energy to lower forms (Odum, 1983).

All models are abstractions of reality and merely descriptions of the conditions of the state variables within the system as they appear to the human mind. For this reason it is worthwhile giving some consideration to the mindset (set of mind states) available to an observer attempting to understand the nature of grass seed dormancy. Weinberg (1975) has aptly summarized the constriction placed on human perceptions of reality as a principle of invariance: 'We understand change only by observing what remains invariant and permanence only by what is transformed.' Change is thus perceived by comparing a state that is remembered on one occasion with what is different on the next occasion. Discriminating change with the human mind thus requires at the absolute minimum a consecutive pair of comparisons.

The task in modelling is thus to partition the dynamic seed (or the world) into sets of stable sub-states that can be differentiated from their changing surroundings by paired, or multiple, comparisons made in the mind. The next section outlines some of the difficulties associated with thinking about, and then modelling, the behaviour of entities such as a dormant seed, or a dormant root, if they are partitioned from their surroundings as though they were 'black boxes' generating behavioural patterns discernible by an intelligent observer.

5.2 A 'black box' perspective of seed behaviour

Weinberg (1975) demonstrated that limiting the scope of observations of behaviour of a biological system such as a seed to either a macro or micro level, in order to partition the state variables by boundaries, requires an observer to either simplify or exclude most of the 'reality'. For example, focusing on a seed seen as an entity defined by its boundary the seed coat separates the seed from its environment but simultaneously lumps everything within the seed into a 'black box' category. Discriminating by focusing down within the seed to the level of a single plastid within a root cell lumps and excludes the rest of the seed and environment into a large 'black box'. Because the mind is incapable of dealing with more than a limited amount of information, lumping into states, defined by boundaries that appear to limit interaction, reduces mental effort. In this sense the mind is incapable of dealing with the complexity of everything interacting with everything else and it reduces complexity to sub-units with boundaries that indicate minimal levels of interaction.

Decomposing a system into sub-systems through discrimination of boundaries carries with it the corollary that the explanation of behaviour of each of the sub-systems will be distinctly different from the behaviour of the whole system. The problem of choosing a satisfactory explanation of dormancy from among a set of explanations derived by looking at the system at different levels is illustrated by the following hypothetical example. Suppose that root behaviour is chosen as the indicator of the physiological state commonly called grass seed dormancy. If the root remains unchanged it is described as dormant and if it elongates it is said to germinate. In order to 'explain' this change from one state to another root behaviour must first be decomposed into sub-states that can be recognized by particular parameters. If the scope (S) of the observations is limited to three parameters: oxygen uptake (O), water status (W) and root length (L) then

$$S = \{O, W, L\}$$

If the range of each parameter is designated by numbers that do not denote measure but simply the number of possible states the hypothetical example might have the following ranges:

O = (1, 2) where 1 = on, 2 = off.
W = (1, 2) where 1 = the same, 2 = different.
L = (1, 2, 3, 4, 5, 6) where 1 = the same, 2 = very,very short, 3 = very short, 4 = short, 5 = long, 6 = very long.

Table 5.1. *Potential states of root behaviour in a dormant grass seed, estimated by three parameters.*

Parameters			
Oxygen uptake	Water status	Length	Name of state
1	1	1	a
1	1	2	b
1	1	3	c
1	1	4	d
1	1	5	e
1	1	6	f
1	2	1	g
1	2	2	h
1	2	3	i
1	2	4	j
1	2	5	k
1	2	6	l
2	1	1	m
2	1	2	n
2	1	3	o
2	1	4	p
2	1	5	q
2	1	6	r
2	2	1	s
2	2	2	t
2	2	3	u
2	2	4	v
2	2	5	w
2	2	6	x

An observer monitoring the root continuously with these ranges for each parameter will see certain combinations from the complete set of 24 possible combinations listed in Table 5.1. The behaviour of the root is described by the sequence of combinations of the three parameters measured for as long as the observer retains interest.

By monitoring the actual states the observer might see the following pattern:

a, n, g, s, a, n, g, s, a, n, g, s ......

When these states are mapped as a sequence of consecutive pairs then *State* a, n, g, s, *is followed by* n, g, s, a, as a short and repeated cycle that goes on indefinitely, see Fig. 5.3.

If a sequence is confined to two observations then it represents a pair of choices from the complete set of 24 states. The number of all possible pairs is the product of the set containing 24^2 (576) possible sequences of length two. Similarly 14 000 sequences of length three and 300 000 of length four, growing combinatorially to an extremely large number with a sequence of ten (24^{10}). If the sequence of observed states is repeated and the cycle is composed of a set with a small number of states the system is described as highly constrained (Weinberg, 1975). The most constrained system would be 'a' moving to 'a' in an indefinite repetitive progression. By contrast the most unconstrained system would have a cycle that would appear to be so random as to have indefinite length. In a system that is highly constrained the sequence of observed states appears to be highly repetitive to the observer and thus reflects stability. If the system is completely unconstrained an observer would see the sequences changing continuously in a random (stochastic) manner.

Intuitively it might seem that state 'a' is somewhat analagous to a root of a dormant seed and state 'x' analagous to a germinating root. The quest then is to understand how state 'a' can become changed to state 'x' and what is the constraining influence that orders pairs of states into a sequence a, n, g, s, a ..... that is repeated.

If the root is influenced by the observer, or the natural environment, the sequence may change so that instead of being a, n, g, s, a... it becomes b, m, h, a (Fig. 5.4).

Fig. 5.3. A short behavioural cycle of a primary root.

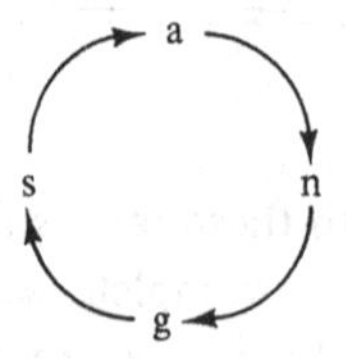

Fig. 5.4. Short and long cycles of root behaviour.

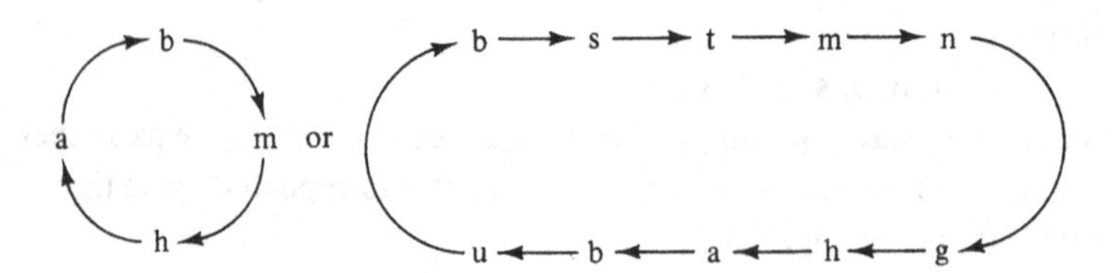

Table 5.2. *State variables for a single parameter, root length.*

State	Root length	Corresponding sub-states
A	1	(a, g, m, s)
B	2	(b, h, n, t)
C	3	(c, i, o, u)
D	4	(d, j, p, v)
E	5	(e, k, q, w)
F	6	(f, l, r, x)

In each case moving from one cycle to another must require an external input unless the system is self-determining. If the system is self-determining the observer would see the change in cycle given patience and a constant environment. Once in the new cycle the sequence of states and thus behaviour is a reflection of internal control. The temptation may be to consider the internal constraint on the pattern and length of the cycle as a genetic constraint that determines behaviour.

However, to another observer who measures only the changes in root length and fails to include changes in either water or oxygen states as measures of behaviour, root behaviour will appear to be quite different. Instead of seeing 24 possible states the observer of root length can only recognize six states (Table 5.2). Thus a, g, m, s are said to be 'lumped' into state A and cannot be recognized by observing only root length. The cycle that was previously observed as a, n, g, s is now seen as ABAA. A cycle m, h, a, s would also be recognized as ABAA and thus is indistinguishable from the ABAA that reflects the cycle a, n, g, s. The cycle observed as a single sequence of ten states when three parameters are monitored (i.e. a, b, d, c, k, g, h, j, i, e ....) becomes a shorter cycle of ABDCE repeated twice. These contrasting views of the behaviour of the same 'realities' are illustrated in (Fig. 5.5).

If root length is the only parameter used to observe root behaviour a single observation that indicated state A would not give any indication of the true position in the sequence of changing root length because state A is not always followed by the same state. As examples, in the first two cycles A can be followed by B or A. In the three hypothetical cycles, each with a scope of three parameters that reflect the true internally determined states, any single observation allows prediction of the next or previous state within each cycle once the cycle is recognized.

If an observer chooses to measure only the two parameters root length and oxygen uptake as assessment of root behaviour a third perspective on behaviour is obtained (Table 5.3). In this example the method of measuring root length is crude and discerns three instead of six lengths (ie. very short is indistinguishable from the same; short is indistinguishable from medium; long is indistinguishable from very long).

When the same cycles observed by measuring three parameters are compared with the cycles observed with the parameters oxygen uptake and root length, in a manner that is less discriminating (3 ranges) than the six ranges used in the example shown in Fig. 5.5, behaviour is interpreted differently because of the method adopted by the observer (Fig. 5.6).

Fig. 5.5. Alternative views of root behaviour when one or three parameters are monitored.

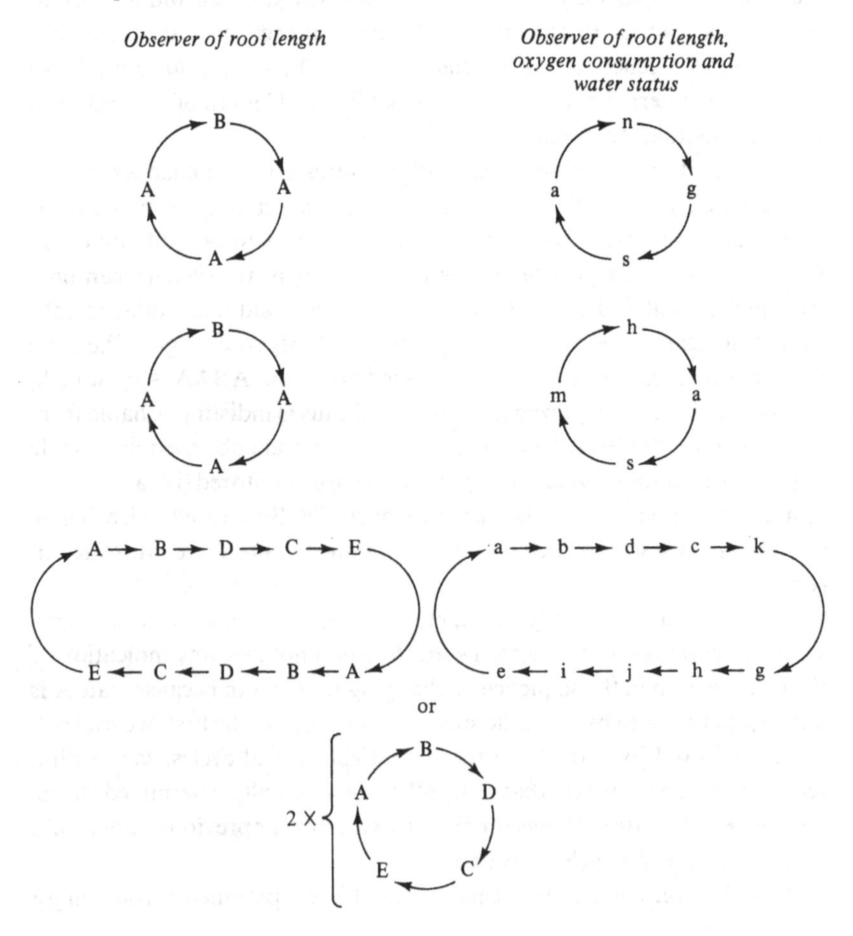

Table 5.3. *Potentially observable states determining root behaviour assessed by two, or three, parameters.*

	Two parameters		
State	Oxygen consumption	Root length	Corresponding states for 3 parameters
S	1	(1, 2)	(a, b, g, h)
T	2	(1, 2)	(m, n, s, t)
U	1	(3, 4)	(c, d, i, j)
V	2	(3, 4)	(o, p, u, v)
W	1	(5, 6)	(e, f, k, l)
X	2	(5, 6)	(q, r, w, x)

The cycles derived from observations of six potential states (A, B, C, D, E, F and S, T, U, V, W, X) are less predictive than the cycles seen from observations of 24 potential states. Discriminating by using more variables to observe behaviour will not help in explanation if behaviour is unconstrained to the extent that the cycles cannot be discerned during the length of time of the observations. Conversely, lumping determinate variables into general categories reduces discrimination, simplifies the description of behaviour and may show patterns that are obscured at the level of greater detail.

The view of reality about the state of dormancy, or germination, is thus related as much to the view adopted by the observer as it is to the degree of constraint, and cyclic or other possible patterns of determinate behaviour, originating from the living system we call a seed. The cyclic patterns of behaviour that change when the 'black box' is interfered with by the observer may not correspond to cycles induced by changes in the natural environment.

Choosing the dimensions of the black box called 'a dormant grass seed' is difficult. Lumping the floret, spike (or panicle) and parent plant together as a single 'dormant system' would seem justified on the basis of knowing that all the parts are involved in influencing the expression of dormancy. On the other hand, focusing on root behaviour alone, because it appears to be the key structure expressing change as the dormant system changes to a non-dormant condition, only permits an explanation of dormancy in terms of the sub-units of a root.

Clearly the 'black box' approach requires the observer to make a choice about the level at which the seed system is entered. The scope of the observations and the ranges used for each parameter are to a great extent defined by the available technology. For example, the search for an explanation of dormancy in terms of promotion and inhibition of germination by naturally occurring growth substances is mainly a reflection of the greatly enhanced ability to monitor these substances in plants using chromatography and mass spectrometry. Changes in growth substances may not, in reality, have any stronger causal relationship to the induction or loss of dormancy than any other molecule in a seed. Nevertheless, discerning the pattern of sequences or cyclic behaviour of a

Fig. 5.6. Potentially observable states determining the behaviour of a germinating root, assessed from several perspectives.

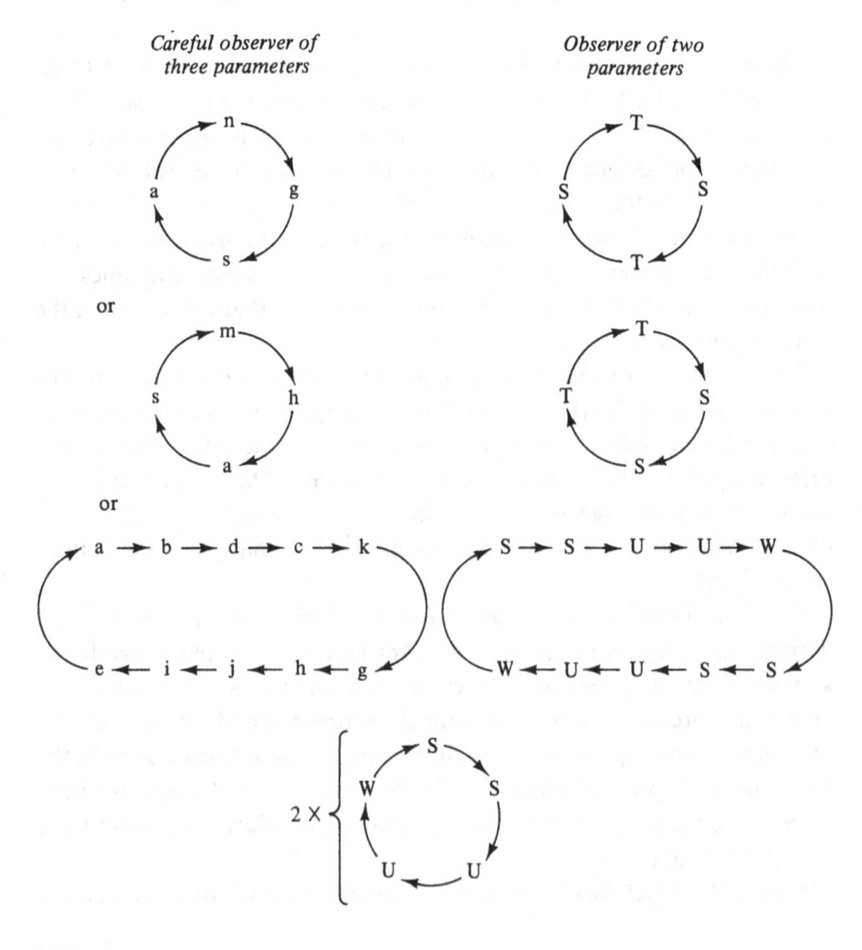

plant growth regulator may allow prediction of onset and termination of the dormant state of an embryonic root.

Limitations to the 'black box' approach to modelling dormancy or germination can be :

(a) Poor choice of the scope and range of observations that might reveal the behavioural patterns of the true state variables determining the general states called dormancy and germination. Historically, the scope has largely been confined to macro observation of primary root elongation in grass seeds. In recent years attention has been focused down to the level of the biochemical system (e.g. respiration) and the molecular level (e.g. growth regulators). To a limited extent the parental genetic contributions to dormancy, particularly the maternal influence, have been evaluated.

(b) Because the glumes, lemma, palea and seed coat constitute major barriers to water availability for the embryo in mature florets, after release from the parent plant, some of the effects of environmental change on embryonic behaviour may be misinterpreted as direct effects when in fact they are indirect. Some of the effects of stratification may simply be on the permeability of the above structures to water and not directly on the embryo. In this sense it is important to distinguish between inert and living components of the seed system.

(c) Physical and technical difficulties associated with timing the measurement of several parameters concomitantly. Biochemical changes are faster than macro changes in cells, tissues and organs. Thus correlating molecular changes with slowly changing root length inevitably leads to fuzziness because of the non-matching time scales.

(d) Interpreting seed behaviour as if it was generated only by a 'black box' exhibiting deterministic behaviour ignores the possibilities that :

   (i) The box (i.e. seed system) eventually grows into a new shape, or size, that then permits new expressions of behaviour.

   (ii) The environment, or observer, may interact with the box and force changes in the sequence of state variables that determine behaviour. The new sequence may change at a different rate from the previous sequence. Change in states during the phase called germination appears to be much faster than change during the phase called dormancy. Thus it may be easier to observe changes in state variables during dormancy, if sequences change slowly, than during rapid germination. In this sense using external force to terminate dormancy as a means of understanding the nature of dormancy is equivalent to shutting the stable door after the horse has bolted. The observer breaking dormancy with a growth regulator, or temperature change, is no longer observing dormancy but instead observes germination. For the same reasons, observing the state of dormancy in a constant environment should, in theory, reveal more about the causal relationships determining the state than observation in a changing environment.

Support for the 'black box' model approach to understanding seed behaviour during the condition of dormancy comes from some experimental observations. Dormant seeds kept in a constant environment change slowly and lose dormancy indicating internal change. The example of *Avena fatua* indicates that the dormancy is terminated by a prior fixed sequence of metabolic changes characteristic of each genotype. In a number of grass species (Chapters 2 & 3) seeds can enter and re-enter the state of dormancy on several occasions provided the changes in the seed system are not beyond some critical limit. This cyclic pattern measured at the macro level of root elongation possibly reflects cyclic behaviour at the level of the sub-systems in other parts of the seed. Lack of coordination of the sub-systems is observed in seeds that 'germinate' by elongating only the coleoptile, or only the root, or only the scutellum, rather than all three organs simultaneously, or in a definite sequence. Different depths of dormancy, measured as different degrees of persistence of dormancy, may be reflections of either the length of a single set of sequences of internal behaviour, or the extent of repeatability of a shorter cycle. In this respect the experimental evidence from grass seeds has not yet confirmed that secondary dormancy is less persistent than primary dormancy, or *vice versa*.

5.3 An energy flow model for grass seed dormancy

Odum (1982) has demonstrated that any system made up of a group of parts that interact in some kind of process, can be visualized as blocks with connecting lines that indicate their interactions. By standardizing these symbols into a systems language it is possible to diagramatically represent any system with energy units because all natural phenomena involve energy transformations. Because energy is never lost the transformations can be followed as pathways entering and leaving the system. The quantities of energy stored within the plant–seed–environment system vary and are called the *state variables* because they describe the state of the system as it varies. Models with varying degrees of simplicity or complexity can be constructed of seeds attached to or separated from the parent plant. Such models can reflect energy flows at different stages of development and help to clarify distinctions between dormant and non-dormant states.

The simple energy diagram depicting a developing seed (Fig. 5.7) uses the symbolic energy language described by Odum (1983). Because the scope of the diagram is broad, measurement of the flows of energy in units per time will only permit general predictions. For example, increasing the inflow of energy from the sun will lead to more stored energy in the seed structures provided that water and inorganic nutrients are not rate limiting, respiration is curtailed, and change in day length leads to a switch from the vegetative phase to the formation of an inflorescence. The model can be tested using quantitative units representing the qualitative differences expressed in storage, flow, resistance and capacitance in a manner similar to an electrical circuit.

Modelling a dormant seed using this approach requires a refinement in the scope of observations. A shift to a more detailed analysis, specifically to within the seed, creates a more complex model. Testing the model may require monitoring a large number of parameters. Using the model to predict the outcome of changes in one or more variables may need complicated calculations or simulation with an analogue computer. Because seed dormancy is expressed during seed development on the parent plant and also in the autonomous seed following abscission from the parent plant, at least two separate models must be developed to depict these different situations.

A model of an autonomous dormant seed is depicted in Fig. 5.8. A comparison of this model with the simpler model of a dormant seed still attached to the parent plant (Fig. 5.7) draws attention to some obvious differences in the two energy flow systems. On the developing plant, water enters the seed via the vascular elements but in the autonomous state it enters from the soil via successive barriers of glumes, coat, aleurone, endosperm and scutellum to the embryo. The energy potential of the water finally delivered to the embryo has marked diurnal cycles related to the changes in whole plant water potential of the parent plant. Diurnal changes in water potential in the environment of an autonomous seed may be even more extreme in seeds at, or close to, the soil surface in comparison to deeply buried seeds.

Oxygen can be available to the developing seed from either the

Fig. 5.7. A simple energy flow diagram for a developing grass seed. Energy circuit symbols after Odum, 1983. With permission of the publisher, John Wiley & Sons, Inc., New York.

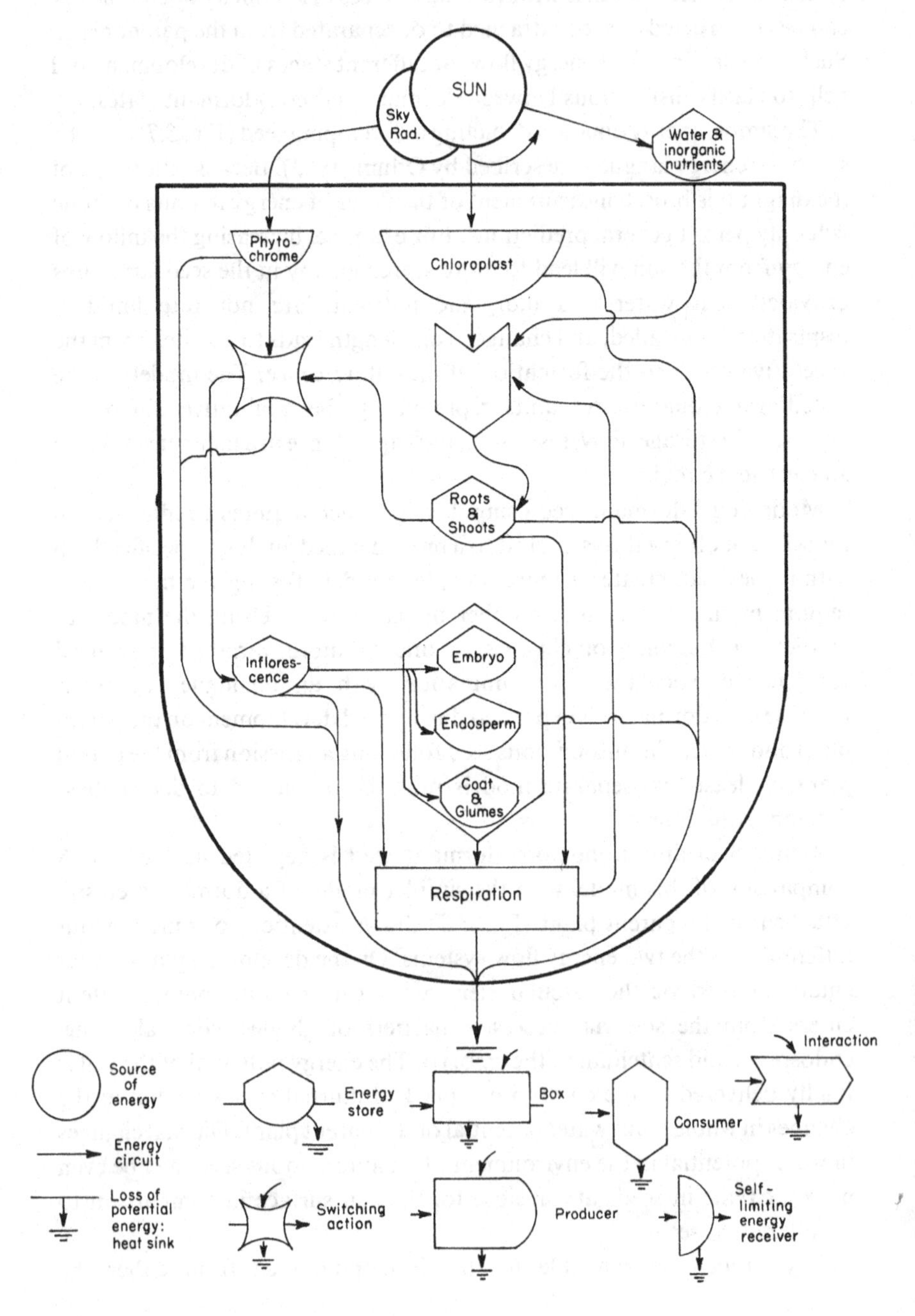

Fig. 5.8. Energy flow diagram for an autonomous grass seed buried close to the soil surface. PGRs = plant growth regulators. Energy circuit symbols after Odum, 1983. With permission of the publisher, John Wiley & Sons, Inc., New York.

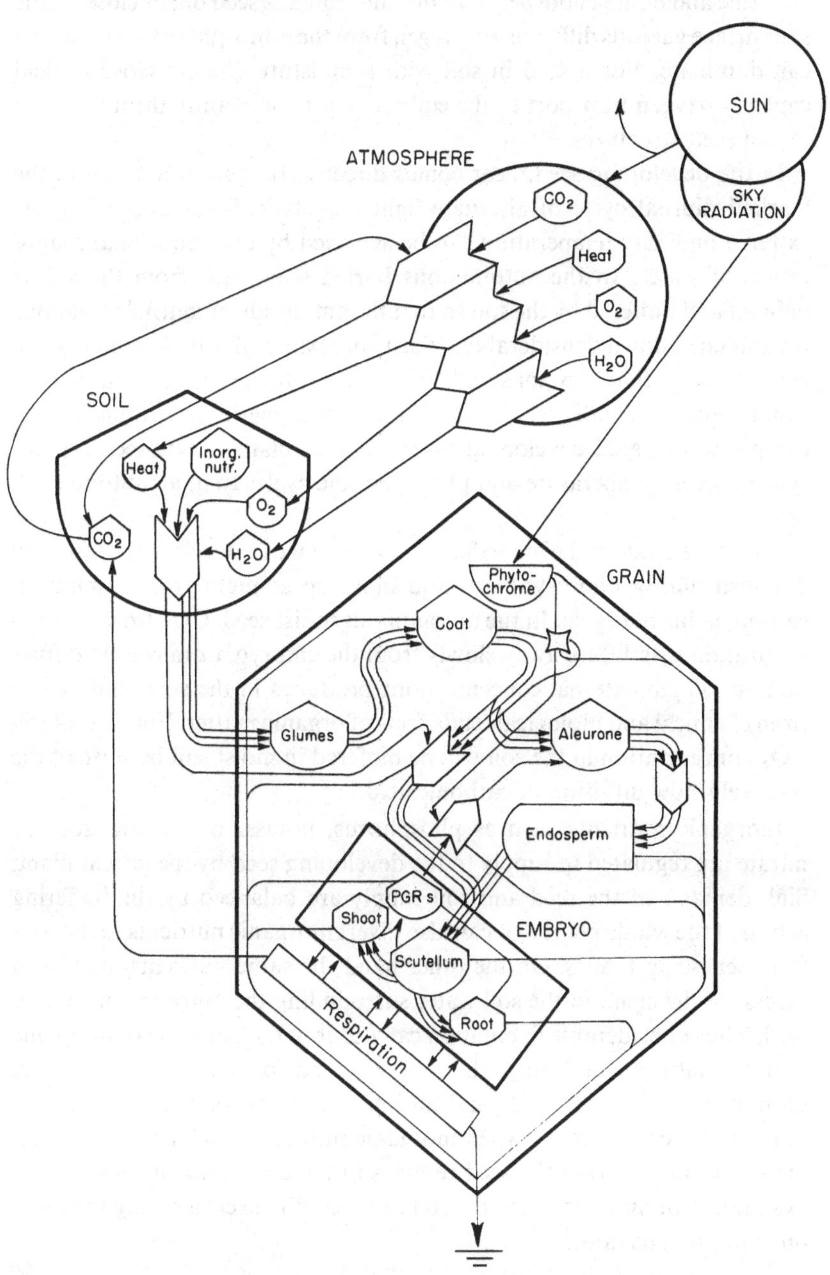

atmosphere or derived internally as a product of photosynthesis. In either case the penetration of oxygen to the embryo is primarily by diffusion across water barriers of the living cells in tissues such as glumes, coat, aleurone and liquid endosperm. In the autonomous seed on, or close to, the soil surface gaseous diffusion of oxygen from the atmosphere to the embryo can dominate. For a seed in soil with a moisture content close to field capacity oxygen transport to the embryo is predominantly through water or wet seed structures.

In the developing seed, heat comes directly from sun radiation in the marked diurnal cycle of alternate light and dark. Some cooling below extreme high day temperature can be achieved by transpirational evaporation of water. In the autonomous buried seed, heat from the sun is indirect and buffered by the soil so that the amplitude of diurnal variations is dampened down considerably. Moist soils will buffer these temperature changes more than dry soils and temperature changes in the soil will lag behind photoperiodic changes in the atmospheric environment. By comparison, a seed developing on the parent plant is exposed to more synchronized temperature and photoperiodic cycles than an autonomous seed.

Carbon dioxide in the immediate environment of a developing grain may diminish during photosynthesis and build up at night from respiration causing a diurnal cycle. In the autonomous moist seed, CO_2 from internal respiration will diffuse away slowly from the embryo, or may even diffuse in, due to high external concentrations produced in the surrounding soil from chemical and biological oxidations of organic matter. Fluctuations in CO_2 concentration in the soil will be buffered in moist soil because of the relatively slow diffusion of carbonic acid.

Inorganic nutrients such as phosphorus, potassium, or ions such as nitrate are regulated in supply to the developing seed by the parent plant. Sink demand of the seed and soil supply are balanced by the buffering action of the whole plant that can also divert inorganic nutrients to the seed from senescing tissues. On the other hand the same nutrients may be in excess, or deficient, in the soil water surrounding the autonomous buried seed. Thus considerable fluctuation can occur in the amounts of inorganic nutrients diffusing in through the seed structures to the embryo, depending on movements of soil water and nutrient status of the soil. Prior to the termination of dormancy when inorganic nutrients, such as phosphorus, are not available from the seed reserves of aleurone and endosperm, the availability of nutrients from the soil reserves may exert a strong influence on embryo behaviour.

In the developing seed the primary source of energy for growth and metabolism of the seed is derived from the photosynthetic and respiratory activity of the parent tissues. To a limited extent, in the early stages of development, photosynthetic activity occurs in the glumes and seed coat. In the later stages of development considerable respiratory activity occurs in association with the production of starch reserves. By contrast, the sole source of energy for metabolism and growth in the autonomous seed is derived from limited reserves stored within the embryo and extensive reserves in the surrounding storage tissues of aleurone and endosperm. The surge in power supply (energy per unit time) essential to initiate rapid germination and early seedling growth comes from the low grade energy of low molecular weight compounds that must first be derived from high grade polymers of carbohydrate, fats, and proteins. The rapid release of this energy pool is governed by complicated hormonal and enzymic feed-back loops to the respiratory and growth functions of the embryo. Autonomous grains with true embryo dormancy appear to require a sequence of changes in respiratory metabolism to achieve a particular state before the power surge from the storage tissues can occur. This sequence of changes can be achieved, or disrupted, in a cyclical manner by inputs from the soil-atmosphere environment.

Timing mechanisms in the developing seed are probably governed by the P and HER sensitivity to changes in photoperiod and quality of light perceived by the whole plant. In the autonomous dormant seed, photoperiodic timing and P activity are governed by the state of hydration of the tissues containing P. In addition the depth of placement of the seed in soil, soil density, compaction and particle size will determine whether light sufficient to activate P can reach the seed.

Some of the differences between the energy flow systems of developing and autonomous dormant seeds can account for contrasting responses to a single environmental factor. For example, low temperatures promote dormancy in the developing seed but break dormancy in the autonomous seed (Chapter 3) indicating that the dormancy states are significantly different.

There are some obvious analogies between the symbolic elements used by Odum (1983) to construct an energy flow diagram, the state variable diagrams of Weinberg (1975) and the holon elements of a system hierarchy (Koestler, 1967). In each case the models attempt to explain behaviour by showing the linkages and dynamic changes in the sub-elements of a system. Explanation of behaviour can be simple, if confined to a macro-system description, or detailed and complex if focused on the micro-system level.

Neither the macro- nor micro-system model can be expected to 'explain' dormancy but they do provide the clarification essential to further experimental observation and the development and testing of working hypotheses. Further progress in understanding grass seed dormancy will be enhanced by constructing and testing models that secure compatibility of data obtained from experimental observations. It is unfortunate that many of the experimental observations of grass seed dormancy from the past have been made on different species, by a multitude of experimental approaches, and at different levels of the seed-environment system. It is not possible therefore to integrate much of this information into a rational model of dormancy. With the advent of the computer it is now possible to construct, and test, theoretical models that can be used to interpret the natural behaviour of seeds as they enter and leave the dormant state.

5.4 Similarity and diversity among dormancy states in grass seeds

The chapter on occurrence of seed dormancy among grass species (Chapter 1) indicated that the trait of seed dormancy is a common feature of the majority of grass species. The adaptive advantage of seed dormancy for survival also confers a competitive advantage that makes many grass species important weeds in cereal and forage crops.In most grass species the lemma, palea and seed coat are important barriers to the uptake of water in the autonomous seed lying on or beneath the soil. The common agricultural practices of scarification, or glume removal, have principal effects in improving germination through increasing the availability of water. The natural removal of these barriers is achieved by cyclical changes in the external environment of the seed. Although it has been argued (Bewley & Black, 1982) that the physical barriers to water uptake are really not part of the mechanism of dormancy it is nevertheless clear that persistence of these barriers leads to various interactions with the caryopsis that intensify the expression of dormancy.

Grass seeds with prolonged dormancy generally express some form of true embryo dormancy. It is rare to find embryo dormancy in genotypes with naked caryopses, indicating a link between the expression of embryo dormancy and the presence of glumes, lemma and palea. The mechanisms for the expression of embryo dormancy appear to be quite complex due to differences in metabolism, hormone sensitivity, and genetic make-up among species with either self- or cross-pollination, or mixtures of both forms of pollination. Even within a single species, such as *Avena fatua*, there can be considerable variation in expression of embryo dormancy. The absence of embryonic dormancy during seed development can be expressed as vivipary in seeds of many species.

The role of endosperm tissues in contributing to the expression of dormancy is not very clear. The conclusions drawn from extensive studies with cereals have been that endosperm utilization is principally a post-germination event. However, there may be a significant role of endosperm in controlling seed dormancy based on the demonstration of genetic linkage between the state of embryo dormancy and non-autonomy of endosperm typified by *A. fatua.*

Studies of the effects of specific environmental factors on dormancy suggest commonalities in response among many grass species. For example, water stress during seed maturation reduces the level of dormancy in many species (Chapter 3, 3.1). Similarly low temperatures during seed development enhance dormancy and high temperatures reduce dormancy in many temperate grass species. In autonomous mature seeds high temperatures favour the induction of secondary dormancy but low temperatures hasten the loss of primary dormancy. Sensitivity to the induction of secondary dormancy appears to be a complex genetic trait that is common among and within many grass species. Based on the example of *A. fatua*, thermally-induced and anaerobically-induced secondary dormancy appear to have separate origins in metabolism and can be expressed distinctly in genetically different lines (Chapter 3). Diversity in temperature-dependent expressions of dormancy is enhanced by structural differences in hulls, coats and reserve tissues among different species.

Light has a major role in breaking seed dormancy in the majority of grass species (Chapter 3). There is an underlying dark dormancy in many species that disappears with time. As long as this dormancy persists, light can be a significant factor in the loss of dormancy through a P mediated stimulus. This P response appears to be universal among grasses, perhaps among seeds of all higher plants. There are subtleties in the P response that act together with the HER light response and confer further diversity in the patterns of response to light within and among grass species. Grass species adapted to different latitudinal zones, particularly in the temperate regions, show variation in photoperiodic induction of dormancy during seed development. Diversity in this photoperiodic response within a species can produce ecotypes with, or without, seed dormancy. There is generally an important interaction between photoperiod and temperature in the determination of this expression of grass seed dormancy. Different genetic races have evolved in each geographic region reflecting different expressions of seed dormancy.

The influence of gases, such as CO_2 and O_2, on seed dormancy is varied and seems to be important mainly in genotypes with true embryo dormancy linked to respiratory metabolism. Potential for considerable genetic diversity in response to gases, within a species, is illustrated by three important species, *Oryza sativa*, *Echinochloa crus-galli*, and *Avena fatua*.

While some plant growth regulators, such as gibberellins and cytokinins, can have marked effects in promoting the loss of dormancy in some species, they do not appear to influence all species of grasses. This may reflect the fact that negative responses, for example to applied gibberellin, may have been due to application of an inappropriate derivative to a genotype with an inability to further synthesize, utilize, or metabolize that particular gibberellin. In the case of inhibitors such as abscisins there may be likewise no universality in response among all grass species. These kinds of responses indicate that there are many different potential ways that endogenous plant growth regulators can be involved in either the induction, maintenance or termination of dormancy in grass seeds.

The response of dormant grass seeds to inorganic substances such as the nitrate ion appears to be quite varied among and within species. The example in *A. fatua* of considerable genetic diversity in this response suggests that diversity in response, rather than commonality in response, to particular chemicals in the environment provides yet another mechanism for adaptation to different ecological niches.

The above, and other evidence, indicates considerable diversity in the expression of dormancy of grasses in different agronomic and ecological situations. The general conclusion can therefore be made that there are many different potential states of dormancy that arise both from the structural and physiological diversity among grasses as well as from the adaptations to external factors of the environment. Thus, despite a great number of commonalities in the response of dormant seeds to particular environmental factors such as water and temperature, it is justifiable to look upon dormancy as being the expression of a unique state within each genotype within a species.

There has been a tendency in the scientific approach to understanding nature to subsume particularities with general rules, or laws. The quest to 'explain' seed dormancy by looking for the generalities that underpin the expression of dormancy has not been successful up to this date. In a sense the common observation that there is great diversity in the expression of dormancy has been marginalized in our thinking in favour of looking for a central powerful and universal explanation. The most universal conclusion that can be drawn from the literature survey (Chapters 1–4) is that the majority of grass species have the potential to achieve a state of dormancy in seeds, given some sequential set of favourable environmental circumstances. The corollary to this conclusion might be that there are as many different ways to achieve this condition of dormancy as there are species and sub-species. It is this very diversity in achieving the dormant state that provides diversity in adaptation within and among species to the great range of ecological niches that grasses occupy on the land surface of the earth. This diversification of the dormant state begins soon after fertilization, extends through the various stages of seed development on the parent plant, and is reflected in the great variation in timing of germination within the population of autonomous seeds found on or under soil.

Significant factors that force achievement of a dormant state in grass seeds are:

(a) The genetic constitution of the seed and type of pollination employed in the parental generation.
(b) Particular barriers to the hydration of the seed created by the presence of glumes, hulls, seed coat, aleurone and endosperm structures.
(c) Structural and metabolic changes induced by temperature changes. Metabolic changes associated with respiration of the embryo are particularly important.
(d) Periodic qualitative changes in the light-dark regime during seed development. Prolonged darkness for autonomous seeds buried beneath the soil.

The same factors also play a significant role in forcing the loss of dormancy and creating diversity in the timing of normal germination.

Past observations of dormancy in grass seeds have been, until recently, largely confined to a narrow range of parameters linked to changes in a single organ, the root. Restricting observation of a complex seed-environment system to the macro level and to a limited range of parameters (i.e. root length) may have unintentionally led to the incorrect conclusion that dormancy is a single state common to all grass seeds. Because other tissues, such as the shoot, scutellum, or reserves are slower to change than the root when dormancy is terminated, they have been considered less important. The discussion in sections 5.1 and 5.2 shows that the true state variables determining dormancy can be missed if an observer chooses to make observations at an inappropriate level in the system. It is therefore possible that the many experiments to understand the nature of grass seed dormancy, using only the on/off state of the root as the indicator of behaviour in the seed system, have provided answers that simply reinforce the conclusion that dormancy is a single condition common to all grass seeds.

However, widening the scope of observations to include other organs and other levels of the seed-environment system reveals a wider range of state variables that contribute to the formation of many different states of grass seed dormancy. The possibility therefore of achieving a single explanation of dormancy in grass seeds diminishes in direct proportion to the intensity with which observations are carried out down to lower and lower levels within the seed-environment system. For similar reasons, a search for 'a general theory' of dormancy embracing all plant species bearing seeds is unlikely to be successful despite some obvious commonalities in behaviour at the macro-level of observation. In this sense every individual seed expresses its own form of dormancy peculiar to each specific environment. Modelling seed behaviour associated with dormancy is thus unlikely to provide a common model for all species, or even sub-species. Nevertheless, models are valuable for delineating the components of a seed-environment system and their interrelationships. Probably every individual seed deserves its own model!

5.5 Conclusion

Several questions were posed in the introduction to this treatise on grass seed dormancy. Summary answers to these questions will serve as a conclusion.

The first question asked what can we conclude about the nature of seed dormancy in a single, well-studied species? A brief answer based on examination of the wild oat (*Avena fatua*) is that dormancy is a genetically controlled lack of synchronism (asynchronism) in the development and functioning of the structural and biochemical components of the seed. While the genetic control is primarily internal, maternal tissues do contribute to this asynchronism during seed development. Prolonged asynchronism in the sub-units of the system delays the transition from seed to seedling. Variations in water, temperature, light, and other environmental factors such as gases and inorganic substances, overcome the asynchronism in a sequential, hierarchically ordered series of steps over time. Thresholds must be attained for action to take place at a given level of physiological sub-units. Termination of the state of asynchronism is perceived by an observer of the macro level as elongation of the primary root and scutellum and soon after by elongation of the coleoptile. Genotypic and phenotypic variation exists within the species for degree of asynchronism. This variation produces variation in the timing of germination within any specific environment. Conversely, different environments force synchronization of the physiological sub-units at different rates for any single genotype. The same environmental variables that can force synchronization may induce asynchronism at some other stage of seed development.

The second question posed in the introduction asked whether there are commonalities between seed dormancy in a single species, such as *A. fatua*, and other grass species. The short answer is a qualified yes. Both the induction during seed development, and maintenance and termination of dormancy in the autonomous seed have many common features among grass species. The similar seed morphologies of many grass species combined with generally similar responses to water, light and temperature produce common features in the expression of seed dormancy. Nevertheless, when any single factor related to dormancy is considered, there are clearly differences both between and among grass species. There is a common tendency for grass seeds to become induced into a dormant state by low temperatures during development on the parent plant, and for dormancy to be diminished by high temperatures and water stress in the late stages of seed development. After separation from the parent plant these same low and high temperatures have opposite effects on dormancy. This suggests that a critical threshold for response to temperature occurs at the time of seed abscission from the parent plant. The response to low temperature during early seed development can be interpreted as a forced asynchronism in the development of physiological sub-units within the seed system. Abscission is the threshold for change in response to temperature that is analagous to shifting from reverse to forward gear in an automobile. Temperature has a synchronizing influence on physiological activity of the seed after abscission from the parent plant.

The response of an autonomous seed to high temperature that leads to secondary dormancy is also a feature common to many grass species. Light also has an important role in overcoming dormancy in a majority of grass seeds. The sequence of genetically determined biochemical blocks found in some genotypes of *A. fatua* that contribute to the expression of dormancy probably occurs in many other grass species. What is certain is that genetic control underpins the expression of seed dormancy in grasses. The specific type of pollination mechanism employed by each species, or sub-species, will also determine the extent of variation in expression of the trait of seed dormancy. *A. fatua* is a particularly good example of a grass with great diversity in the expression of seed dormancy. The diversity is related to the combination of a predominantly selfing mode of pollination with a small percentage of out-crossing that provides for new genetic combinations.

With these particular attributes the species competes well in a specific niche and can also invade new niches.

The third question in the introduction was concerned with whether it is possible to establish definitions of dormancy and draw new conclusions about the effects on dormancy of physiological and environmental conditions. The review of experimental evidence (Chapters 2 and 3), together with the discussion about models (Chapter 4), suggests several general conclusions. Firstly, there is no single state of dormancy into which all seeds of the grass family must enter, remain, and exit. Instead there are multiple states of dormancy arising from the many different combinations of environments and seed morphologies and functions. These multiple states can be aptly subsumed with the term '*holonic asynchronism*'. This term shifts the sense of the expression 'dormancy', from being a positive *presence* of some specific state, to an emphasis on the *absence* of synchronism among a number of component sub-states that have the potential for unification into a single active state. The absence of synchronism in some parts of the seed system is induced by environmental factors. The same environmental factors also play an important role in the achievement of synchronism among the sub-parts.

The apparently antithetical germination response to a single environmental factor such as temperature at different stages of embryo development and maturity may be due to the dominating influence of maternal tissues. An environment that supports maturation of parental tissue, when filtered through those same tissues, becomes an external environment unsuited for the normal progression of development in an embryo. Even after abscission of the seed from the parent plant some parental tissue envelopes the embryo and filters external environmental influences. In this way lack of synchronism in the parts of the embryonic system is sustained until the intervening barriers are removed by the changing environment. This is well demonstrated by the simple device of excising the embryo at any stage of seed development on the parent plant, or from the mature autonomous seed. True embryonic dormancy (expressed as the failure of an excised embryo to germinate) is of short duration, by comparison with dormancy expressed in an entire floret. This temporary lack of synchronism apparent within an excised embryo, arising from the previous forced asynchronism of functional sub-parts of the entire floret, can be sustained by environmental extremes such as high temperatures and anaerobiosis. With 'average' environmental conditions the lack of synchronism is quickly overcome. It seems possible that for any stage of development of a grass plant there is a particular set of environmental circumstances that might be so conducive to lack of synchronism among the functional sub-parts of the system that growth is suspended in the entire plant or particular sub-units of the plant. In this sense seed dormancy has similarities with bud dormancy or other expressions of dormancy located in specific plant organs.

From the considerations of different approaches to modelling a seed-environment system reflecting dormancy, it would seem that the method of monitoring energy flow used by Odum (1983) has merit as an approach to understanding seed dormancy. An experimental method that permits the use of common units of measurement for monitoring all sub-parts in both the living seed sub-system and the environment sub-system should help to overcome many of the problems of interpretation encountered in the past from fragmentary approaches employing different methodologies at different levels. Monitoring energy flow in a seed-environment system will require a new set of experimental techniques. It is to be hoped that such a new approach to investigating the nature of seed dormancy in grasses will provide seed biologists with an enlightened understanding of the true nature of this fascinating aspect of plant behaviour.

Bibliography

Adams, C. E. (1956). Starr millet germination problems. *Seed Technology News*, **55**, 17.

Adkins, S. (1981). Studies on the mechanisms of seed dormancy in *Avena fatua* L. Ph.D. Thesis. University of Reading, England.

Adkins, S., Gosling, P. G. & Ross, J. D. (1980). Glucose-6–phosphate dehydrogenase and 6–phosphogluconic acid dehydrogenase of wild oat seeds. *Phytochemistry*, **19**, 2523–2525.

Adkins, S., Loewen, M. & Symons, S. J. (1986). Variation within pure lines of wild oats (*Avena fatua*) in relation to degree of primary dormancy. *Weed Science*, **34**, 859–864.

(1987). *Weed Science*, **35**, 169–172.

Adkins, S. W., Naylor, J. M. & Simpson, G. M. (1984). The physiological basis of seed dormancy in *Avena fatua*. V. Action of ethanol and other organic compounds. *Physiologia Plantarum*, **62**, 18–24.

Adkins, S. W. & Ross, J. D. (1981a). Studies in wild oat seed dormancy. I. The role of ethylene in dormancy breakage and germination of wild oat seeds (*Avena fatua* L.). *Plant Physiology*, **67**, 358–362.

(1981b). Studies in wild oat seed dormancy. II. Activities of pentose phosphate pathway dehydrogenases. *Plant Physiology*, **68**, 15–17.

Adkins, S. W. & Simpson, G. M. (1988). The physiological basis of seed dormancy in *Avena fatua*. IX. Characterization of two distinct dormancy systems. *Physiologia Plantarum*, **73**, 15–20.

Adkins, S. W., Simpson, G. M. & Naylor, J. M. (1984a). The physiological basis of seed dormancy in *Avena fatua*. III. Action of nitrogenous compounds. *Physiologia Plantarum*, **60**, 227–233.

(1984b). The physiological basis of seed dormancy in *Avena fatua*. IV. Alternative respiration and nitrogenous compounds. *Physiologia Plantarum*, **60**, 234–238.

(1984c). The physiological basis of seed dormancy in *Avena fatua*. VI. Respiration and the stimulation of germination by ethanol. *Physiologia Plantarum*, **62**, 148–152.

(1985). The physiological basis of seed dormancy in *Avena fatua*. VII. Action of organic acids and pH. *Physiologia Plantarum*, **65**, 310–316.

Adkins, S. W., Symons, S. J. & Simpson, G. M. (1987). The physiological basis of seed dormancy in *Avena fatua*. VIII. Action of malonic acid. *Physiologia Plantarum*, **72**, 477–482.

Agrawal, P. K. (1981). Genotypic variation in seed dormancy of paddy and simple methods to break it. *Seed Research*, **9**, 20–27.

Agrawal, P. K. & Kaur, S. (1975). Standardization of the tetrazolium test for ragi (*Eleusine coracana*) seeds. *Seed Science and Technology*, **3**, 565–568.

Agrawal, P. K. & Nanda, J. S. (1969). A note on dormancy in rice. *Riso*, **18**, 325–326.

Ahring, R. M. (1963). Methods of handling introductions of grass seed belonging to the tribe Andropogoneae. *Crop Science*, **3**, 102.

Ahring, R. M., Dunn, N. L. & Harlan, J. R. (1962). Effect of various treatments in breaking seed dormancy in Sand Lovegrass, *Eragrostis trichodes* (Nutt.) Wood. *Crop Science*, **3**, 131–133.

Ahring, R. M., Eastin, J. D. & Garrison, C. S. (1975). Seed appendages and germination of two Asiatic bluestems. *Agronomy Journal*, **67**, 321–325.

Ahring, R. M. & Harlan, J. R. (1961). Germination characteristics of some accessions of *Bothriochloa ischaemum* (L.) Keng. *Oklahoma Experiment Station. Technical Bulletin* T-89. 19pp.

Ahring, R. M. & Todd, G. W. (1978). Seed size and germination of hulled and unhulled bermudagrass seeds. *Agronomy Journal*, **70**, 667–670.

Aisien, A. O. & Palmer, G. H. (1983). The sorghum embryo in relation to the hydrolysis of the endosperm during germination and seedling growth. *Journal of the Science of Food and Agriculture*, **34**, 113–121.

Akamine, E. K. (1944). Germination of Hawaiian range grass seeds. *Hawaii Agricultural Experiment Station. Technical Bulletin* No. 2. 60pp.

Akazawa, T. & Miyata, S. (1982). Biosynthesis and secretion of α-amylase and other hydrolases in germinating cereal seeds. *Essay in Biochemistry*, **18**, 40–78.

Al-Aish, M. & Brown, W. V. (1958). Grass germination responses to *iso*-propyl-phenyl carbamate and classification. *American Journal of Botany*, **45**, 16–23.

Albrecht, K. A., Oelke, E. A. & Brenner, M. L. (1979). Abscisic acid levels in the grain of wild rice. *Crop Science*, **19**, 671–676.

Allan, R. E., Vogel, O. A. & Craddock, J. C. (1961). Effect of gibberellic acid upon seedling emergence of slow and fast emerging wheat varieties. *Agronomy Journal*, **53**, 30–32.

Allard, R. W. (1965). Genetic systems associated with colonizing ability in predominantly self-pollinated species. In *The Genetics of Colonizing Species*, ed. H. G. Baker & G. L. Stebbins, pp. 49–76. New York: Academic Press.

Amemiya, A., Akemine, H. & Toriyama, K. (1956a). Studies on the embryo culture in rice plant. 1. Cultural conditions and growth of immature embryo in rice plant. *National Institute for Agricultural Science. Series D.*, **6**, 1–40.

(1956b). Studies on the embryo culture in rice plant. 2. The first germinative stage and varietal differences in growth response of cultivated embryos in rice plant. *National Institute for Agricultural Science. Series D*, **6**, 41–60.

Amen, R. D. (1965). Seed dormancy in the alpine rush, *Luzula spicata* L. *Ecology*, **46**, 361–364.

(1968). A model of seed dormancy. *Botanical Review*, **34**, 1–31.

(1970). Nature of seed dormancy and germination in the salt marsh grass (*Distichlus spicata*). *New Phytologist*, **69**, 1005–1013.

Andersen, A. M. (1938). Comparison of methods used in germinating seeds of *Poa compressa*. *Proceedings. International Seed Testing Association*, **10**, 307–315.

(1941). The effect of different temperatures on the germination of freshly harvested and mature seeds of *Axonopus compressus*. *Proceedings. Association of Official Seed Analysts*, **33**, 99–102.

(1944). Germination of freshly harvested seed of western grown Astoria bentgrass. *Proceedings. Association of Official Seed Analysts*, **35/36**, 138–146.

(1947a). Some factors influencing the germination of seed of *Poa compressa* L. *Proceedings. Association of Official Seed Analysts*, **37**, 134–43.

(1947b). The effect of alternating temperatures, light intensities, and moistening agents of the substratum on the germination of freshly harvested seed of Oregon grown ryegrass (*Lolium* spp.). *Proceedings. Association of Official Seed Analysts*, **37**, 152–161.

(1953). Germination of buffel grass, *Pennisetum ciliare* (L.) (Link) seed. *Association of Official Seed Analysts News Letter*, **27**, 36–37.

(1955). A germination study of Merion Kentucky bluegrass with special reference to the interfering fungi. *Proceedings. Association of Official Seed Analysts*, **45**, 94–101.

(1958). A preliminary study of dormancy in brown top and cattail millets. *Proceedings. Association of Official Seed Analysts*, **48**, 85–92.

(1961). A study of dormant and firm seeds of brown-top millet. *Proceedings. Association of Official Seed Analysts*, **51**, 92–98.

(1962). Effect of gibberellic acid, kinetin-like substance, ceresan and phenacridane chlorite on the germination of *Panicum ramosum* seeds. *Proceedings. International Seed Testing Association*, **27**, 730–741.

(1963). Germination of seed of Texas needlegrass (*Stipa leuchotricha*). *Association of Official Seed Analysts News Letter*, **37**, 18.

Andersen, A. M. & Drake, V. C. (1944). Preliminary study of crested wheatgrass exhibiting delayed germination. *Proceedings. Association of Official Seed Analysts*, **35/36**, 146–152.

Andersen, S. (1965). The germination of freshly harvested seed of ripe and unripe barley and oats. *Euphytica*, **14**, 91–96.

Andrew, M. H. & Mott, J. J. (1983). Annuals with transient seed banks. The population biology of indigenous sorghum species of tropical northwest Australia. *Australian Journal of Ecology*, **8**, 265–276.

Andrews, C. J. (1967). The initiation of dormancy in developing seed of *Avena fatua* L. Ph.D. Thesis. University of Saskatchewan, Saskatoon, Canada. 139pp.

Andrews, C. J. & Burrows, V. D. (1972). Germination response of dormoat seeds to low temperature and gibberellin. *Canadian Journal of Plant Science*, **52**, 295–303.

(1974). Increasing winter survival of dormoat seeds by a treatment inducing secondary dormancy. *Canadian Journal of Plant Science*, **54**, 565–571.

Andrews, C. J. & Simpson, G. M. (1969). Dormancy studies in seed of *Avena fatua*. 6. Germinability of the immature embryo. *Canadian Journal of Botany*, **47**, 1841–1849.

Andronescu, D. I. (1919). Germination and further development of the embryo of *Zea mays* separated from the endosperm. *American Journal of Botany*, **6**, 443–452.

Anghel, G. & Raianu, M. (1959). Germination of wild oats (*Avena fatua, A. ludoviciana*) in the laboratory and in the field. *Analele Institutului de Cercatari Agronomice, Seria C, Fiziologie, Genetica, Ameliorare, Protectia Plantelor si Technologie Agricola*, **27**, 83–95.

Anonymous. (1954). Dormancy and some other problems in seed testing. *New Zealand Journal of Agriculture*, **88**, 207–209.

Anonymous. (1956). Comparative germination tests on certain *Agropyron* spp. *Association of Official Seed Analysts News Letter*, **30**, 39–46.

Anonymous. (1962). Germination inhibitions in native grasses (*Lasirus hirsutis, Cenchrus ciliaris, Cenchrus setigereus*). *Agricultural Research*, **12**, 58–59.

Anonymous. (1983). The shedding mechanism of wild-oat seed. Hormonal control of enzyme production in aleurones of the wild oat. DNA synthesis in early germination. *10th Annual Report, Weed Research Organization, Agricultural and Food Research Council, United Kingdom*, pp. 103–104.

Anton de Triquell, A. (1986). Grass gametophytes: their origin, structure, and relationship with the sporophyte. In *Grass Systematics and Evolution*, ed. T. R. Soderstrom, K. W. Hilu, C. S. Campbell & M. E. Barkworth, pp. 11–36. Washington, D.C.: Smithsonian Institute Press.

Arai, M. & Chisake, H. (1961). Ecological studies on *Alopecurus aequalis* Sobol., a noxious weed in winter cropping. 7–8. On the primary dormancy of the seed. *Proceedings. Crop Science Society of Japan*, **29**, 428–32.

Arai, M. & Miyahara, M. (1960). Physiological and ecological studies on *Echinochloa crus-galli*. 1. On primary dormancy of seed. *Proceedings. Crop Science Society of Japan*, **29**, 130–133.

(1962). Physiological and ecological studies on barnyard grass (*Echinochloa crus-galli* Beauv. var. oryzicola Ohwi). II. On the primary dormancy of the seed. (2) On the dormancy broken of the seed in the soil. *Proceedings. Crop Science Society of Japan*, **31**, 73–79.

(1963). Physiological and ecological studies on barnyard grass (*Echinochloa crus-galli* Beauv. var. oryzicola Ohwi). V. On the germination of the seed. *Proceedings. Crop Science Society of Japan*, **31**, 362–366.

Arber, A. (1934). *The Gramineae. A study of cereal, bamboo and grass.* Cambridge: Cambridge University Press.

Arnon, I. (1972). *Crop Production in Dry Regions.* Vol. 1. Background and Principles. London: Leonard Hill Books.

Arora, N. & Bannerjee, S. K. (1978). Seed testing procedure for finger millet (*Eleusine coracana*). *Seed Research*, **6**, 158–160.

Aspinall, D. J. (1965). Effects of soil moisture stress on the growth of barley. III. Germination of grain from plants subjected to water stress. *Journal of the Institute of Brewing*, **72**, 174–176.

Atterberg, A. (1907). Die nachreife des Getreides. *Landwirtschaftliches Versuchungen*, **67**, 129–143.

Atwood, W. M. (1914). A physiological study of the germination of *Avena fatua. Botanical Gazette*, **57**, 386–414.

Augsten, H. (1956). Wachstumsversuche mit isolierten Weizen-Embryonen. *Planta (Berlin)*, **48**, 24–46.

Avery, G. S. (1930). Comparative anatomy and morphology of embryos and seedlings of maize, oats and wheat. *Botanical Gazette*, **89**, 1–4.

Babu, V. R. & Joshi, M. C. (1970). Studies on physiological ecology of *Borreria articularis* D. a common weed of bajra (*Pennisetum typhoides*) fields. *Tropical Ecology*, **11**, 126–139.

Baker, L. O. & Leighty, D. H. (1958). Germination studies with wild oat seed. *Proceedings. Western Weed Control Conference*, **16**, 69–74.

Banting, J. D. (1962). The dormancy behaviour of *Avena fatua* L. in cultivated soil. *Canadian Journal of Plant Science*, **42**, 22–39.

(1966). Studies on the persistence of *Avena fatua. Canadian Journal of Plant Science*, **46**, 129–140.

(1974). *Growth Habit and Control of Wild Oats.* Publication 1531. Canada Department of Agriculture, Ottawa. 34pp.

Banting, J. D. & Gebhardt, J. P. (1979). Germination, after ripening, emergence, persistence and control of Persian darnel (*Lolium persicum*). *Canadian Journal of Plant Science*, **59**, 1037–1046.

Banting, J. D., Molberg, E. S. & Gebhardt, J. P. (1973). Seasonal emergence and persistence of green foxtail. *Canadian Journal of Plant Science*, **53**, 369–376.

Barralis, G. (1965). La germination des folles avoines. *Annales des Epiphyties (Paris)*, **16**, 295–314.

Barrett, S. C. H. & Wilson, B. F. (1983). Colonizing ability in the *Echinochloa crus-galli* complex barnyard grass. 2. Seed biology. *Canadian Journal of Botany*, **61**, 556–562.

Barton, K. A., Roe, C. H. & Khan, A. A. (1971). Inhibition and germination: influence of hard seed coats on RNA metabolism. *Physiologia Plantarum*, **25**, 402–406.

Baskin, J. M. & Baskin, C. C. (1967). Germination and dormancy in cedar glade plants. I. *Aristida longespica* and *Sporobolus vaginiflorus. Journal of the Tennessee Academy of Science*, **42**, 132–133.

(1981). Ecology of germination and flowering in the weedy winter annual grass *Bromus japonicus. Journal of Range Management*, **34**, 369–372.

(1983). Seasonal changes in the germination responses of fall panicum to temperature and light. *Canadian Journal of Plant Science*, **63**, 973–979.

(1985). The annual dormancy cycle in buried weed seeds a continuum. *BioScience*, **35**, 492–498.

Baskin, J. M., Schank, S. C. & West, S. H. (1969). Seed dormancy in 2 species of *Digitaria* from Africa. *Crop Science*, **9**, 584.

Bibliography

Bass, L. N. (1948). Germination of freshly harvested oats. *Proceedings. Association of Official Seed Analysts*, **38**, 47–52.

(1950a). Dormancy in redtop. *Association of Official Seed Analysts News Letter*, **24**, 42.

(1950b). Effect of wave length bands of filtered light on germination of seeds of Kentucky bluegrass (*Poa pratensis*). *Proceedings. Iowa Academy of Science*, **57**, 61–71.

(1951). Effect of light intensity and other factors on germination of seeds of Kentucky bluegrass (*Poa pratensis* L.). *Proceedings. Association of Official Seed Analysts*, **41**, 83–86.

(1955). Determining the viability of western wheatgrass seed lots. *Proceedings. Association of Official Seed Analysts*, **45**, 102–104.

(1959). Comparison of germination percentages obtained for highland bentgrass seed tested at different temperature alternations. *Proceedings. Association of Official Seed Analysts*, **49**, 73–76.

Bekendam, J. (1975). Report of the working group on the application of gibberellic acid in routine germination testing to break dormancy of cereal seed. *Seed Science and Technology*, **3**, 92–93.

Belderok, B. (1961). Studies on dormancy in wheat. *Proceedings. International Seed Testing Association*, **26**, 697–760.

(1962). Histochemical determination of bound disulfide groups in barley and wheat with regard to germ dormancy. *Berichte Getreidechemiker-Tagung, Detmold*, **1962**, 21–25.

(1968). Seed dormancy problems in cereals. *Field Crop Abstracts*, **212**, 203–211.

Berg, T. (1982). Seed dormancy in local populations of *Phalaris arundinacea* L. *Acta Agriculturae Scandinavica*, **32**, 405–409.

Berrie, A. M. M., Don, R., Buller, D., Alam, M. & Parker, W. (1976). The occurrence and function of short chain fatty-acids in plants. *Plant Science Letters*, **6**, 163–173.

Berrie, A. M. M., Buller, D., Don, R. & Parker, W. (1979). Possible role of volatile fatty acids and abscisic acid in the dormancy of oats. *Plant Physiology*, **63**, 758–764.

Bespalova, Z. G. & Borisova, I. V. (1979). Germination capacity and sprouting characteristics of caryopses of feather grass (*Stipa poaceae*). *Botanicheskii Zhurnal (Leningrad)*, **64**, 1081–1090.

Best, E. P. H. (1978). Growth substances and dormancy in *Ceratophyllum demersum*. *Physiologia Plantarum*, **45**, 399–406.

Bewley, J. D. & Black, M. (1978). *Physiology and Biochemistry of Seeds in Relation to Germination*. Vol. 1. Development and Growth. Berlin: Springer-Verlag.

(1982). *Physiology and Biochemistry of Seeds in Relation to Germination*. Vol. 2. Viability, Dormancy and Environmental Control. Berlin: Springer-Verlag.

Bews, J. W. (1929). *The World's Grasses*. New York: Russell and Russell.

Bhupathi, P., Selvaraj, J. A. & Ramaswamy, K. R. (1983). A note on seed dormancy in *Cenchrus ciliaris* L. *Seed Research (New Delhi)*, **11**, 232–234.

Binrad, L. (1958). Resultats de quelques essais sur la germination de *Panicum maximum*. *Agricultura (Louvain)*, **6**, 305–310.

Bishop, L. R. (1944). Memorandum on barley germination. *Journal of the Institute of Brewing*, **50**, 166–185.

(1948). Notes on the dormancy of varieties of Scotch common barley. *Brewers Guild Journal*, **34**, 427–428.

(1958). Barley dormancy as a varietal characteristic. *Journal of the Institute of Brewing*, **64**, 484–488.

Black, M. (1959). Dormancy studies in seed of *Avena fatua*. 1. The possible role of germination inhibitors. *Canadian Journal of Botany*, **37**, 393–402.

Black, M., Butler, J. & Hughes, M. (1987). Control and development of dormancy in cereals. In *4th International Symposium, Pre-harvest Sprouting in Cereals*, ed. D. J. Mares, pp. 379–392. Boulder, Colorado: Westview Press.

Black, M. & Naylor, J. M. (1957). Control of dormancy in wild oats. *Research Report*,

Canadian National Weed Committee, Western Section, p. 130.

(1959). Prevention of the onset of dormancy by gibberellic acid. *Nature (London)*, **184**, 468–469.

Blanchard, M. (1957). Contribution a l'etude du phenomene de dormance chez *Hordeum vulgare* L. *Proceedings. International Seed Testing Association*, **22**, 192–195.

Bloch, F. & Morgan, A. I. (1967). Germination inhibition in wheat and barley during steeping, and α-amylase development in presence of gibberellic acid. *Cereal Chemistry*, **44**, 61–69.

Blowers, D. P. & Trewavas, A. J. (1987). Autophosphorylation of plasma membrane bound calcium calmodulin dependent protein kinase from pea seedlings and modification of catalytic activity by autophosphorylation. *Biochemical and Biophysical Research Communications*, **143**, 691–696.

Bor, N. L. (1960). *The Grasses of Burma, Ceylon, India and Pakistan*. Oxford: Pergamon Press.

Born, V. W. H. (1971). Green foxtail: seed dormancy, germination and growth. *Canadian Journal of Plant Science*, **51**, 53–59.

Bose, S., Ghosh, B. & Sircar, S. M. (1977). Note on the role of cytokinin in the primary dormancy of rice seed. *Indian Journal of Agricultural Science*, **47**, 634–636.

Bouharmont, J. (1961). Embryo culture of rice on sterile medium. *Euphytica*, **10**, 283–293.

Bowman, H. F. (1975). *The effects of foliar applied gibberellic acid on the dormancy of wild oat seeds*. M.Sc. Thesis. Montana State University, Bozeman, USA.

Boyce, K. G. (1973). Seed dormancy in tall fescue (*Festuca arundinacea* Schreb.): Acquisition, effect on metabolic process and relief by temperature and growth regulators. *Dissertation Abstracts International. Section B. Sciences and Engineering*, **33**, 5615–5616.

Brad, I., Laszlo, I. & Valuta, G. (1959). Physiological and biochemical study of the behaviour of some varieties of winter barley during dormancy. *Analele. Institutului de Cercetari Agronomice. Academia Republicii Populare Romine. Series C*, **27**, 35–45.

Brecke, B. J. & Duke, W. B. (1980). Dormancy, germination and emergence characteristics of fall panicum (*Panicum dichotomiflorum*) seed. *Weed Science*, **28**, 683–685.

Briggs, D. E. (1962). Development of enzymes by barley embryos *in vitro*. *Journal of the Institute of Brewing*, **68**, 470–475.

(1963). Effects of gibberellic acid on barley germination and its use in malting: A review. *Journal of the Institute of Brewing*, **69**, 244–248.

(1973). Hormones and carbohydrate metabolism in germinating cereal grains. In *Biosynthesis and its Control in Plants*, ed. B. V. Milborrow, pp. 219–277. London: Academic Press.

Brown, A. J. (1909). The selective semi-permeability of the covering of seeds of *Hordeum vulgare*. *Proceedings. Royal Society of London. Series B.*, **81**, 82–93.

(1912). The influence of temperature on the absorption of water by seeds of *Hordeum vulgare* in relation to the temperature coefficient of chemical change. *Proceedings. Royal Society of London. Series B.*, **85**, 546–553.

Brown, A. J. & Tinker, F. (1916). Selective permeability: the absorption of phenol and other solutions by the seeds of *Hordeum vulgare*. *Proceedings. Royal Society of London. Series B.*, **89**, 373–379.

Brown, C. M. & Aryeetey, A. N. (1973). Maternal control of oil content in oat (*Avena sativa* L.). *Crop Science*, **13**, 120–121.

Brown, E., Stanton, T. R., Wiebe, G. A. & Martin, J. H. (1948). Dormancy and the effect of storage on oats, barley and sorghum. *United States Department of Agriculture. Technical Bulletin 953*, 30pp. Washington: United States Government.

Brown, H. T. & Morris, G. H. (1890). The germination of some of the Gramineae. *Journal of the Chemical Society*, **57**, 458–528.

Brown, N. A. C. & Bridglall, S. S. (1987). Preliminary studies of seed dormancy in *Datura stramonium*. *South African Journal of Botany*, **53**, 107–109.

Brown, R. (1931). The absorption of water by seeds of *Lolium perenne* (L.) and certain other

Gramineae. *Annals of Applied Biology*, **18**, 559–573.

Brown, R. F. (1982). Seed dormancy in *Aristida armata*, Australian Journal of Botany, **30**, 67–74.

Bruns, V. F. (1965). The effects of fresh water storage on the germination of certain weed seeds. *Weeds*, **13**, 38–39.

Bulard, C. (1960). Aseptic cultures of barley embryos separated from the grain: action of gibberellic acid. *Comptes Rendus. Academie des Sciences*, **250**, 3716–3718.

Bulgakova, Z. P. (1951). Biology of the dormancy period in seeds of some tree species. *Byulleten' Moskovskogo Obshchestva Ispytateley Prirody Byull Otdel Biol*, **57**, 77–81.

Buller, D. C. & Grant Reid, J. S. (1977). Do fatty-acids inhibit gibberellin-induced amylolysis – a reply. *Nature (London)*, **270**, 193–194.

Buller, D. C., Parker, W. & Grant Reid, J. S. (1976). Short chain fatty-acids as inhibitors of gibberellin induced amylolysis in barley endosperm. *Nature (London)*, **260**, 169–170.

Buraas, T. & Skinnes, H. (1984). Genetic investigations on seed dormancy in barley (*Hordeum vulgare*). *Hereditas*, **101**, 235–244.

(1985). Development of seed dormancy in barley, wheat and Triticale under controlled conditions. *Acta Agriculturae Scandinavica*, **35**, 233–244.

Burrows, V. D. (1964). Seed dormancy, a possible key to high yield of cereals. *Agricultural Institute Review*, **19**, 33–35.

(1970). Yield and disease-escape potential of fall-sown oats possessing seed dormancy. *Canadian Journal of Plant Science*, **50**, 371–377.

Burton, G. (1969). Breaking seed dormancy in seeds of pearl millet, *Pennisetum typhoides*. *Crop Science*, **9**, 659–654.

Burton, G. W. (1939). Scarification studies on southern grass seeds. *Journal of the American Society of Agronomy*, **31**, 179–187.

Butler, L., Helms, K. & Ogle, D. (1983). Establishing an official blowing method and germination method for kleingrass (*Panicum coloratum*). *Association of Official Seed Analysts News Letter*, **57**, 40–45.

Button, E. F. (1959). Effect of gibberellic acid on laboratory germination of creeping red fescue (*Festuca rubra*). *Agronomy Journal*, **51**, 60–61.

Cairns, A. L. P. (1982). Induction of dormancy and gibberellic acid insensitivity in *Avena fatua* seeds by saccarides. In *Proceedings. 4th National Weeds Conference. South Africa*, ed. H. A. Vandeventer & M. Mason, pp. 75–81.

(1984). *The physiology of seed dormancy and germination in* Avena fatua *L.*, Ph.D. Thesis, University of Stellenbosch, South Africa.

Cairns, A. L. P. & de Villiers, O. T. (1980). Effect of aluminium phosphide fumigation on the dormancy and viability of *Avena fatua* seed. *South African Journal of Science*, **76**, 323.

(1982). Effects of various saccharides on gibberellic acid sensitivity of *Avena fatua* seeds. In *Proceedings. 3rd International Symposium on Pre-harvest Sprouting in Cereals*, ed. J. E. Kruger & D. E. Laberge, pp. 66–71. Boulder, Colorado: Westview Press.

(1986a). Breaking dormancy of *Avena fatua* seeds by treatment with ammonia. *Weed Research*, **26**, 191–197.

(1986b). Physiological basis of dormancy breaking in wild oat (*Avena fatua* L.) by ammonia. *Weed Research*, **26**, 365–374.

Cameron-Mills, V. & Duffus, C. M. (1977). The in vitro culture of immature barley embryos on different culture media. *Annals of Botany*, **41**, 1117–1127.

Canode, C. L., Horning, E. V. & Maguire, J. D. (1963). Seed dormancy in *Dactylis glomerata*. *Crop Science*, **3**, 17–19.

Cantoria, M. & Gacutan, M. V. C. (1972). Germination behaviour of *Gynura crepidioides*, *Portulaca oleracea* and *Dactyloctenium aegyptium*. *Plant Physiology*, **49**, Supp. p.14.

Cardwell, V. B., Oelke, E. A. & Elliott, W. A. (1978). Seed dormancy mechanisms in wild rice

(*Zizania aquatica*). *Agronomy Journal*, **70**, 481–488.

Casey, J. E. (1947). Apparent dormancy in sorghum seed. *Association of Official Seed Analysts News Letter*, **21**, 34–36.

Cavers, P. B. & Benoit, D. L. (1987). Seed banks in agricultural land. *American Journal of Botany*, **74**, 635.

Chancellor, R. J., Catizone, P. & Peters, N. C. B. (1976). Breaking seed dormancy in *Avena fatua* L. *Proceedings. Symposium Status, Biology and Control of Grassweeds in Europe*, Vol. 1, pp. 95–102. Paris: UNESCO.

Chancellor, R. J. & Parker, C. (1972). The effects of plant growth-regulatory chemicals on seed germination. *Proceedings. 11th British Weed Control Conference*, pp. 772–777. Oxford: Weed Research Organization.

Chancellor, R. J., Parker, C. & Teferedegn, T. (1971). Stimulation of dormant weed seed germination by 2–chloroethylphosphonic acid. *Pesticide Science*, **2**, 35–37.

Chandraratna, M. F., Fernando, L. H. & Wattegedera, C. (1952). Seed dormancy in rice. *Tropical Agriculturist*, 108, 261–264.

Chandrasekariah, S. R., Govindappa, T. & Kulkarni, G. N. (1976). Pattern of dormancy breaking in seeds of few paddy varieties. *Current Science (Bangalore, India)*, **45**, 463–464.

Chang, C. W. (1963). Comparative growth of barley embryos *in vitro* and *in vivo*. *Bulletin of the Torrey Botanical Club*, **90**, 385–391.

Chang, S. C. (1943). Length of dormancy in cereal crops and its relationship to after-harvest sprouting. *Journal of the American Society of Agronomy*, **35**, 482–490.

Chang, T. T. & Yen, S. T. (1969). Inheritance of grain dormancy in four rice crosses. *Botanical Bulletin. Academia Sinica*, **10**, 1.

Chaudhary, T. & Ghildyal, B. P. (1969). Germination response of rice seeds to constant and alternating temperatures. *Agronomy Journal*, **61**, 328–330.

Chaudri, I. I., Khalifa, M. M. & Rahman, F. S. A. (1975). Investigations on germination ecology of some dune plants of Tripoli area Libya. *Libyan Journal of Science*, **5**, 1–6.

Cheam, A. H. (1986). Seed production and seed dormancy in wild radish (*Raphanus raphanistrum* L.) and some possibilities for improving control. *Weed Research*, **26**, 405–413.

Chen, F. S., MacTaggart, J. M. & Elofsen, R. M. (1982). Chemical constituents in wild oat (*Avena fatua*) and their effects on seed germination. *Canadian Journal of Plant Science*, **62**, 155–161.

Chen, L-C., Chang, W-L. & Chiu, F-T. (1980). Varietal differences of pre-harvest sprouting and its relationship with grain dormancy. *National Science Council Monographs*, **8**, 151–161.

Chen, S. S. C. & Chang, J. (1972). Does gibberellic acid stimulate seed germination via amylase synthesis? *Plant Physiology*, **49**, 441–442.

Chen, S. S. C. & Park, W. (1973). Early actions of gibberellic acid on the embryo and endosperm of *Avena fatua* seeds. *Plant Physiology*, **52**, 174–176.

Chen, S. S. C. & Varner, J. E. (1969). Metabolism of ^{14}C-maltose in *Avena fatua* seeds during germination. *Plant Physiology*, **44**, 770–774.

(1970). Respiration and protein synthesis in dormant and after-ripened seeds of *Avena fatua*. *Plant Physiology*, **46**, 108–112.

Chin, H. F. & Raja Harum, R. M. (1979). Ecology and physiology of *Eleusine indica* seeds. *Proceedings. 7th Asian-Pacific Weed Science Society Conference*, pp. 313–315. Sydney, Australia.

Ching, T. M. & Foote, W. H. (1961). Post-harvest dormancy in wheat varieties. *Agronomy Journal*, **53**, 183–186.

Chippindale, H. G. (1933). The effect of soaking in water on the 'seeds' of *Dactylis glomerata* L. *Annals of Botany*, **47**, 841–849.

Bibliography

Chirco, E. M., Goodman, J. R. & Clark, B. E. (1979). Germination of deertongue (*Panicum clandestinum* L.) seeds. *Association of Official Seed Analysts News Letter*, **53**, 40–42.

Christie, L. A. (1956). *A study of the control of dormancy in wild oats*. M.Sc. Thesis. University of Saskatchewan, Saskatoon, Canada. 80pp.

Chu, C-C., Sweet, R. D. & Ozbun, J. L. (1978). Some germination characteristics in common lambsquarters (*Chenopodium album*). *Weed Science*, **26**, 255–258.

Clark, D. C. & Bass, L. N. (1970). Germination experiments with seeds of Indian ricegrass, *Oryzopsis hymenoides* (Roem. and Schult.) Ricker. *Proceedings. Association of Official Seed Analysts*, **60**, 226–239.

Clark, L. E. (1967). Seed dormancy in sorghums. *Dissertation Abstracts*, **28**, 437B.

Clark, L. E., Collier, J. W. & Langston, R. (1967). Dormancy in *Sorghum bicolor* (L.) Moench. I. Relationship to seed development. *Crop Science*, **7**, 497–501.

(1968). Dormancy in *Sorghum bicolor* (L.) Moench. II. Effects of pericarp and testa. *Crop Science*, **8**, 155–158.

Clifford, H. T. (1986). Spikelet and floral morphology. In *Grass Systematics and Evolution*, ed. T. R. Soderstrom, K. W. Hilu, C. S. Campbell & M. E. Barkworth, pp. 21–30. Washington, D.C.: Smithsonian Institute.

Clutter, M. E. (1978). *Dormancy and Developmental Arrest*. New York: Academic Press.

Cobb, R. D. & Jones, L. G. (1962). Germinating dormant seeds of *Avena fatua*. *American Journal of Botany*, **49**, 658–659.

Cocks, P. S., Boyce, K. G. & Kloot, P. M. (1976). The *Hordeum murinum* complex in Australia. *Australian Journal of Botany*, **24**, 651–662.

Cocks, P. S. & Donald, C. M. (1973). The germination and establishment of two annual pasture grasses (*Hordeum leporinum* Link. and *Lolium rigidum* Gaud.) *Australian Journal of Agricultural Research*, **24**, 1–10.

Coffman, F. A. (1961). A method of obtaining more F_1 plants from hybrid oat seed. Treatment to ensure germination. *Crop Science*, **1**, 378.

Coffman, F. A. & Stanton, T. (1940). Dormancy in fatuoid and normal oat kernels. *Journal of the American Society of Agronomy*, **32**, 459–466.

Cohn, M. A., Boullion, K. J. & Chiles, L. A. (1987). Structure-activity studies of dormancy-breaking chemicals. *Plant Physiology*, **83**, Supp. p. 39.

Cohn, M. A. & Butera, D. L. (1982). Seed dormancy in red rice (*Oryza sativa*). 2. Response to cytokinins. *Weed Science*, **30**, 200–205.

Cohn, M. A., Butera, D. L. & Hughes, J. A. (1983). Seed dormancy in red rice (*Oryza sativa*). 3. Response to nitrite, nitrate and ammonium ions. *Plant Physiology*, **73**, 381–384.

Cohn, M. A. & Castle, L. (1983). Nitrogen dioxide as a dormancy breaking agent. *Plant Physiology*, 72, Supp. p. 55.

Cohn, M. A., Chiles, L. A., Hughes, J. A. & Boullion, K. J. (1987). Seed dormancy in red rice (*Oryza sativa*). 6. Monocarboxylic acids: A new class of pH-dependent germination stimulants. *Plant Physiology*, **84**, 716–719.

Cohn, M. A. & Hughes, J. A. (1981). Seed dormancy in red rice (*Oryza sativa*). 1. Effect of temperature on dry after-ripening. *Weed Science*, **29**, 402–404.

Colbry, V. L. (1953). Factors affecting the germination of reed canary grass. *Proceedings. Association of Official Seed Analysts*, **55**, 50–53.

Collins, E. J. (1918). The structure of the integumentary system of the barley grain in relation to localised water absorption and semi-permeability. *Annals of Botany*, **32**, 381–414.

Colosi, J. C. & Cavers, P. B. (1983). Biotype by environmental effects on winter seed survival. *American Journal of Botany*, **70**, p. 46.

Côme, D., Lecat, S. & Corbineau, F. (1987). Role of glumellae in oat (*Avena sativa* L.) seed dormancy. *International Botanical Congress Abstracts*, **17**, 139.

Côme, D., Lenoir, C. & Corbineau, F. (1984). The dormancy of cereals and its elimination. *Seed Science and Technology*, **12**, 629–640.

Conn, J. S. & Farris, M. L. (1987). Seed viability and dormancy of 17 weed species after 21 months in Alaska, USA. *Weed Science*, **35**, 524–528.

Conover, D. G. & Geiger, D. R. (1984a). Germination of Australian channel millet (*Echinochloa turnerana*) seeds. 1. Dormancy in relation to light and water. *Australian Journal of Plant Physiology*, **11**, 395–408.

(1984b). Germination of Australian millet (*Echinochloa turnerana*) seeds. 2. Effects of anaerobic conditions, continuous flooding and low water potential. *Australian Journal of Plant Physiology*, **11**, 409–418.

Corbineau, E. & Côme, D. (1980). Some dormancy characteristics of barley caryopses (*Hordeum vulgare*) cv. Sonja. *Comptes Rendus Hebdomedaires des Seances. Academie des Sciences*, **290**, 547–550.

Corbineau, F. & Côme, D. (1982). Seed dormancy variation of 2 barley varieties (*Hordeum vulgare*) during maturation and dry storage. *Comptes Rendus Hebdomedaires des Seances. Academie des Sciences*, **294**, 967–970.

Corbineau, F., Lecat, S. C. & Côme, D. (1986). Dormancy of 3 cultivars of oat seeds (*Avena sativa* L.). *Seed Science and Technology*, **14**, 725–735.

Corns, W. G. (1960). Effects of gibberellin treatments on germination of various species of weed seeds. *Canadian Journal of Plant Science*, **40**, 47–51.

Coukos, C. J. (1944). Seed dormancy and germination in some native grasses. *Journal of the American Society of Agronomy*, **36**, 337–345.

Courtney, A. D. (1968). Seed dormancy and field emergence in *Polygonum aviculare*. *Journal of Applied Ecology*, **5**, 675–684.

Crabb, D. (1971). Changes in the response to gibberellic acid of barley endosperm slices during storage. *Journal of the Institute of Brewing*, **77**, 522–528.

Crescini, F. & Spreafico, L. (1953). Cultivated plants versus weeds. I. Researches on the biology of the germination of the achenes of *Artemisia vulgaris* L. *Annali della Sperimentazione Agraria (Rome)*, **7**, 1597–1610.

Crocker, W. (1906). Role of the seed coats in delayed germination. *Botanical Gazette*, **42**, 265–291.

(1916). Mechanics of dormancy in seeds. *American Journal of Botany*, **3**, 99–120.

Crocker, W. & Barton, L. V. (1957). *Physiology of Seeds*. Waltham, Mass. USA: Chronica Botanica.

Cross, D. O. (1936). A rare occurrence. Germination of Kikuyu grass seed. *Agricultural Gazette of New South Wales*, **47**, 485.

Cseresnyes, Z. & Bude, A. (1976). Research work concerning the duration of dormancy with the determination method for the seed germination of some varieties of barley and 2–row barley. *Analele, Institutului de Cercetari Cereale Plante Teh-Fundulea*, **41**, 483–489.

Cullinan, B. (1947). Germinating certain southern grass seeds. *Proceedings. Association of Official Seed Analysts*, **37**, 162–163.

Cuming, A. C. & Osborne, D. J. (1978a). Membrane turnover in imbibed and dormant embryos of the wild oat (*Avena fatua* L). 1. Protein turnover and membrane replacement. *Planta (Berlin)*, **139**, 209–217.

(1978b). Membrane turnover in imbibed and dormant embryos of the wild oat (*Avena fatua* L). 2. Phospholipid turnover and membrane replacement. *Planta (Berlin)*, **39**, 219–226.

Cumming, B. G. (1957). Interaction of light and dormancy in the wild oat (*Avena fatua* L.). *Research Report. National Weed Committee, Western Section. Agriculture Canada*, p.133.

Cumming, B. G. & Hay, J. (1958). Light and dormancy in wild oats (*Avena fatua* L.). *Nature (London)*, **182**, 609–610.

Curran, P. L. & McCarthy, H. V. (1986). Dormancy studies on commercial seed lots of the barley cultivar Igri. *Seed Science and Technology*, **14**, 657–576.

Cussans, G. W. (1976). Population dynamics of wild oats in relation to systematic control. *Report. Agricultural Research Council, Weed Research Organization*, **1974–75**, 47–56.

Daletskaia, T. V. & Nikolaeva, M. G. (1987). Fusicoccin action on deep-dormant seed germination. *Doklady Akademii Nauk SSSR*, **293**, 510–512.

Dan, V. (1971). Germination dormancy and sensitivity of barley to the water: theoretical and practical aspects. *Industria Alimentaria (Bucaresti)*, **22**, 312–314.

Danjo, T. & Inosaka, M. (1960). On the tissue of scutellum concerned with absorption of nutrients of endosperm in rice and oat seeds. *Proceedings. Crop Science Society of Japan*, **29**, 100–102.

Darmency, H. & Aujas, C. (1986). Polymorphism for vernalization requirement in a population of *Avena fatua*. *Canadian Journal of Botany*, **64**, 730–733.

(1987). Character inheritance and polymorphism in a wild oat (*Avena fatua*) population. *Canadian Journal of Botany*, **65**, 2352–2356.

Dashek, W. V., Singh, B. N. & Walton, D. C. (1977). Metabolism and localization of abscisic acid in barley aleurone layers. *Plant Physiology*, **59**, Supp., 76.

Dathe, W., Schneider, G. & Sembdner, G. (1978). Endogenous gibberellins and inhibitors in caryopses of rye. *Phytochemistry*, **17**, 963–966.

Datta, S. C., Evenari, M. & Gutterman, Y. (1970). The heteroblasty of *Aegilops ovata* L. *Israel Journal of Botany*, **19**, 463–483.

Datta, S. C., Gutterman, Y. & Evenari, M. (1972). The influence of the origin of the mother plants on yield and germination of their caryopses in *Aegilops ovata*. *Planta (Berlin)*, **105**, 155–164.

Davis, D. G. (1962). *Studies of phyto-toxic properties of extracts from mature seeds of wild oats, Avena fatua L.* M.Sc. Thesis. North Dakota State University. USA.

De Toledo, F. F., Marcos Filho, J., Silvarolla, M. B. & Batista Neto, J. F. (1981). Maturation and dormancy of *Paspalum notatum* seeds. *Revista de Agricultura (Brazil)*, **56**, 83–91.

Delouche, J. C. (1956). Dormancy in seeds of *Agropyron smithii*, *Digitaria sanguinalis* and *Poa pratensis*. *Iowa State College Journal of Science*, **30**, 348–349.

(1958). Germination of Kentucky bluegrass harvested at different stages of maturity. *Proceedings. Association of Official Seed Analysts*, **48**, 81–84.

(1961). Effect of gibberellin and light on germination of centipedegrass seed (*Eremochloa ophiuroides*). *Proceedings. Association of Official Seed Analysts*, **51**, 147–150.

Delouche, J. C. & Bass, L. N. (1954). Effect of light and darkness upon the germination of seeds of western wheatgrass (*Agropyron smithii* L.) *Proceedings. Association of Official Seed Analysts*, **44**, 104–113.

Delouche, J. C. & Nguyen, N. T. (1964). Methods for overcoming seed dormancy in rice. *Proceedings. Association of Official Seed Analysts*, **54**, 41–49.

Deore, B. P. & Solomon, S. (1981). Seed dormancy in rice. *Maharashtra Vidnyan Mandir Patrika*, **16**, 23–28.

(1982). Inheritance of dormancy in rice. *Journal of Maharashtra Agricultural University*, **7**, 220–223.

Depauw, R. M. & McCaig, T. N. (1983). Recombining dormancy and white seed color in a spring wheat cross. *Canadian Journal of Plant Science*, **63**, 581–589.

Derera, N. F., Bhatt, G. M. & McMaster, G. J. (1977). On the problem of pre harvest sprouting of wheat. *Euphytica*, **26**, 299–308.

Deunff, Y. L. (1983). Mise en evidence de l'influence benefique de la'alcool ethylique en solution aquese sur la levee de dormance des orges. *Comptes Rendus Hebdomadaires des Sciences. Academie des Sciences. Serie III.*, **296**, 433–436.

Dighe, R. S. & Patil, V. N. (1985). Evaluation of dormancy behaviour of paddy varieties recommended for Vidarbha region of Maharashtra State, India. *Punjabrao Krishi Vidyapeeth Research Journal*, **9**, 23–28.

Doggett, H. (1970). *Sorghum*, London: Longmans, Green & Co. Ltd.

Don, R. (1979). Use of chemicals, particularly gibberellic acid, for breaking cereal seed dormancy. *Seed Science and Technology*, **7**, 355–367.

Donald, W. W. & Zimdahl, R. L. (1987). Persistence, germinability and distribution of jointed goatgrass (*Aegilops cylindrica*) seed in soil. *Weed Science*, **35**, 149–154.

Dore, J. (1955). Dormancy and viability of padi seed. *Malayan Agricultural Journal*, **38**, 163–173.

Drake, V. C. (1948). Germination of certain dormant oat and barley samples. *Association of Official Seed Analysts News Letter*, **22**, 13–16.

(1949). Germination of rescue grass seed as affected by temperature, substrata, light, and removal of glumes. *Association of Official Seed Analysts News Letter*, **23**, 42–46.

Drennan, D. S. H. & Berrie, A. M. M. (1961). Physiological studies of germination in the genus Avena. I. The development of amylase activity. *New Phytologist*, **61**, 1–9.

Duke, S. O. (1978). Significance of fluence response data in phytochrome-mediated seed germination. *Photochemistry and Photobiology*, **28**, 383–388.

Dunwell, J. M. (1981a). Dormancy and germination in embryo of *Hordeum vulgare* L. Effect of dissection, incubation temperature and hormone application. *Annals of Botany*, **48**, 203–213.

(1981b). Influence of genotype and environment on growth of barley embryos *in vitro*. *Annals of Botany*, **48**, 535–542.

Dure, L. S. (1960). Gross nutritional contributions of maize endosperm and scutellum to germination growth of maize axis. *Plant Physiology*, **35**, 919–935.

Dyck, P. L., Noll, J. S. & Czarnecki, E. (1986). Heritability of RL-4137 type of dormancy in two populations of random lines of spring wheat (*Triticum aestivum*). *Canadian Journal of Plant Science*, **66**, 855–862.

Eckerson, S. A. (1913). A physiological and chemical study of after-ripening. *Botanical Gazette*, **55**, 286–299.

Edelman, J., Shibko, A. J. & Keys, A. J. (1959). The role of the scutellum of cereal seedlings in the synthesis and transport of sucrose. *Journal of Experimental Botany*, **10**, 178–189.

Edwards, D. C. (1933). 'Hard' seeds in *Panicum coloratum*, Stapf. *Nature (London)*, **132**, 208.

Edwards, M. M. (1976). Dormancy in seeds of charlock (*Sinapis arvensis* L.) – early effects of gibberellic acid on synthesis of amino acids and their proteins. *Plant Physiology*, **58**, 626–630.

Eggens, J. L. & Ormrod, D. P. (1982). Creeping bentgrass, Kentucky bluegrass and annual bluegrass seed germination response to elevated temperature. *HortScience*, **17**, 624–625.

Egley, G. H. & Chandler, J. M. (1983). Longevity of seeds after 5.5 years in the Stoneville 50 year buried seed study. *Weed Science*, **31**, 264–270.

Ehara, K. & Abe, S. (1952). Studies on the wild Japanese barnyard millet as a weed in the lowland rice-field. VI. Physiological investigations on the germination. *Proceedings. Crop Science Society of Japan*, **21**, 61–62.

Eira, M. T. S. (1983). Comparison of methods for overcoming seed dormancy in Andropogon grass. *Revista Brasileira de Sementes*, **5**, 37–49.

Elliott, B. B. & Leopold, A. C. (1953). An inhibitor of germination and of amylase activity in oat seeds. *Physiologia Plantarum*, **6**, 65–77.

Ellis, R. H., Hong, T. D. & Roberts, E. H. (1983). Procedures for the safe removal of dormancy from rice seed. *Seed Science and Technology*, **11**, 77–112.

(1985a). *Handbooks for Genebanks. No. 2. Handbook of Seed Technology for Genebanks.* Vol. 1. Principles and Methodology. Rome: International Board for Plant Genetic Resources.

(1985b). *Handbooks for Genebanks. No. 3. Handbook of Seed Technology for Genebanks.* Vol. 2. Compendium of Specific Germination Information and Test Recommendations. Rome: International Board for Plant Genetic Resources.

(1986). The response of seeds of *Bromus sterilis* L. and *Bromus mollis* L. to white light of varying photon flux density and photoperiod. *New Phytologist*, **104**, 485–496.

(1987). Comparison of cumulative germination and rate of germination of dormant and

aged barley seed lots at different constant temperatures. *Seed Science and Technology*, **15**, 717–727.
Ellis, R. H. & Roberts, E. H. (1980). The influence of temperature and moisture on seed viability period in barley (*Hordeum distichum*). *Annals of Botany*, **45**, 31–38.
Emal, J. G. & Conard, E. C. (1973). Seed dormancy and germination in Indiangrass as affected by light, chilling, and certain chemical treatments. *Agronomy Journal*, **65**, 383–385.
Erasmus, D. J. & Van Staden, J. (1983). Germination of *Setaria chevalieri* caryopses. *Weed Research*, **23**, 225–229.
Ermilov, G. B. (1961). Effect of gibberellic acid upon the germination ability of corn sprouts. *Akademiya Nauk SSSR. Izvestiya. Seriya Biologiya*, **1**, 33–39.
Essery, R. E. & Pollock, J. R. A. (1957). Studies in barley and malt. X. Note on the behaviour of desiccation and heating on the germinative behaviour of dormant barley. *Journal of the Institute of Brewing*, **63**, 221–222.
Evans, G. R. & Tisdale, E. W. (1972). Ecological characteristics of *Aristida longiseta* and *Agropypron spicatum* in west-central Idaho. *Ecology*, **53**, 137–142.
Evans, L. T. (1964). Reproduction. In *Grasses and Grassland*, ed. C. Barnard, pp. 126–153. London: MacMillan & Co. Ltd.
Evans, R. A. & Young, J. A. (1975). Enhancing germination of dormant seeds of downy brome. *Weed Science*, **23**, 354–357.
Fawcett, R. S. & Slife, F. W. (1975). Germination stimulation properties of carbamate herbicides. *Weed Science*, **23**, 419–424.
Febles, G. & Harty, R. (1973). Effect of light and alternate temperatures on the germination of *Digitaria didactyla*. *Cuban Journal of Agricultural Science*, **7**, 233–236.
Febles, G. & Padilla, C. (1970). The effect of *Rhizobium melilotti*, scarification and temperature in breaking dormancy in common guinea grass seed (*Panicum maximum* Jacq.). *Revista Cubana de Ciencia Agricola*, **4**, 71–78.
(1971). Effect of temperature on germination of guinea grass seed (*Panicum maximum* Jacq.). *Revista Cubana de Ciencia Agricola*, **5**, 77–87.
Feekes, W. (1938). The tendency to germination in the ear of about 60 varieties either in cultivation or undergoing trials in Holland. *Report of the Technical Committee on Wheat*, **11**, 211–237. Gebroeders Hoistemi, Groningen, Holland.
Fendall, R. K. & Canode, C. L. (1971). Dormancy-related growth inhibitors in seeds of orchardgrass *Dactylis glomerata* L. *Crop Science*, **11**, 727.
Fendall, R. K. & Carter, J. F. (1965). New-seed dormancy of green needlegrass (*Stipa viridula* Trin.). I. Influence of the lemma and palea on germination, water absorption and oxygen uptake. *Crop Science*, **5**, 533–536.
Fernandez, D. B. (1980). Some aspects on the biology of *Pennisetum polystachyon* (L.) Schult. *Philippine Journal of Weed Science*, **7**, 1–10.
Fischbeck, G. & Reimer, L. (1971). Dormancy and water sensitivity of 1971 barley. *Brauwelt.*, **111**, 1183–1185.
Fischer, M. L., Stritzke, J. F. & Ahring, R. M. (1982). Germination and emergence of little barley (*Hordeum pusillum*). *Weed Science*, **30**, 624–628.
Fischnich, O., Grahl, A. & Thielebein, M. (1962). Demonstration of primary dormancy in a winter barley variety. *Naturwissenschaften*, **49**, 41.
Fischnich, O., Thielebein, M. & Grahl, A. (1961). Sekundare Keimruhe bei Getreide. *Proceedings. International Seed Testing Association*, **26**, 89–114.
Fishman, S., Erez, A. & Couvillon, G. A. (1987). The temperature-dependence of dormancy breaking in plants – mathematical analysis of a 2-step model involving cooperative transition. *Journal of Theoretical Biology*, **124**, 473–483.
Fitzgerald, P. H. (1959). Germination induced by excision of the endosperm of immature wheat grains. *New Zealand Journal of Agricultural Research*, **2**, 735–774.
Foley, M. E. (1987). The effect of wounding on primary dormancy in wild oat (*Avena fatua*) caryopses. *Weed Science*, **35**, 180–184.

Foral, A. & Bosak, Q. (1971). Postharvest dormancy and sprouting in selected European spring barley varieties. *Rostlinna Vyroba*, **17**, 427–434.
Forst, R. A. & Cavers, P. B. (1975). The ecology of pigweeds (*Amaranthus*) in Ontario, Canada. 1. Interspecific and intraspecific variation in seed germination among local collections of *Amaranthus powellii* and *Amaranthus retroflexus*. *Canadian Journal of Botany*, **53**, 1276–1284.
Forward, B. F. (1949). Studies of germination in oats. *Proceedings. Association of Official Seed Analysts*, **39**, 83–84.
Frank, A. B. & Larson, K. S. (1970). Influence of oxygen, sodium hypochlorite, and dehulling on germination of green needlegrass (*Stipa viridula* Trin.). *Crop Science*, **10**, 679–682.
Frankland, B. (1981). Germination in shade. In *Plants and the Daylight Spectrum*, ed. H. Smith, pp. 187–204. London: Academic Press.
Froud-Williams, R. J. (1981). Germination behaviour of *Bromus* species and *Alopecurus myosuroides*. In *Grass Weeds in Cereals in the United Kingdom, Proceedings of the Association of Applied Biologists, Conference, Warwick, 1981*, 31–40.
Froud-Williams, R. J. & Chancellor, R. J. (1986). Dormancy and seed-germination of *Bromus catharticus* and *Bromus commutatus*. *Seed Science and Technology*, **14**, 439–450.
Froud-Williams, R. J., Drennan, D. S. H. & Chancellor, R. J. (1984). The influence of burial and dry storage upon cyclic changes in dormancy, germination and response to light in seeds of various weeds. *New Phytologist*, **96**, 473–482.
Froud-Williams, R. J. & Ferris, R. (1987). Germination of proximal and distal seeds of *Poa trivialis* L. from contrasting habitats. *Weed Research*, **27**, 245–250.
Froud-Williams, R. J., Hilton, J. R. & Dixon, J. (1986). Evidence for an endogenous cycle of dormancy in dry stored seeds of *Poa trivialis*. *New Phytologist*, **102**, 123–132.
Fryer, J. R. (1922). The influence of light and of fluctuating temperatures on the germination of *Poa compressa* (L.). *Scientific Agriculture (Ottawa)*, **2**, 225–230.
Fuerst, E. P., Upadhyaya, M. K., Simpson, G. M., Naylor, J. M. & Adkins, S. W. (1983). A study of the relationship between seed dormancy and pentose phosphate pathway activity in *Avena fatua*. *Canadian Journal of Botany*, **61**, 667–670.
Fujii, T. (1963). On the anaerobic process involved in the photoperiodically induced germination of *Eragrostis* seed. *Plant and Cell Physiology*, **4**, 357–359.
Fujii, T. & Yokohama, Y. (1965). Physiology of light-requiring germination in *Eragrostis* seeds. *Plant and Cell Physiology*, **6**, 135–145.
Fukuyama, T., Takahashi, R. & Hayashi, J. (1973). Studies on seed dormancy of post-harvest barley. II. Differential characteristics in seed dormancy by variety, year and grade of barley. *Nogaku Kenkyu*, **54**, 185–198.
Fulbright, T. E., Redente, E. F. & Wilson, A. M. (1983). Germination requirements of green needlegrass (*Stipa viridula*). *Journal of Range Management*, **36**, 390–394.
Fulwider, J. R. & Engel, R. E. (1959). The effect of temperature and light on germination of seed of goosegrass, *Eleusine indica*. *Weeds*, **7**, 359–361.
Fursov, O. V., Kurbakov, O. I. & Darkanbaev, T. B. (1984). Enzymes of maize starch hydrolysis. *Fiziologiya i Biokhimiya Kul'turnykh Rastenii*, **16**, 430–437.
Furya, S. & Kataoka, T. (1983). The effect of temperature and soil moisture on innate dormancy-breaking in *Echinochloa* spp. and temperature and light conditions for their germination. *Aspects of Applied Biology*, **4**, 55–62.
Fykse, H. (1970a). Studium vedkomande spiring, dormans og levetid for frø av floghavre. *Statens plantevern, Ugrasbiologisk avdeling*. Melding nr. 80, 120pp. Scientific Report, Agricultural College of Norway, Vollebekk, Norway.
(1970b). Spiring og frøkvile hos floghavre. *Jord og Avling*, **13**, 20–23.
(1974). Investigations of *Sonchus arvensis*. 2. Distribution in Norway, growth and dormancy partially compared with related species. *Forskning og Forsøk i Landbruket*, **25**, 389–412.
Gaber, S. D., Abdalla, F. H. & Mahdy, M. T. (1974). Treatments affecting dormancy in sweet

sorghum seed. *Seed Science and Technology*, **2**, 305–316.
Gaber, S. D. & Roberts, E. H. (1969). Water-sensitivity in barley seeds. II. Association with micro-organism activity. *Journal of the Institute of Brewing*, **75**, 303–314.
Gain, E. & Jungelson, A. (1915). Sur les graines de maies issus de la vegetation d'embryons libres. *Comptes Rendus*, **160**, 142–144.
Garber, R. J. & Quisenberry, K. S. (1923). Delayed germination and the origin of false wild oats. *Journal of Heredity*, **14**, 267–274.
Garcia-Huidroba, J., Monteith, J. L. & Squire, G. R. (1982). Time, temperature and germination of pearl millet (*Pennisetum typhoides* S. and H.). I. Constant temperature. *Journal of Experimental Botany*, **33**, 288–296.
Gaspar, S., Fazekas, J. & Petho, A. (1973). Effects of gibberellic acid (GA_3) and prechilling on breaking dormancy in cereals. *Seed Science and Technology*, **3**, 555–563.
Gaspar, T., Wyndaele, R., Bouchet, M. & Ceulemans, E. (1977). Peroxidase and α-amylase activities in relation to germination of dormant and non-dormant wheat. *Physiologia Plantarum*, **40**, 11–14.
Gassner, G. (1911). Keimuntersuchungen mit *Chloris ciliata*. *Berichte. Deutsche Botanische Gesellschaft*, **29**, 708–722.
Geng, S. & Barnett, F. L. (1969). Effects of various dormancy-reducing treatments on seed germination and establishment of Indiangrass, *Sorghastrum nutans* L. *Crop Science*, **9**, 800.
George, D. W. (1972). Effect of season and location on post-harvest dormancy in wheat. *Nevada Agricultural Experiment Station, Series T*, **14**, 14.
Gfeller, F. & Svejda, F. (1960). Inheritance of post-harvest seed dormancy and kernel colour in spring wheat lines. *Canadian Journal of Plant Science*, **40**, 1–6.
Ghosh, B. N. (1962). Agro-meteorological studies on rice. I. Influence of climatic factors on dormancy and viability of paddy seeds. *Indian Journal of Agricultural Science*, 32, 235–241.
Gianfagna, A. J. & Pridham, A. M. S. (1951).Some aspects of dormancy and germination of crabgrass seed, *Digitaria sanguinalis* Scop. *Proceedings. American Society of Horticultural Science*, **58**, 291–297.
Giles, B. E. & Lefkovitch, L. F. (1984). Differential germination in *Hordeum spontaneum* from Iran and Morocco. *Zeitschrift fuer Pflanzenzuechtung*, **92**, 234–238.
Gill, N. T. (1938). The viability of weed seeds at various stages of maturity. *Annals of Applied Biology*, **25**, 447–456.
Goedert, C. O. (1984). *Seed dormancy of tropical grasses and implications for the conservation of genetic resources*, Ph.D. Thesis, University of Reading, England.
Goedert, C. O. & Roberts, E. H. (1986). Characterization of alternating temperature regimes that remove seed dormancy in seeds of *Brachiaria humidicola*. *Plant and Cell Environment*, **9**, 521–526.
Gonzalez, Y. & Torriente, O. (1983). Effect of KNO_3 on dormancy breaking of *Panicum maximum* cv. Likoni. 1. Storage at ambient temperature. *Journal y Pastas y Forrajes*, **6**, 59–72.
Goodsell, S. F. (1957). Germination of dormant sorghum seed. *Agronomy Journal*, **49**, 387–389.
Gordon, E. M. (1951). Light and temperature sensitiveness in germinating seed of timothy (*Phleum pratense* L.). *Scientific Agriculture (Ottawa)*, **31**, 71–84.
Gordon, I. L. (1979a). Selection against sprouting damage in wheat: a synopsis. In *Proceedings. 5th International Wheat Genetics Symposium*, vol. 2, pp. 954–962. Delhi, India.
(1979b). Alpha-amylase, grain redness and flavanols. *Australian Journal of Agricultural Research*, **30**, 387–402.
Gosling, P. G., Butler, R. A., Black, M. & Chapman, J. M. (1981). The onset of germination

ability in developing wheat. *Journal of Experimental Botany*, **32**, 621–627.
Gould, F. W. (1968). *Grass Systematics*. New York: McGraw Hill.
Grabe, D. F. (1955). Germination responses of smooth bromegrass seed. *Proceedings. Association of Official Seed Analysts*, **45**, 68–71.
Grahl, A. (1965). Induction of secondary dormancy in wheat by means of high temperature and the effect of the humidity of the seed bed. *Proceedings. International Seed Testing Association*, **30**, 787–801.
(1969). Embryo and dormancy in barley. *Landwirtschaftliche Forschung. Sonderhft*, **24**, 41–49.
(1970). Einfluss de Keimungstemperatur und Statifikation auf die Keimruhe von Getreide. *Proceedings. International Seed Testing Association*, **35**, 427–438.
(1975). Weather conditions before maturity and dormancy in wheat in relation to prediction of sprouting. *Seed Science and Technology*, **3**, 815–826.
(1984). Relation between intensity and duration of primary dormancy in wheat. *Seed Science and Technology*, **12**, 687–695.
Grahl, A., Thielebein, M. & Fischnich, O. (1962). Germination of barley caryopses during primary dormancy in relation to their position in the spike. *Naturwissenschaften*, **49**, 41–42.
Gramshaw, D. (1972). Germination of annual ryegrass seeds (*Lolium rigidum* Gaud.) as influenced by temperature, light, storage environment and age. *Australian Journal of Agricultural Research*, **23**, 779–787.
Gramshaw, D. & Stern, W. R. (1977). Survival of annual ryegrass (*Lolium rigidum*) seed in a Mediterranean type environment. 2. Effects of short-term burial on persistence of viable seed. *Australian Journal of Agricultural Research*, **28**, 93–101.
Gray, R. A. (1958). Breaking the dormancy of peach seeds and crabgrass seeds with gibberellins. *Plant Physiology*, **33**, 40–41.
Green, J. G. & Helgeson, E. A. (1957). The developmental morphology of wild oats. *Proceedings. North Central Weed Control Conference*, **14**, 5.
Griffeth, W. L. & Harrison, C. M. (1954). Maturity and curing temperatures and their influence on germination of Reed canary grass seed. *Agronomy Journal*, **46**, 163–167.
Griffiths, D. J. (1961). The influence of different daylengths on ear emergence and seed setting in oats. *Journal of Agricultural Science*, **57**, 279–288.
Griffiths, C. M., MacWilliam, I. C. & Reynolds, T. (1967). Comparative effect of gibberellins and their derivatives on germination and malting of barley. *Journal of the Institute of Brewing*, **73**, 189–193.
Grigg, D. B. (1974). *The Agricultural Systems of the World. An Evolutionary Approach*. Cambridge: Cambridge University Press.
Grilli, I., Lioi, L., Anguillesi, M. C., Meletti, P. & Floris, C. (1986). Metabolism in seed ripening protein and polyadenylate RNA pattern in developing embryos of *Triticum durum*. *Journal of Plant Physiology*, **124**, 321–330.
Grime, J. P. (1981). The role of seed dormancy in vegetation dynamics. *Annals of Applied Biology*, **98**, 555–558.
Grime, J. P., Mason, G., Curtis, A. V., Redman, J., Band, S. R., Mowforth, M. A. G., Neal, A. M. & Shaw, S. (1981). A comparative study of germination characteristics in a local flora. *Journal of Ecology*, **69**, 1017–1059.
Griswold, S. M. (1936). Effect of alternate moistening and drying on germination of seeds of western range plants. *Botanical Gazette*, **98**, 243–269.
Gritton, E. T. & Atkins, R. E. (1963a). Germination of sorghum seed as affected by dormancy. *Agronomy Journal*, **55**, 169–174.
(1963b). Germination of grain sorghum as affected by freezing temperatures. *Agronomy Journal*, **55**, 139–142.
Groves, R. H., Hagon, M. W. & Ramakrishnan, P. S. (1982). Dormancy and germination of

seed of 8 populations of *Themeda australis*. *Australian Journal of Botany*, **30**, 373–386.
Gupta, K. C. (1973). Factors influencing dormancy in seeds of Crowfoot grass. *Biochemie und Physiologie der Pflanzen*, **164**, 582–587.
Gupta, R., Bhandal, I. S. & Malik, C. P. (1985). A possible mechanism of abscisic acid response in rice (*Oryza sativa* L.) seed germination. *Journal of Biological Research*, **5**, 18–24.
Gupta, R. K. & Saxena, S. K. (1980). Ecological studies on *Eleusine compressa* – a potential grass for sheep pasturage in the arid zone. *Annals of Arid Zone*, **19**, 1–14.
Gutormson, T. J. & Wiesner, L. E. (1987). Methods for the germination of beardless wildrye (*Elymus triticoides* Buckl.). *Journal of Seed Technology*, **11**, 1–6.
Guzevskii, A. E. (1947). Presowing heating of seeds in water. *Soviet Agronomy*, **1947**, 67–70.
Haagen-Smit, A. J., Siu, R. & Wilson, G. (1945). A method for culturing of excised, immature corn embryos *in vitro*. *Science*, **101**, 234.
Haber, A. H. (1962). Effects of indoleacetic acid on growth without mitosis and on mitotic activity in absence of growth by expansion. *Plant Physiology*, **37**, 18–26.
Hacker, J. B., Andrew, M. M., McIvor, J. G. & Mott, J. J. (1984). Evaluation in contrasting climates of dormancy characteristics of seed of *Digitaria milanjiana*. *Journal of Applied Ecology*, **21**, 961–969.
Haferkamp, M. R., Jordan, G. L. & Matsuda, K. (1977). Pre-sowing seed treatments, seed coats and metabolic activity of Lehmann lovegrass seeds. *Agronomy Journal*, **69**, 527–530.
Hagemann, M. G. & Ciha, A. J. (1987). Environmental x genotype effects on seed dormancy and afterripening in wheat. *Agronomy Journal*, **79**, 192–196.
Haggquist, M-L., Petterson, A. & Liljenburg, C. (1984). Growth inhibitor in oat grains. I. Leakage of inhibitory factors from oat grains during imbibition and the effect of these on germination and growth. *Physiologia Plantarum*, **61**, 75–80.
Hagon, M. W. (1976). Germination and dormancy of *Themeda australis* Danthonia spp., *Stipa bigeniculata* and *Bothriochloa macra*. *Australian Journal of Botany*, **24**, 319–327.
Haight, J. C. & Grabe, D. F. (1972). Wetting and drying treatments to improve the performance of orchardgrass seed. *Proceedings. Association of Official Seed Analysts*, **62**, 135–148.
Halfron, E. (1979). *Theoretical Systems Ecology. Advances and Case Studies*. New York: Academic Press.
Halstead, E. H. & Vicario, B. T. (1969). Effect of ultrasonics on the germination of wild rice (*Zizania aquatica*). *Canadian Journal of Botany*, **47**, 1638–1640.
Hanssen, K. B. & Nicholls, E. B. (1965). Investigations into techniques for the germination of *Panicum maximum* Jacq. *Proceedings. International Seed Testing Association*, **30**, 715–722.
Hardesty, B. & Elliott, F. C. (1956). Differential post-ripening effects among seeds from the same parental wheat spike. *Agronomy Journal*, **48**, 406–409.
Harlan, H. V. & Pope, M. N. (1922). The germination of barley seeds harvested at different stages of growth. *Journal of Heredity*, **13**, 72–75.
Harlan, J. R. (1956). *Theory and Dynamics of Grassland Agriculture*. New York: D. Van Nostrand Co. Ltd.
Harmer, R. & Lee, J. A. (1978). The germination and viability of *Festuca vivipara* L. Sm. plantlets. *New Phytologist*, **81**, 745–51.
Harper, L. W. (1970). Dormancy in Japanese millet (*Echinochloa crus-galli* var. Frumentacea (Roxb.) wight) seed. *Proceedings. Association of Official Seed Analysts*, **60**, 132–137.
Harradine, A. R. (1980). The biology of African feather grass (*Pennisetum macrourum* Trin.) in Tasmania. 1. Seedling establishment. *Weed Research*, **20**, 165–169.
Harrington, G. T. (1916). Germination and viability tests of Johnson grass seed. *Proceedings. Association of Official Seed Analysts*, **9**, 24–28.

(1923). Forcing the germination of freshly harvested wheat and other cereals. *Journal of Agricultural Research*, **23**, 79–100.

Harrington, G. T. & Crocker, W. (1923). Structure, physical characteristics, and composition of the pericarp and integument of Johnson grass seed in relation to its physiology. *Journal of Agricultural Research*, **23**, 193–222.

Harrington, J. B. (1932). The comparative resistance of wheat varieties to sprouting in the stook and windrow. *Scientific Agriculture (Ottawa)*, **12**, 635–644.

(1949). Testing cereal varieties for dormancy. *Scientific Agriculture (Ottawa)*, **29**, 538–550.

Harrington, J. B. & Knowles, P. F. (1940). Dormancy in wheat and barley varieties in relation to breeding. *Scientific Agriculture (Ottawa)*, **20**, 355–364.

Hart, J. W. & Berrie, A. M. M. (1966). The germination of *Avena fatua* under different gaseous environments. *Physiologia Plantarum*, **19**, 1020–1025.

(1967). Relationship between endogenous levels of malic acid and dormancy in grain of *Avena fatua* L. *Phytochemistry*, **7**, 1257–1260.

Harty, R. L. & Butler, J. E. (1975). Temperature requirements for germination of green panic, *Panicum maximum* var. trichoglume, during the after-ripening period. *Seed Science and Technology*, **3**, 529–536.

Harty, R. L., Hopkinson, J. M., English, B. H. & Alder, J. (1983). Germination, dormancy and longevity in stored seed of *Panicum maximum*. *Seed Science and Technology*, **11**, 341–352.

Haun, C. R. (1956). *Dormancy and germination studies of the wild oat (Avena fatua)*. M.Sc. Thesis, Montana State College, Bozemann, Montana, USA. 53pp.

Hawkes, J. B., Williams, J. T. & Hanson, J. (1976). *A Bibliography of Plant Genetic Resources*. Rome: International Board for Plant Genetic Resources.

Hawton, D. (1979). Temperature effects on *Eleusine indica* and *Setaria anceps* grown in association. *Weed Research*, **19**, 279–284.

Hawton, D. & Drennan, D. S. H. (1980). Studies on the longevity and germination of seed of *Eleusine indica* and *Crotolaria goreensins*. *Weed Research*, **20**, 217–223.

Hay, J. R. (1960). Experiments on the mechanism of dormancy in wild oats. *Abstracts. Weed Society of America Meeting*. p. 34.

(1962). Experiments on the mechanism of induced dormancy in wild oats, *Avena fatua* L. *Canadian Journal of Botany*, **40**, 191–202.

(1967). Induced dormancy in wild oats. In *Physiologie, Oekologie und Biochemie der Keimung*, ed. H. Boriss, pp. 319–323. Greifswald: Ernst-Moritz-Arndt Universität.

Hay, J. R. & Cumming, B. G.. (1959). A method for inducing dormancy in wild oats (*Avena fatua* L.). *Weeds*, **7**, 34–40.

Hayashi, M. (1976). Studies on dormancy and germination of rice seed. 4. Location and identification of inhibitors and auxin within dormant seed organs. *Japanese Journal of Tropical Agriculture*, **19**, 156–161.

(1979). Studies on dormancy and germination in rice seed. 6. Chromatographic identification of a germination inhibitor in rice seed. *Japanese Journal of Tropical Agriculture*, **23**, 1–5.

(1980). Studies on dormancy and germination of rice seed. 9. The effects of oxygen and moisture on the release of rice seed dormancy and inactivation of inhibitors in the dormant seed. *Kagoshima Daigaku Nogakubu Gakujutsu Hokoku*, **30**, 1–9.

(1987). Relationship between endogenous germination inhibitors and dormancy in rice seeds. *Japanese Agricultural Research Quarterly*, **21**, 153–161.

Hayashi, M. & Hidaka, Y. (1979). Studies on dormancy and germination of rice seed. 8. Temperature treatment effects on rice seed dormancy and hull tissue degeneration in rice seed during the ripening period and after harvest. *Bulletin. Faculty of Agriculture, Kagoshima University*, **1979**, 21–32.

Hayashi, M. & Himeno, M. (1973). Studies on the dormancy and germination of rice seed. 2.

Relationship between seed dormancy and growth substances in rice. *Japanese Journal of Tropical Agriculture*, **16**, 270–275.

Hayashi, M. & Matsuo, T. (1983). Dormancy and germination of rice seed. 11. Ascertainment of the relation between the endogenous gibberellin-like substances and seed dormancy. *Kagoshima Daigaku Nogakubu Gakujutsu Hokoku*, **33**, 1–6.

Hayashi, M. & Morifuji, N. (1972). Studies on the dormancy and germination of rice seed. 1. The influences of temperature and gaseous conditions on dormancy and germination in rice seeds. *Japanese Journal of Tropical Agriculture*, **16**, 115–120.

Hayashi, M. & Tanaka, T. (1979). Studies on germination of rice seed. 7. Assay of endogenous germination inhibitors by germination of excised embryos and ascertainment of the relationship between the germination inhibitors and degree of seed dormancy. *Bulletin. Faculty of Agriculture, Kagoshima University*, **1979**, 11–20.

Heichel, G. H. & Day, P. R. (1972). Dark germination and seedling growth in monocots and dicots of different photosynthetic efficiencies in 2% and 20.9% oxygen. *Plant Physiology*, **49**, 280–283.

Heise, A. C. (1941). Germination of green foxtail seeds. *Proceedings. Association of Official Seed Analysts*, **33**, 43–44.

Heit, C. E. (1945). Viability testing of extremely dormant seeds by embryo excision. *New York Agricultural Experiment Station, Annual Report*, **1943/44**, 46.

(1948). Report of subcommittee on dormancy in seeds. *Proceedings. Association of Official Seed Analysts*, **38**, 25–26.

Helgeson, E. A. & Davis, D. G. (1963). Studies of the phytotoxic properties of some extracts from wild oats. *Proceedings. North Dakota Academy of Sciences*, **17**, 66.

Helgeson, E. A. & Green, J. G. (1957). Phytotoxic properties of wild oat extracts. *Abstracts. North Central Weed Control Conference*, **14**, 38–39.

(1958). Gibberellic acid vs. wild oats. *North Dakota Agricultural Experiment Station Bimonthly Report*, B20, 7–8.

Hemberg, T. (1958). Auxins and growth inhibiting substances in maize kernels. *Physiologia Plantarum*, **11**, 284.

Hendricks, S. B. & Taylorson, R. B. (1974). Promotion of seed germination by nitrate, nitrite, hydroxylamine, and ammonium salts. *Plant Physiology*, **54**, 304–309.

(1976). Variation in germination and amino-acid leakage of seeds with temperature related to membrane phase change. *Plant Physiology*, **58**, 7–11.

(1978). Dependence of phytochrome action in seeds on membrane organization. *Plant Physiology*, **61**, 17–19.

(1980). Reversal by pressure of seed germination promoted by anaesthetics. *Planta (Berlin)*, **149**, 108–111.

Hewett, P. D. (1958). Effects of heat and loss of moisture on the dormancy of barley. *Nature (London)*, **181**, 424–425.

(1959). Effect of heat and loss of moisture on the dormancy of wheat and some interactions with 'Morganna D'. *Nature (London)*, **183**, 1600.

Hilton, J. R. (1982). An unusual effect of the far-red absorbing form of phytochrome: Photoinhibition of seed germination in *Bromus sterilis* L. *Planta (Berlin)*, **155**, 525–28.

(1984a). The influence of light and potassium nitrate on the dormancy and germination of *Avena fatua* L. (wild oat) seed and its ecological significance. *New Phytologist*, **96**, 31–34.

(1984b). The influence of dry storage temperature on the response of *Bromus sterilis* seeds to light. *New Phytologist*, **98**, 129–134.

(1987). Photoregulation of germination in freshly-harvested and dried seeds of *Bromus sterilis* L. *Journal of Experimental Botany*, **38**, 286–292.

Hilton, J. R. & Bitterli, C. J. (1983). The influence of light on the germination of *Avena fatua* wild oat seed and its ecological significance. *New Phytologist*, **95**, 325–334.

Hilton, J. R., Froud-Williams, R. J. & Dixon, J. (1984). A relationship between phytochrome photoequilibrium and germination of seeds of *Poa trivialis* L. from contrasting habitats. *New Phytologist*, **97**, 375–379.

Hilton, J. R. & Owen, P. D. (1985a). Phytochrome regulation of extractable cytochrome oxidase activity during early germination of *Bromus sterilis* L. and *Lactuca sativa* L. Grand Rapid seeds. *New Phytologist*, **100**, 163–171.

(1985b). Light and dry storage influences on the respiration of germinating seeds of five species. *New Phytologist*, **99**, 523–531.

Hilton, J. R. & Thomas, B. (1987). Photoregulation of phytochrome synthesis in germinating embryos of *Avena sativa* L. *Journal of Experimental Botany*, **38**, 1704–1712.

Hilu, K. W. & De Wet, J. M. J. (1980). Effect of artificial selection on grain dormancy in *Eleusine* (Gramineae). *Systematic Botany*, **5**, 54–60.

Hoffman, O. L. (1961). Breaking wild oat (*Avena fatua*) dormancy with gases at high pressure. *Weeds*, **9**, 493–496.

Holm, L., Pancho, J. V., Herberger, J. P. & Plucknett, D. L. (1979). *A Geographical Atlas of World Weeds*. New York: John Wiley and Sons.

Holm, R. E. & Miller, M. R. (1972a). Hormonal control of weed seed germination. *Weed Science*, **20**, 209–212.

(1972b). Weed seed germination responses to chemical and physical treatments. *Weed Science*, **20**, 150–153.

Hooley, R. (1984a). Gibberellic acid controls specific acid- phosphatase isozymes in aleurone cells and protoplasts of *Avena fatua* L. *Planta (Berlin)*, **161**, 355–360.

(1984b). Gibberellic acid controls the secretion of acid phosphatase in aleurone layers and isolated aleurone protoplasts of *Avena fatua*. *Journal of Experimental Botany*, **35**, 822–828.

Hooley, R. & Zwar, J. A. (1986). Hormonal regulation of α-amylase gene expression. *Journal of Cell Biochemistry*, Supp. **10B**, 22.

Hopkins, C. Y. (1936). Thermal death point of certain weed seeds. *Canadian Journal of Research*, C-D, **14**, 178–183.

Hsiao, A. I. & Quick, W. A. (1983). The induction and breakage of seed dormancy in wild oats. In *Wild Oat Symposium Proceedings '83*, ed. H. Smith, pp. 173–185. Regina, Saskatchewan: Canadian Plains Research Centre.

(1984). Actions of sodium hypochlorite and hydrogen peroxide on seed dormancy and germination of wild oats, *Avena fatua* L. *Weed Research*, **24**, 411–419.

(1985). Wild oats (*Avena fatua* L.) seed dormancy as influenced by sodium hypochlorite, moist storage and gibberellin A_3. *Weed Research*, **25**, 281–288.

Hsiao, A. I. & Simpson, G. M. (1971). Dormancy studies in seed of *Avena fatua*. VII. The effects of light and variation in water regime on germination. *Canadian Journal of Botany*, **49**, 1347–1357.

Huang, W. Z. & Hsiao, A. I. (1987a). Factors affecting seed dormancy and germination of Johnsongrass (*Sorghum halapense* L. Pers.). *Weed Research*, **27**, 1–12.

(1987b). Factors affecting seed dormancy and germination of *Paspalum distichum*. *Weed Research*, **27**, 405–415.

Hubac, C. (1966). Dormancy of immature caryopses of wheat cv. Fylgia. *Annales des Sciences Naturelles. Botanique et Biologie Vegetale*, **7**, 475–504.

Hughes, R. M. (1979). Effects of temperature and moisture stress on germination and seedling growth of four tropical species. *Journal of the Australian Institute of Agricultural Science*, **45**, 125.

Hurtt, W. & Taylorson, R. B. (1978). Greenhouse studies of chemical effects on wild oat seed shedding and germination. *Proceedings. Northeastern Weed Science Society*, **32**, 114.

Hyde, E. O. C. (1935). Observations on the germination of newly harvested Algerian oats. *New Zealand Journal of Agriculture*, **51**, 361–367.

Hylton, L. O. & Bass, L. N. (1961). Germination of sixweeks fescue. *Proceedings. Association of Official Seed Analysts*, **51**, 118–124.

Ikeda, M. (1963). Studies on the viviparous germination of rice seed. *Bulletin. Faculty of Agriculture, Kagoshima University*, **13**, 89–115.

Ikehashi, H. (1973). Studies on the environmental and varietal differences of germination habits in rice seeds with special reference to plant breeding. *Journal of the Central Agricultural Experiment Station, Konosu*, **19**, 1–60.

(1975). Dormancy formation and subsequent changes of germination habits in rice seeds. *Japanese Agricultural Research Quarterly*, **9**, 8–12.

Imam, A. G. & Allard, R. W. (1965). Population studies in predominantly self-pollinating species. VI. Genetic variability between and within natural populations of wild oats from differing habitats in California. *Genetics*, **51**, 49–62.

Inoue, K., Hayashi, T. & Yamazaki, K. (1970). Control of barnyard grass by induction of dormancy breaking. 1. Dormancy breaking of barnyard grass seeds by respiration inhibitors. *Nippon Dojo-Hiryogaku Zasshi*, **41**, 377–382.

International Rice Research Institute. (1968). Seed dormancy. *IRRI Reporter*, **4**, 1–4.

Isikawa, S., Fujii, T. & Yokohama, Y. (1961). Photoperiodic control of the germination of *Eragrostis* seeds. *Botanical Magazine (Tokyo)*, **74**, 14–18.

Ito, H., Kitahara, M. & Kawamura, M. (1931). Differences in the amylolytic activity of the germinating grains of 15 strains of *Setaria italica* Kunth. *Kagami Bulletin. Imperial College of Agriculture, Gihu*, **16**, 1–33.

Jacobsen, J. V. (1983). Regulation of protein synthesis in aleurone cells by gibberellin and abscisic acid. In *Biochemistry and Physiology of the Gibberellins*. ed. A. Crozier, pp. 159–187. New York: Praeger Publishers.

Jacobsohn, R. & Andersen, R. N. (1968). Differential response of wild oat lines to diallate, triallate, and Barban. *Weed Science*, **16**, 491–494.

Jain, J. C., Quick, W. A. & Hsiao, A. I. (1983). ATP synthesis during water imbibition in caryopses of genetically dormant and non-dormant lines of wild oat (*Avena fatua* L.). *Journal of Experimental Botany*, **34**, 381–387.

Jain, S. K. (1982). Variation and adaptive role of seed dormancy in some annual grassland species. *Botanical Gazette*, **143**, 101–106.

Jain, S.K. & Marshall, D. R. (1967). Population studies in predominantly self-pollinating species. X. Variation in natural populations of *Avena fatua* and *A. barbata*. *American Naturalist*, **101**, 19–33.

Jain, S. K. & Rai, K. N. (1977). Natural selection during germination in wild oat (*Avena fatua*) and California Burclover (*Medicago polymorpha* var. vulgaris) populations. *Weed Science*, **25**, 495–498.

Jalote, S. & Vaish, C. P. (1976). Dormancy behaviour of paddy varieties in U.P. *Seed Research (New Delhi)*, **4**, 187–190.

Jana, S., Acharya, S. N. & Naylor, J. M. (1979). Dormancy studies in seed of *Avena fatua*. 10. Inheritance of germination behaviour. *Canadian Journal of Botany*, **57**, 1663–1667.

Jana, S. & Naylor, J. M. (1982). Adaptation to herbicide tolerance in populations of *Avena fatua*. *Canadian Journal of Botany*, **60**, 1611–1617.

Jana, S. & Thai, K. M. (1987). Patterns of changes of dormant genotypes in *Avena fatua* populations under different agricultural conditions. *Canadian Journal of Botany*, **65**, 1741–1745.

Jana, S., Upadhyaya, M. K. & Acharya, S. N. (1988). Genetic basis of dormancy and differential response to sodium azide in *Avena fatua* seeds. *Canadian Journal of Botany*, **66**, 635–641.

Jansson, G. (1959). Chemically induced water-sensitivity in barley seed. *Arkiv für Kemi*, **14**, 279–289.

(1960). Breaking of water sensitivity of barley seeds by treatment with chemicals. *Arkiv für Kemi*, **15**, 439–450.

Jennings, R. W., Collins, N. A., Bettis, R. B. & Biswas, P. K. (1968). Effects of several chemical

stimulants and inhibitors on seed germination and oxygen consumption of selected weed species. *Abstracts. Weed Science Society of America*, **1968**, 23–24.
Johnson, L. P. V. (1935a). *Physiological and genetical studies on delayed germination in Avena*. Ph.D. Thesis, State College of Washington, Pullman, Washington, USA.
(1935b). General preliminary studies on the physiology of delayed germination in *Avena fatua*. *Canadian Journal of Research. C*, **13**, 283–300.
(1935c). The inheritance of delayed germination in hybrids of *Avena fatua* and *A. sativa*. *Canadian Journal of Research. C*, **13**, 367–387.
Johnston, M. E. H. (1972). Report of the working group for the germination of tropical and sub-tropical seeds. *Proceedings. International Seed Testing Association*, **37**, 355–359.
(1981). Report of the germination committee working group on tropical and sub-tropical seeds. 1977–1980. *Seed Science and Technology*, **9**, 137–140.
Johnston, M. E. H. & Miller, J. G. (1964). Investigation into techniques for the germination of *Paspalum dilatatum*. *Proceedings. International Seed Testing Association*, **29**, 145–148.
Johnston, M. E. H. & Tattersfield, J. G. (1971). A preliminary report on germination techniques for *Panicum maximum* Jacq. *Proceedings. International Seed Testing Association*, **36**, 115–121.
Jones, C. A. (1985). *C4 Grasses and Cereals. Growth, Development and Stress Response*. New York: John Wiley and Sons.
Jones, J. F. & Hall, M. A. (1981). The effect of ethylene on quantitative and qualitative aspects of respiration during the breaking of dormancy of *Spergula arvensis* seeds. *Annals of Botany*, **48**, 291–300.
Jong, F. S. (1984). *The regulation of endosperm hydrolysis in Avena fatua*. M.Sc. Thesis, University of Saskatchewan, Saskatoon, Saskatchewan, Canada. 76pp.
Jordan, D. (1977). A decade of wild oats. *Span*, **20**, 21–24.
Jordan, J. L. (1981). Seed dormancy in Pennsylvania smart weed and barnyard grass. *Dissertation Abstracts. International B.*, **42**, 1256–1257.
Jordan, L. S. & Jordan, J. L. (1982). Effects of pre-chilling on *Convolvulus arvensis* seed coat and germination. *Annals of Botany*, **49**, 421–424.
Jorgenson, E. M., O'Sullivan, P. A. & Vanden Born, W. H. (1974). Response of four wild oat seed color categories to barban, benzoylprop-ethyl and AC 84777. *Research Report. National Weed Committee of Canada, Western Section*, 487–488. Ottawa: Agriculture Canada.
Joubert, D. C. & Small, J. G. C. (1982). Seed-germination and dormancy of *Stipa trichotoma*, Nasella tussock. 1. Effect of dehulling, constant temperatures, light, oxygen, activated charcoal and storage. *South African Journal of Botany*, **1**, 142–146.
Judkins, W. P. (1946). The influence of kernel size, age, location in the panicle, and variety of oat, on the variability of the *Avena* test. *American Journal of Botany*, **33**, 181–184.
Juliano, B. O. (1980). Properties of the rice caryopsis. In *Rice Production and Utilization*, ed. B. S. Luh, pp. 403–438. Westport, Conn. USA: Avi Pub. Co. Inc.
Juliano, B. O. & Chang, T-T. (1987). Pre-harvest sprouting in rice. In *4th International Symposium. Pre-harvest Sprouting in Cereals*, ed. D. J. Mares, pp. 34–42. Boulder, Color. USA: Westview Press.
Juntilla, O. (1977). Dormancy in dispersal units of various *Dactylis glomerata* populations. *Seed Science and Technology*, **5**, 463–472.
Juntilla, O., Landgraff, A. & Nilsen, A. J. (1978). Germination of *Phalaris arundinacea* seeds. *Acta Horticulturae (The Hague)*, **83**, 163–166.
Justice, O. L. & Bass, L. (1978). *Principles and Practices of Seed Storage*. United States Department of Agriculture Handbook, No. 506. Washington, D.C.
Justice, O. L. & Whitehead, M. D. (1946). Seed production, viability, and dormancy in the nutgrasses, *Cyperus rotundus* and *C. esculentus*. *Journal of Agricultural Research*, **73**, 303–318.
Kahre, L., Kolk, H. & Fritz, T. (1965). Gibberellic acid for breaking of dormancy in cereal

seed. *Proceedings. International Seed Testing Association*, **30**, 887–891.

Kahre, L., Kolk, H. & Wiberg, H. (1962). Note on dormancy-breaking in seeds (cereals and timothy). *Proceedings. International Seed Testing Association*, **27**, 679–683.

Kamalavalli, D., Amin, J. V., Pathak, C. H., Sen, D. H. & Bansal, R. P. (1978). The cyanide induced early germination of sorghum. In *Environment and Physiological Ecology of Plants*, ed. D. N. Sen & R. P. Bansal, pp. 67–76. Dehra Dun, India: Bishen Singh & Mahendra Pal Singh.

Karl, R. & Rudiger, W. (1982). Naturliche Hemmstoffe von Keimung und Wachstum. 1. Ausarbeitung eines quantitaven Biotests und Anwendung auf Extrakte aus Spelzen von *Avena sativa* L. *Zeitschrift für Naturforschung. Section C. Biosciences*, **37**, 793–801.

Karssen, C. M. (1982). Seasonal patterns of dormancy in weed seeds. In *Physiology and Biochemistry of Seed Development, Dormancy and Germination*, ed. A. A. Khan, pp. 243–270. Amsterdam, Holland: Elsevier Biomedical Press.

Katayama, T. C. & Nakagama, A. (1972). Studies on the germination behaviour of teff seeds (*Eragrostis abyssinica* Schrad.) with the emphasis on storage conditions. *Japanese Journal of Tropical Agriculture*, **16**, 97–105.

Kaufman, P. B., Ghoshesh, N. S., Nakosteen, L., Pharis, R. P. & Durley, R. C. (1976). Analysis of native gibberellins in internode, nodes, leaves, and inflorescence of developing *Avena* plants. *Plant Physiology*, **58**, 131–134.

Kearns, V. & Toole, E. H. (1938). Temperature and other factors affecting the germination of the seed of fescues. *Proceedings. International Seed Testing Association*, **10**, 337–341.

Kendrick, R. E. & Frankland, B. (1983). *Phytochrome and Plant Growth.* Studies in Biology no. 68. London: Edward Arnold Publishers Ltd.

Keng, J. & Foley, M. E. (1987). Effect of gibberellin on protein and non-structural carbohydrate in dormant *Avena fatua* caryopses. *Plant Science (Limerick)*, **51**, 37–41.

Kennedy, R. A., Fox, T. C. & Siedow, J. N. (1987). Activities of isolated mitochondrial enzymes from aerobically and anaerobically germinated barnyard grass (*Echinochloa*) seedlings. *Plant Physiology*, **85**, 474–480.

Kennedy, R. A., Rumpho, M. E. & Vanderzee, D. (1983). Germination of *Echinochloa crus-galli* barnyard grass seeds under anaerobic conditions. Respiration and response to metabolic inhibitors. *Plant Physiology*, **72**, 787–794.

Kent, N. & Brink, R. A. (1947). Growth *in vitro* of immature *Hordeum* embryos. *Science*, **106**, 547–548.

Key, J. M. (1987). Dormancy in the wild and weedy relatives of modern cereals. In *4th International Symposium. Pre-harvest Sprouting in Cereals*, ed. D. J. Mares, pp. 414–24. Boulder, Colo., USA: Westview Press.

Keys, C. E. (1949). Observations on the seed and germination of *Setaria italica* (L). *Transactions. Kansas Academy of Science*, **52**, 474–477.

Khan, A. A. (1969). Cytokinin-inhibitor antagonism in the hormonal control of α-amylase synthesis and growth in barley seed. *Physiologia Plantarum*, **22**, 94–103.

Khan, A. A., Tolbert, M. A. & Mitchell, E. D. (1964). Partial chemical and physiological characterization of a dormancy factor from bulbs of immature wheat. *Plant Physiology*, **39**, Supp. p. 28.

Khan, A. A. & Waters, E. C. (1969). Hormonal control of postharvest dormancy and germination in barley seeds. *Life Sciences*, **8**, 729–736.

Kiewnick, L. (1964). Experiments on the influence of seedborne and soilborne microflora on the viability of wild oat seeds (*Avena fatua* L.). II. Experiments on the influence of microflora on the viability of seeds in the soil. *Weed Research*, **4**, 31–43.

Kijima, K. & Takei, K. (1971). Germination test of tropical and sub-tropical grasses. 1. On the germination of coloured guineagrass (*Panicum coloratum*). *Journal of the Japanese Society of Grassland Science*, **17**, 170–175.

Kilcher, M. R. & Lawrence, T. (1960). Quality of intermediate wheatgrass seed, with and without hulls. *Canadian Journal of Plant Science*, **40**, 482–486.

Kim, C. M. (1954). Effect of soil moisture on the germination and early development of Bengal grass (*Setaria italica*) and barn grass (*Panicum crusgalli frumentaceum*). *Ecological Review*, **13**, 271–276.

Kinch, R. C. (1963). A method of inducing rapid germination of western wheatgrass. *Proceedings. Association of Official Seed Analysts*, **53**, 55–57.

King, L. J. (1952). Germination and chemical control of the giant foxtail grass. *Contributions of the Boyce Thompson Institute*, **16**, 469–487.

King, R. W. (1976). Abscisic acid in developing wheat grains and its relationship to grain growth and maturation. *Planta (Berlin)*, **132**, 43–51.

(1983). The physiology of pre-harvest sprouting – a review. In *3rd International Symposium. Pre-harvest Sprouting in Cereals*, ed. J. E. Kruger & D. E. Laberge, pp. 11–21. Boulder, Colo., USA: Westview Press.

Kirk, J. & Courtney, A. D. (1972). A study on the survival of wild oats *Avena fatua* seeds burie, in farmyard manure and fed to bullocks. *Proceedings. 11th British Weed Control Conference*, Vol. 1. 226–233.

Kiseleva, N. A. (1956). On a second state of dormancy in seeds of wild oat (*Avena fatua*) and chess brome (*Bromus secalinus*). *Agrobiologia*, **2**, 130–134.

Kleine, R. (1929). Uber die Keimung mit *Poa fertilis*. *Pflanzenbau*, **5**, 211–213.

Kneebone, W. R. (1960). Size of caryopses in Buffalo grass (*Buchloe dactyloides* (Nutt.) Engelm.) as related to their germination and longevity. *Agronomy Journal*, **52**, 553.

Koch, W. (1968a). Environmental factors affecting the germination of some annual grasses. *Proceedings. 9th British Weed Control Conference*, pp. 14–19.

(1968b). Influence of oxygen and carbon dioxide on the germination of weed seeds. *Saatgutwirtschaft*, **9**, 283–285.

(1974). *Einfluss von Umweltfaktoren auf die Samenphase annueller Unkrauter insbesondere unter dem Gesichtspunkt de Unkrautbekampfung*. Arbeiten Der Universität Hohenheim, Landwirtschaftliche Hochschule. Stuttgart: Verlag Eugen Ulmer. 204pp.

Koestler, A. (1967). *The Ghost in the Machine*. London: Hutchinson and Co. Ltd.

Kohout, V. (1977). The degree of mortality of wild oat seeds kept on the surface of the soil. *Proceedings. European Weed Research Symposium, Methods of Weed Control and their Integration*, pp.195–201. Uppsala, Sweden.

Kohout, V. & Loudova, H. (1981). Differences in dormancy of weed seeds of the genera *Echinochloa* and *Setaria*. *Sbornik UVTI (Ustav Vedeckotechnickych Informaci) Ochrana Rostlin*, **17**, 145–150.

Kohout, V. & Pulkrabek, J. (1977). Prispevek ke studii dormance a vzchazivosti obilek ovsa hlucheho. In *Sbornik AF Vysoke Skoly Zemedelske Praze*, **1977**, 155–164.

Koller, D.. (1972), Environmental control of seed germination. In *Seed Biology*, ed. T. T. Kozlowski, pp. 1–101. New York: Academic Press.

Koller, D. & Negbi, M. (1957). Hastening the germination of *Panicum antidotale Retz.*, *Bulletin. Research Council of Israel, Sec. D*, **5**, 225–238.

Koller, D. & Roth, N. (1963). Germination regulating mechanisms in some desert seeds. 7. *Panicum turgidum* (Gramineae). *Israel Journal of Botany*, **12**, 64–73.

Kollman, G. E. & Staniforth, D. W. (1972). Hormonal aspects of seed dormancy in Yellow Foxtail. *Weed Science*, **20**, 472–477.

Kommedahl, T., DeVay, J. E. & Christensen, C. M. (1958). Factors affecting dormancy and seedling development in wild oats. *Weeds*, **6**, 12–18.

Kono, Y., Takeuchi, S., Kawarada, A. & Ota, Y. (1975). Anaerobic respiration in the dormancy of rice seed. *Proceedings. Crop Science Society of Japan*, **44**, 194–198.

Konzak, C. F., Randolph, L. F. & Jensen, N. F. (1951). Embryo culture of barley species hybrids. *Journal of Heredity*, **42**, 125–134.

Koshimuzu, T. (1936). On the relation between the ripening stages of the maize-seed and its germination. *Botanical Magazine (Tokyo)*, **50**, 504–513.

Koves, E. (1957). Paper chromatographic investigations of ether-soluble germination and growth inhibitors in oat husks. *Acta Biologica (Szeged)*, **3**, 179–187.

Krishnaswamy, P. (1952). Storage and germination of millet seeds. *Madras Agricultural Journal*, **39**, 485–490.

Kropac, Z. (1980). Distribution of *Avena fatua* in Czechoslovakia. *Folia Geobotanica et Phytotaxonomica (Praha)*, **15**, 259–307.

Kropac, Z., Havranek, T. & Dobry, J. (1986). Effect of duration and depth of burial on seed survival of *Avena fatua* in arable soil. *Folia Geobotanica et Phytotaxonomica (Praha)*, **21**, 249–262.

Kucera, C. L. (1966). Some effects of gibberellic acid on grass seed germination. *Iowa State Journal of Science*, **41**, 137–143.

Kuhnel, W. (1965). Ecological studies on the occurrence of wild oats in Oderbruch region. *Nachrichtenblatt. Deutscher Pflanzenschutzdienst (Berlin)*, **19**, 145–149.

Kumari, J., Thomas, T. P. & Sen, D. N. (1987). Breaking seed dormancy in som grasses of Indian arid zone. *Geobios (Johdpur)*, **14**, 131–133.

Kuo, W. H. J. & Chu, C. (1979). Mechanisms of seed dormancy: A review of pentose phosphate pathway hypothesis. *K'o Hsueh Nung Yeh (Taipei)*, **27**, 71–77.

(1982). Hull peroxidase in a prophylactic dormancy mechanism of rice *Oryza sativa* grains. *Plant Physiology*, **69**, Supp. p. 4.

(1983). No relation between catalase and dormancy of rice *Oryza sativa* L. grains. *Journal of the Agricultural Association of China (Taiwan)*, **123**, 13–20.

(1985). Prophylactic role of hull peroxidase in the dormancy mechanism of rice *Oryza sativa* grain. *Botanical Bulletin. Academia Sinica (Taipei)*, **26**, 59–66.

Kurth, H. (1965). Untersuchungen uber die Keimungsphysiologie des Wildhafers (*Avena fatua* L.) und zu seiner Bekampfung mit Herbiziden aus der Reihede Chlorierten Aliphatischen Carbonsauren. *Nachrichtenblatt. Deutscher Pflanzenschutzdienst (Berlin)*, **19**, 29–35.

Kyurdzhiyeva, V. (1956). Several problems in the biology of rice seeds germination. *Sbornik Nauchnykh-Issled. Rabot Stud. Stavropol'sk Sel'skokhozyaistvennyi Institut*, **1956**, 7–10.

Kurth, H. (1967). The germinative behaviour of weeds. *SYS Reporter*, (3), 6–11.

LaCroix, L. J., Naylor, J. M. & Larter, E. N. (1962). Factors controlling embryo growth and development in barley (*Hordeum vulgare* L.). *Canadian Journal of Botany*, **40**, 1515–1523.

Lagreze-Fossat, A. (1856). De la reproduction de la folle-avoine. *Moniteur des Comices*, **1856**, 346–348.

Lal, R. & Reed, W. B. (1980). The effect of microwave energy on germination and dormancy of wild oat seeds. *Canadian Agricultural Engineering*, **22**, 85–88.

Landgraff, A. & Juntilla, O. (1979). Germination and dormancy of reed canary grass seeds (*Phalaris arundinacea*). *Physiologia Plantarum*, **45**, 96–102.

Lang, G. A., Early, J. D., Martin, G. C. & Darnell, R. L. (1987). Endo-, para-, and ecodormancy: physiological terminology and classification for dormancy research. *HortScience*, **22**, 371–377.

Larcher, W. (1983). *Physiological Ecology*. 3rd edn. Berlin: Springer-Verlag.

Larondelle, Y., Corbineau, F., Dethier, M., Côme, D. & Hers, H-G. (1987). Fructose 2,6–biphosphate in germinating oat seeds. A biochemical study of seed dormancy. *European Journal of Biochemistry*, **166**, 605–610.

Larson, A. H., Harvey, R. B. & Larson, J. (1936). Length of the dormant period in cereal seeds. *Journal of Agricultural Research*, **52**, 811–836.

LaRue, C. D. & Avery, G. S. (1938). The development of the embryo of *Zizania aquatica* in the seed and in artificial culture. *Bulletin. Torrey Botanical Club*, **65**, 11–21.

Lascorz, M. & Drapon, R. (1987a). Activity of three oxido-reduction enzymes from oat seed hulls during ripening. Possible inhibitory role towards germination. *Plant Physiology and Biochemistry*, **25**, 117–123.

(1987b). Lipoxygenase system in oat hulls in relation to dormancy. *Phytochemistry*, **26**, 349–351.

Laude, H. M. (1949). Delayed germination of California oatgrass, *Danthonia californica. Agronomy Journal*, **41**, 404–408.

(1956). Germination of some freshly harvested seed of some western range species. *Journal of Range Management*, **9**, 126–129.

Lawson, G. W. (1986). *Plant Ecology in West Africa, Systems and Processes*. Chichester & New York: John Wiley and Sons.

Lecat, S. (1987). *Quelques aspects metabolique de la dormance des semences d'Avoine (Avena sativa* L.). Etude plus particuliere de l'action des glumelles. Ph.D. Thesis, Universite Pierre et Marie Curie, Paris, France.

Leggett, H. W. & Banting, J. D. (1958). *Wild Oats in Western Canada. Research for Farmers*. Canada Department of Agriculture, Publication 3, pp.10–11. Ottawa.

Lehle, F. R., Staniforth, D. W. & Stewart, C. R. (1978). Caryopsis dormancy in *Setaria lutescens* endosperm starch susceptibility to amylase digestion. *Plant Physiology*, **61**, Supp. p. 16.

(1983). Lipid mobilization in dormant and nondormant caryopses of yellow foxtail (*Setaria lutescens*). *Weed Science*, **31**, 28–36.

Lehle, F. R., Stewart, C. R. & Staniforth, D. W. (1981). Ultrastructure of dormancy release in caryopses of *Setaria lutescens* Gramineae. *Plant Physiology*, **67**, Supp. p. 39.

Lenoir, C., Corbineau, F. & Côme, D. (1986). Barley (*Hordeum vulgare*) cultivar Sonja seed dormancy as related to glumella characteristics. *Physiologia Plantarum*, **68**, 301–307.

Leshem, Y. (1977). Germination inhibition in wheat by ethylene producing compounds ethrel and al-sol. *Plant Physiology*, **59**, Supp. p. 37.

Leshem, Y. Y. (1978). An exogenous chemical imposition of ephemeral dormancy in cereal seed. *Journal of Agricultural Science*, **90**, 459–462.

Letham, D. S. & Palni, L. M. S. (1983). The biosynthesis and metabolism of cytokinins. In *Annual Review of Plant Physiology*, ed. R. W. Briggs, R. L. Jones & V. Walbot, vol. 34, pp. 163–197. Palo Alto, California: Annual Reviews Inc.

Lewis, J. (1961). The influence of water level, soil depth and type on the survival of crop and weed seeds. *Proceedings. International Seed Testing Association*, **26**, 68–85.

Lindauer, L. L. (1972). Germination ecology of *Danthonia sericea* M. populations. *Dissertation Abstracts*, **32**, 5061B-5062B.

Linnington, S., Bean, E. W. & Tyler, B. F. (1979). The effects of temperature upon seed germination in *Festuca pratensis* var. Appennina. *Journal of Applied Ecology*, **16**, 933–938.

Lisman, J. E. (1985). A mechanism for memory storage insensitive to molecular turnover: A bistable autophosphorylating kinase. *Proceedings. National Academy of Science, USA.*, **82**, 3055–3057.

Lodge, G. M. & Whalley, R. D. B. (1981). Establishment of warm season and cool season native perennial grasses on the northwest slopes of New South Wales, Australia 1. Dormancy and germination. *Australian Journal of Botany*, **29**, 111–120.

Lohaus, E., Blos, I., Schafer, W. & Rudiger, W. (1982). Naturliche Hemstoffe von Keimung und Wachstum. II. Isolierung und Struktur von Hemstoffen aus *Avena sativa* L. *Zeitschift für Naturforschung C*, **37**, 802–811.

Lucanus, P. (1862). Uber den Einfluscher Reife unde der Nachreife auf die Keimungs- und Vegetationskraft der Roggen korner. *Landwirtschaftliche Versuchstation*, **4**, 253–263.

Ludwig, H. (1971). Dormancy in seeds of Gramineae and its problem in seed testing with special regard to treatment with gibberellic acid. *Proceedings. International Seed Testing Association*, **36**, 289–305.

Lush, W. M. & Groves, R. H. (1981). Germination, emergence and surface establishment of wheat (*Triticum aestivum*) cv. Egret and rye grass (*Lolium rigidum*) cv. Wimmera in response to natural and artificial dehydration cycles. *Australian Journal of Agricultural Research*, **32**, 731–740.

Lute, A. M. (1938). Germination characteristics of wild oats. *Proceedings. Association of Official Seed Analysts*, **1930–38**, 70–73.

MacLean, D. & Grof, B. (1968). Effect of seed treatments on *Brachiaria mutica* and *B. ruziziensis*. *Queensland Journal of Agricultural Science*, **25**, 81–83.

MacLeod, A. M. (1967a). The physiology of malting – a review. *Journal of the Institute of Brewing*, **73**, 146–162.

(1967b). Gibberellic acid and malting. *Wallerstein Laboratories. Communications*, **30**, 85–95.

(1969). The utilization of cereal seed reserves. *Scientific Progress*, **57**, 99–112.

MacLeod, A. M. & Palmer, G. H. (1966). The embryo of barley in relation to modification of the endosperm. *Journal of the Institute of Brewing*, **72**, 580–589.

Mai-Tran-Ngoc-Tieng & Nguyen-Thi-Ngoc-Lang. (1971). Seed dormancy in a rice variety. *Proceedings. Pacific Science Congress*, **1**, 67.

Maiti, S., Purkait, A. & Chatterjee, B. N. (1981). Seed dormancy in deenanath grass (*Pennisetum pedicellatum*). *Forage Research*, **7**, 97–99.

Major, R. L. & Wright, L. N. (1974). Seed dormancy characteristics of sideoats gramagrass, (*Bouteloua curtipendula* Michx. Torr.). *Crop Science*, **14**, 37–40.

Major, W. & Roberts, E. H. (1968). Dormancy in cereal seeds. I. The effects of oxygen and respiratory inhibitors. *Journal of Experimental Botany*, **19**, 77–89.

Mangelsdorf, P. C. (1930). The inheritance of dormancy and premature germination in maize. *Genetics*, **15**, 462–494.

Manson, J. M. (1932). Weed survey of the Prairie provinces. *Dominion of Canada, National Research Council, Report 26*,pp. 34. Ottawa.

Mapelli, S., Lombardi, L. & Rocchi, P. (1984). Gibberellin and abscisic acid effects on the activity of hydrolytic enzymes in de-embryonated barley seeds. *Plant Growth Regulation*, **2**, 31–40.

Marais, G. F. & Kruis, W. J. G. (1983). Pre-harvest sprouting – the South African situation for seed dormancy in barley. In *3rd International Symposium. Pre-harvest Sprouting in Cereals*, eds. J. E. Kruger & D. E. Laberge, pp. 267–273. Boulder, Color. USA: Westview Press.

Marbach, I. & Mayer, A. M. (1979). Germination, utilization of storage materials and potential for cyanide release in cultivated and wild sorghum. *Physiologia Plantarum*, **47**, 100–104.

Marshall, D. R. & Jain, S. K. (1968). Interference in pure and mixed populations of *Avena fatua* and *A. barbata*. *Journal of Ecology*, **57**, 252–271.

(1969). Genetic polymorphism in natural populations of *Avena fatua* and *A. barbata*. *Nature (London)*, **221**, 276–278.

Martin, C. C. (1975). Role of glumes and gibberellic acid in dormancy of *Themeda triandra* spikelets. *Physiologia Plantarum*, **33**, 171–176.

Maruyama, K. (1980). Methods of selection against pre harvest sprouting of rice during rapid generation advance. *Japanese Journal of Breeding*, **30**, 344–350.

Mastovsky, J. & Karel, V. (1961). A study of the possibility of shortening dormancy of barley and reducing its water sensitivity. *American Brewer*, **94**, 41–44.

Mather, H. J. & Greaney, F. J. (1949). Wild oat control by cultural methods. *Line Elevator Farm Service. Circular 11*, 6pp. Winnipeg, Manitoba, Canada.

Mathews, A. C. (1947). Observations on methods of increasing germination of *Panicum anceps* Michx. and *Paspalum notatum* Flugge. *Journal of the American Society of Agronomy*, **39**, 439–442.

Matumura, M. & Nakajima, N. (1981). Fundamental studies on artificial propagation by seeding useful grasses in Japan. VIII. Some observations concerning the seed propagation of the dwarf bamboo, *Sasa senanensis* Rehd. *Research Bulletin. Faculty of Agriculture, Gifu University*, **45**, 289–297.

Matumura, M., Yukimura, T. & Shinoda, S. (1983). Fundamental studies on artificial propagation by seeding useful wild grasses in Japan. IX. Seed fertility and germinability of the intraspecific 2 types of *Chigaya alang-alang, Imperata cylindrica* cv. Koenigii. *Journal Japanese Society of Grassland Science*, **28**, 395–404.

Mayer, A. M. & Poljakoff-Mayber, A. (1963). *The Germination of Seeds*. Oxford, England: Pergamon Press.

Maynard, M. L. (1963). Effects of wetting and drying on germination of crested wheatgrass seed. *Journal of Range Management*, **16**, 119–121.

McClure, F. A. (1966). *The Bamboos. A Fresh Perspective*. Cambridge, Mass. USA: Harvard University Press.

McCrate, A. J., Nielsen, M. T., Paulsen, G. M. & Heyne, E. G. (1982). Relationship between sprouting in wheat (*Triticum aestivum*) and embryo response to endogenous inhibition. *Euphytica*, **31**, 193–200.

McDonald, M. B. & Khan, A. A. (1977). Factors determining germination of Indian rice-grass seed. *Agronomy Journal*, **69**, 558–563.

(1983). Acid scarification and protein synthesis during seed germination. *Agronomy Journal*, **75**, 111–114.

McIntyre, G. I. & Hsiao, A. I. (1983). The role of water in the mechanism of seed dormancy in wild oats (*Avena fatua* L.). In *Wild Oat Symposium '83*, ed. A.E. Smith, pp. 187–200. Regina, Saskatchewan, Canada: Canadian Plains Research Centre.

(1985). Seed dormancy in *Avena fatua*. II. Evidence of embryo water content as a limiting factor. *Botanical Gazette*, **146**, 347–352.

McKeon, G. M. (1985). Pasture seed dynamics in a dry monsoonal climate. II. The effect of water availability, light and temperature on germination speed and seedling survival of *Stylosanthes humulis* and *Digitaria ciliaris*. *Australian Journal of Ecology*, **10**, 149–163.

McKeon, G. M., Rose, C. W., Kalma, J. D. & Torsell, B. W. R. (1985). Pasture seed dynamics in a dry monsoonal climate. I. Germination and seed bed environment of *Stylosanthes humulis* and *Digitaria ciliaris*. *Australian Journal of Ecology*, **10**, 135–147.

McWha, J. A., de Ruiter, J. & Jameson, E. P. (1976). The nature of dormancy in wild oats. *Proceedings. 29th New Zealand Weed and Pest Control Conference*, pp. 102–105.

McWha, J. A. & Jackson, D. L. (1976). Some growth promotive effects of abscisic acid. *Journal of Experimental Botany*, **27**, 1004–1008.

Merry, J. (1942). Studies on the embryo of *Hordeum sativum*. II. The growth of the embryo in culture. *Bulletin. Torrey Botanical Club*, **69**, 360–372.

Metz, R. (1970). The spread of wild oat (*Avena fatua*) caryopses and possibilities for farm hygiene measures for destroying or eliminating wild oat seeds. *Nachrichtenblatt. Deutscher Pflanzenschutzdienst (Berlin)*, **24**, 85–88.

Metzger, J. D. (1983). Role of endogenous plant-growth regulators in seed dormancy of *Avena fatua*. 2. Gibberellins. *Plant Physiology*, **73**, 791–795.

Metzger, J. D. & Sebesta, D. K. (1982). Role of endogenous growth regulators in seed dormancy of *Avena fatua*. *Plant Physiology*, **70**, 1480–1485.

Middendorf, F. G. (1938). Cytology of dormancy in *Phaseolus* and *Zea*. *Botanical Gazette*, **100**, 485–499.

Mikkelsen, D. S. & Sinah, M. N. (1961). Germination inhibition in *Oryza sativa* and control by preplanting soaking treatments. *Crop Science*, **1**, 332–335.

Milby, T. H. & Johnson, F. L. (1987). Germination of downy brome from southern Kansas, central Oklahoma and North Texas, USA. *Journal of Range Management*, **40**, 534–536.

Miller, P. M., Ahrens, J. F. & Stoddard, E. M. (1965). Stimulation of crabgrass seed germination by 1,2–dibromo-3–chloropropane and ethylene dibromide. *Weeds*, **13**, 13–14.

Miller, S. D., Nalewaja, J. D. & Mulder, C. E. G. (1982). Morphological and physiological variation in wild oat. *Agronomy Journal*, **74**, 771–775.

Ministry of Agriculture, Fisheries and Food-Agricultural Development and Advisory Service. (1974). *Wild Oats*. Advisory Leaflet 452. Tolcarne Drive, Pinner, Middlesex, United Kingdom.

Misra, P. K. & Misro, B. (1970). Seed dormancy in African cultivated rice (*Oryza glaberrima* Steud.). *Indian Journal of Agricultural Science*, **40**, 13.

Mitchell, F. S., Caldwell, F. & Hampson, G. (1958). Influence of the enclosing protective tissues on the metabolism of barley grain. *Nature (London)*, **181**, 1270–1271.

Mitra, A. K., Mukherji, D. K. & Mukherjee, P. (1975). *Proceedings. Indian Science Congress Association*, **62**, 69.

Mitsui, T., Akazawa. T., Christeller, J. T. & Tartakoff, A. M. (1985). Biosynthesis of rice seed α-amylase: Two pathways of amylase secretion by the scutellum. *Archives of Biochemistry and Biophysics*, **241**, 179–186.

Miyamoto, T., Tolbert, N. E. & Everson, E. H. (1961). Germination inhibitors related to dormancy in wheat seeds. *Plant Physiology*, **36**, 739–746.

Miyata, S., Okamoto, K., Watanabe, A. & Akazawa, T. (1981). Enzymic mechanism of starch breakdown in germinating rice (*Oryza sativa*) cultivar Kinmaze seeds. 10. *In vivo* and *in vitro* synthesis of α-amylase in rice seed scutellum. *Plant Physiology*, **68**, 1314–1318.

Mohamed, H. A., Clark, J. A. & Ong, C. K. (1985). The influence of temperature during seed development on the germination characteristics of millet seeds. *Plant Cell and Environment*, **8**, 361–362.

Mondrus-Engle, M. (1981). Tetra ploid perennial Teosinte (*Zea perennis*) seed dormancy and germination. *Journal of Range Management*, **34**, 59–61.

Moore, D. J. & Fletchall, O. H. (1963). Germination-regulating mechanisms of giant foxtail (*Setaria faberii*). *Research Bulletin. Missouri Agricultural Experiment Station*, No. 829.

Moore, G. M. & Hoffman, W. D. (1964). Dormancy of the seed of *Panicum ramosum* (Browntop millet). *Association of Official Seed Analysts News Letter*, **38**, 24–26.

Moormann, B. (1942). Untersuchungen uber Keimruhe bei Hafer unde Gerste. *Kuehn-Arkiv*, **56**, 41–79.

Morgan, S. F. & Berrie, A. M. M. (1970). Development of dormancy during seed maturation in *Avena ludoviciana* winter wild oat. *Nature (London)*, **228**, 1225.

Morinaga, T. (1926). The favorable effect of reduced oxygen supply upon the germination of certain seeds. *American Journal of Botany*, **13**, 159–166.

Morishima, H. & Oka, H. I. (1959). Variation in seed dormancy of wild rice species. *Annual Report. National Institute of Genetics*, **10**, 55–56.

Morrison, I. N. & Dushnicky, L.. (1982). Structure of the covering layers of the wild oat (*Avena fatua*) caryopses. *Weed Science*, **30**, 352–359.

Morrow, L. A. & Gealy, D. (1983). Growth characteristics of wild oat (*Avena fatua*) in the Pacific Northwest. *Weed Science*, **31**, 226–229.

Mott, J. J. (1974). Mechanisms controlling dormancy in the arid zone grass *Aristida contorta* L. 1. Physiology and mechanisms of dormancy. *Australian Journal of Botany*, **22**, 635–645.

(1978). Dormancy and germination in 5 native grass species from savannah woodland communities of the northern territory Australia. *Australian Journal of Botany*, **26**, 621–632.

(1980). Germination and establishment of the weeds *Sida acuta* and *Pennisetum pedicellatum* in the Northern Territory. *Australian Journal of Experimental Agriculture and Animal Husbandry*, **20**, 463–469.

Mott, J. J. & Tynan, P. W. (1974). Mechanisms controlling dormancy in arid zone grass (*Aristida contorta*). 2. Anatomy of hull. *Australian Journal of Botany*, **22**, 647–653.

Mouton, J. A. (1960). La dormance chez *Orysa sativa* L.; revision bibliographiqe. *Journal d'Agriculture Tropicale et de Botanique Applique*, **77**, 588–590.

Mukherjee, A. & Chatterji, V. N. (1970). Photoblastism in some of the desert grass seeds. *Annals of Arid Zone*, **9**, 104–113.

Mullverstedt, R. (1961). *Investigations on several questions concerning cultural weed control, with particular regard to weed seed germination in relation to oxygen*. Dissertation, Landwirtschaft Hochschule, Stuttgart-Hohenheim. 75pp.

(1963a). Untersuchungen uber die Ursachen des Vermehrten Auflaufens von Unkrauten nach Mechanischen Unkrautbekampfungsmassnahmen (Nachauflauf). *Weed Research*, **3**, 298–303.

(1963b). Untersuchungen uber die Keimung von Unkrautsamen in Abhangigkeit vom Sauerstoffpartialdruck. *Weed Research*, **3**, 154–163.

Munerati, M. O. (1925). Existe-t-il une apres maturation chez les cereales recemment recoltees? *Comptes Rendus. Academie de Sciences (Paris)*, **181**, 1081–1083.

Munn, M. T. (1946). Germinating freshly harvested winter barley and wheat. *Proceedings. Association of Official Seed Analysts*, **36**, 151–152.

Murdoch, A. J. & Roberts, E. H. (1982). Biological and financial criteria of long-term strategies for annual weeds. *Proceedings. 1982 British Crop Protection Conference on Weeds*, pp. 741–748.

Myers, A. (1963). Germination of *Phalaris* seed. *Agricultural Gazette of New South Wales*, **74**, 635–637.

Na, J. H. & Lee, Y. B. (1984). Studies on the dormancy effects of plant growth substances for breaking the seed dormancy of *Raphanus sativus* L. *Nongop Kisul Yongu Pogo Chungnam Taehakkyo*, **11**, 85–93.

Nakajima, T. & Moroshima, H. (1958). Studies on embryo culture in plants. II. Embryo culture interspecific hybrids in *Oryza*. *Japanese Journal of Breeding*, **8**, 105–110.

Nakamura, S. (1962). Germination of grass seeds. *Proceedings. International Seed Testing Association*, **27**, 710–729.

(1963). Short communication on dormancy of rice seed. *Proceedings. International Seed Testing Asociation*, **28**, 57–59.

Nakamura, S., Watanabe, S. & Ichihara, J. (1960). Effect of gibberellin on the germination of agricultural seeds. *Proceedings. International Seed Testing Association*, **25**, 433–439.

Narziss, L., Reicheneder, E., Duerr, P, & Eder, J. (1980). Trials on the breaking of dormancy and water sensitivity by physical methods. 2. Effects of warm or cold treatment on barleys of the 1979 harvest. *Brauwissenschaft*, **33**, 295–303.

Navasero, E. P., Baun, L. C. & Juliano, B. O. (1975). Grain dormancy, peroxidase activity and oxygen uptake in *Oryza sativa*. *Phytochemistry*, **14**, 1899–1902.

Naylor, J. M. (1978). Studies on the genetic control of germination in wild oats. *Proceedings. Wild Oat Action Committee Seminar '78*, pp. 73–74. Ottawa: Agriculture Canada.

(1983). Studies on the genetic control of some physiological processes in seeds. *Canadian Journal of Botany*, **61**, 3561–3567.

Naylor, J. M. & Fedec, P. (1978). Dormancy studies in seed of *Avena fatua*. 8. Genetic diversity affecting response to temperature. *Canadian Journal of Botany*, **56**, 2224–2229.

Naylor, J. M.. & Jana, S. (1976). Genetic adaptation for seed dormancy in *Avena fatua*. *Canadian Journal of Botany*, **54**, 306–312.

Naylor, J. M., Sawhney, R. & Jana, S. (1980). Genotype-environment interaction in the regulation of seed dormancy in wild oats (*Avena fatua* L.). *Weed Science. Abstracts*, **1980**, p. 87.

Naylor, J. M. & Simpson, G. M. (1961a). Bioassay of gibberellic acid using excised embryos of *Avena fatua* L. *Nature (London)*, **92**, 679–680.

(1961b). Dormancy studies in seeds of *Avena fatua*. 2. A gibberellin-sensitive inhibitory

mechanism in the embryo. *Canadian Journal of Botany*, **39**, 281–295.
Naylor, R. E. L. & Abdalla, A. F. (1982). Variation in germination behaviour. *Seed Science and Technology*, **10**, 67–76.
Negbi, M. (1984). The structure and function of the scutellum of the Gramineae. *The Botanical Journal of the Linean Society*, **88**, 205–222.
Negbi, M. & Koller, D. (1964). Dual action of white light in the photocontrol of germination of *Oryzopsis miliacea*. *Plant Physiology*, **39**, 247–253.
Neill, S. J., Horgan, R. & Rees, A. F. (1987). Seed development and vivipary in *Zea mays* L. *Planta (Berlin)*, **171**, 358–364.
Nelson, A. & MacLagan, J. F. A. (1935). Factors in the germination of *Aira flexuosa*. *Edinburgh Royal Botanical Garden, Notes*, **18**, 251–266.
Nelson, J. R. & Wilson, A. M. (1969). Influence of age and awn removal on dormancy of Medusahead seeds. *Journal of Range Management*, **22**, 289.
Nicholls, P. B. (1979). Induction of sensitivity to gibberellic acid in developing wheat caryopses – effect of rate of desiccation. *Australian Journal of Plant Physiology*, **6**, 229–240.
(1986). Induction of sensitivity to gibberellic acid in wheat and barley caryopses: Effect of dehydration, temperature and the role of the embryo during caryopsis maturation. *Australian Journal of Plant Physiology*, **13**, 785–794.
Nielsen, E. L., Dickerson, J. G. & Smith, D. C. (1959). Strain and seed treatment as factors in germination and seedling growth of smooth bromegrass. *Phytopathology*, **49**, 8–12.
Nieser, O. (1924). Beitrage zur Kenntnis de Keimungsphysiologie von *Anthoxanthum puelli*, *Festuca ovina* and *Aera flexuosa*. *Archiv für Botanik*, **6**, 275–312.
Nieto-Hatem, J. (1963). Seed dormancy in *Setaria lutescens*. *Dissertation Abstracts. Ser. B.*, **24**, 1360–1361.
Niffenegger, D. & Schneiter, A. A. (1963). A comparison of methods of germinating green needlegrass seed. *Proceedings. Association of Official Seed Analysts*, **53**, 67–73.
Nilolaeva, M. G. (1969). *Physiology of Deep Dormancy in Seeds*. Leningrad: Izdatel'stvo 'Nauka'. (Translation. Israel Program for Scientific Translations, Jerusalem. 1969: For the National Science Foundation, Washington, D.C.).
Nishiyama, I. & Inamori, Y. (1966). Length of dormant period in seeds of *Avena* species. *Japanese Journal of Breeding*, **16**, 73–76.
Norris, R. F. & Schoner, C. A. (1980). Yellow foxtail (*Setaria lutescens*) biotype studies – dormancy and germination. *Weed Science*, **28**, 159–163.
Norstog, K. & Klein, R. M. (1972). Development of cultured barley embryos. 2. Precocious germination and dormancy. *Canadian Journal of Botany*, **50**, 1887–1894.
Oakes, A. J. (1959). Germination of elephant grass (*Pennisetum purpureum* Schum.). *Journal of the Agricultural University of Puerto Rico*, **43**, 140.
Obeid, M. & Tagelseed, M. (1976). Factors affecting dormancy and germination of seeds of *Eichhornia crassipes* (Mart.) solms from Nile. *Weed Research*, **16**, 71–80.
Obhlidlova, L. & Hradilik, J. (1973). Effects of ethrel (2–chloroethylphosphonic acid) and kinetin (6–furfurylaminopurine) on the dormancy of barley embryos. *Rostinna Vyroba*, **19**, 905–916.
(1975). Simulation of dormancy in barley embryos and its reversal. *Acta Universitatis Agriculturae (Brno). Facultas Agronomica*, **23**, 855–862.
Odgaard, P. (1972). Wild oat *Avena fatua*. II. The influence of climate, soil and other site-dependent factors. *Tiddskrift for Planteavl*, **76**, 132–144.
Odum, H. T. (1983). *Systems Ecology. An Introduction*. New York: John Wiley and Sons.
Oelke, E. A. & Albrecht, K. A. (1978). Mechanical scarification of dormant wild rice seed. *Agronomy Journal*, **70**, 691–694.
(1980). Influence of chemical seed treatments on germination of wild rice seeds. *Crop Science*, **20**, 595–598.

Ogawara, K. & Hayashi, J. (1964). Dormancy studies in *Hordeum spontaneum* seeds. *Berichte des Ohara Instituts für Landwirtschaftliche Biologie*, **12**, 159–188.

Oka, H. I. & Tsai, K. H. (1955). Dormancy and longevity of rice seed with regard to their variations among varieties. *Japanese Journal of Breeding*, **5**, 22–26.

Okada, T. & Ochi, M. (1986). Studies on fall panicum seed for better germination. I. Discovery of soaking treatment at atmospheric temperature as a dormancy-breaking method. *Bulletin. National Grassland Research Institute*, **34**, 26–35.

Okada, T., Ochi, M. & Ohta, K. (1982). Seed treatment to secure high germination percentage of fall panicum seed, soaking at room temperature. *Journal of the Japanese Society of Grassland Science*, **28**, 119–120.

Oke, H. G. (1952/53). Germination, longevity and dormancy tests of the seed of *Dichanthium annulatum* Stapf. and *Bothriochloa intermedia* (Br.) A. Camus. *Poona Agricultural College Magazine*, **43**, 155–159.

Okigbo, B. N. (1964a). Studies of seed germination in star grasses: I. The effect of nitrate and alternating temperatures. *Journal. West African Science Association*, **8**, 141–158.

(1964b). Studies of seed germination in star grasses. II. Effect of mechanical and acid scarification. *Journal. West African Science Association*, **8**, 159–166.

Opsomer, J. E. & Bronckers, F. (1958). Stimulation de la germination, des semences de *Panicum maximum*. *Academie Royale des Sciences Colon. Bulletin des Seances*, **4**, 330–334.

Ortega-Delagado, M. L. Vega-Vasquez, O., Kruger, J. E. & Laberge, D. E. (1983). Carbohydrates and germination inhibitors during corn seed maturation and germination. In *3rd International Symposium. Pre-harvest Sprouting in Cereals*, ed. J. E. Kruger & D. E. Laberge, pp. 181–187. Boulder, Color. USA: Westview Press.

Osborne, D. J. (1980). Senescence in seeds. In *Senescence in Plants*. ed. K. V. Thimann, pp. 13–37. Boca Raton, Florida, USA: CRC Press.

O'Toole, J. J. & Cavers, P. B. (1981). Variations in numbers of dormant viable seeds from different soil samples taken from a single arable field. *Proceedings. International Botanical Congress*, **13**, 111.

Ovesnov, A. M. & Ovesnov, S. A. (1972). Ecology of germination of stipa grass seeds. *Soviet Journal of Ecology*, **2**, 220–224.

Palmer, G. H. (1970). Response of cereal grains to gibberellic acid. *Journal of the Institute of Brewing*, **76**, 378.

Pammel, L. H. & King, C. M. (1925). Germination of some pines and other trees. *Proceedings. Iowa Academy of Science*, **32**, 123–132.

Pannangpetch, K. & Bean, E. W. (1984). Effects of temperature on germination in populations of *Dactylis glomerata* from NW Spain and central Italy. *Annals of Botany*, **53**, 633–639.

Parasher, V. & Singh, O. S. (1984). Physiology of anaerobiosis in *Phalaris minor* and *Avena fatua*. *Seed Research (New Delhi)*, **12**, 1–7.

(1985). Mechanisms of anoxia induced secondary dormancy in canary grass *Phalaris minor* Retz.) and wild oat (*Avena fatua* L.). *Seed Research (New Delhi)*, **13**, 91–97.

Paterson, J. G. (1976). The distribution of *Avena* species naturalized in Western Australia. *Journal of Applied Ecology*, **13**, 257–264.

Paterson, J. G., Boyd, W. J. & Goodchild, N. A. (1976). Vernalization and photoperiod requirement of naturalized *Avena fatua* and *Avena barbata* Pott. ex Link. in Western Australia. *Journal of Applied Ecology*, **13**, 257–264.

Paterson, J. G., Goodchild, N. A. & Boyd, W. J. R. (1976). Effect of storage temperature, storage duration and germination temperature on dormancy of *Avena fatua* L. and *Avena barbata* Pott. ex Link. *Australian Journal of Agricultural Research*, **27**, 373–379.

Paul, A. K. & Mukherji, S. (1977). Germination behaviour and metabolism of rice seeds under waterlogged conditions. *Zeitschift für Pflanzenphysiologie*, **82**, 117–124.

Peers, F. G. (1958). Germination inhibitory substances in oat husks. *West African Journal of Biological Chemistry*, **2**, 9–14.

Pejka, H. (1971). Investigations on the ecology and control of wild oat (*Avena fatua* L.) in the Wroclaw Voivodeship. *Pamietnik Pulawski*, **46**, 83–119.

Pelton J. (1956). A study of dormancy in eighteen species of high altitude Colorado plants. *Butler University Botanical Studies*, **18**, (1).

Pemadasa, M. A. & Lovell, P. H. (1975). Factors controlling germination of some dune annuals. *Journal of Ecology*, **63**, 41–60.

Percival, F. W. & Bandurski, R. S. (1976). Esters of indole-3–acetic acid from *Avena seeds*. *Plant Physiology*, **58**, 60–67.

Pernes, J., Rene, J., Rene-Chaume, R., Savidan, Y. & Souciet, J. L. (1975). Problems associated with the reproduction of *Panicum maximum* by seeds. CAH ORSTOM Ser. Biol. **10**, 127–133.

Peters, N. C. B. (1978). *Factors influencing the emergence and competition of* Avena fatua *L. with spring barley*. Ph.D. Thesis. University of Reading, England. 232pp.

(1982a). The dormancy of wild oat seed (*Avena fatua* L.) from plants grown under various temperature and soil moisture conditions. *Weed Research*, **22**, 205–212.

(1982b). Production and dormancy of wild oat (*Avena fatua*) seed from plants grown under soil water stress. *Annals of Applied Biology*, **100**, 189–196.

(1985). Competitive effects of *Avena fatua* L. plants derived from seeds of different weights. *Weed Research*, **23**, 305–311.

(1986). Factors affecting seedling emergence of different strains of *Avena fatua*. *Weed Research*, **26**, 29–38.

Peters, N. C. B., Chancellor, R. J. & Drennan, D. S. H. (1975). Dormancy of seed from wild oat plants sprayed with sub-lethal levels of herbicides and gibberellic acid. *Proceedings. European Weed Research Society, Symposium on the Status and Control of Grassweeds in Europe*, **1**, 87–94.

Peters, R. A. & Yokum H. C. (1961). Progress report on a study of the germination and growth of yellow foxtail (*Setaria glauca* (L.) Beauv.). *Proceedings. Northeast Weed Control Conference*, **15**, 350–355.

Petersen, H. I. (1949). Nogle Ukrudts planters Udbredelse og Betydning i Danmark. *Tiddskrift for Planteavl*, **52**, 468–483.

Petrasovits, I. (1958). Germination studies in rice. *Novenytermeles*, **7**, 27–36.

Phaneendranath, B. R. (1977a). *Dormancy and viability of Kentucky Bluegrass* (Poa pratensis L.) *seed as affected by stage of maturity, storage conditions and other treatments*. Ph.D. Thesis. Rutgers State University, USA. 167 pp.

(1977b). Effects of accelerated ageing and dry heat treatment on dormancy and viability of freshly harvested Kentucky bluegrass. *Journal of Seed Technology*, **2**, 11–17.

Phaneendranath, B. R., Duell, R. W. & Funk, C. R. (1978). Dormancy of Kentucky bluegrass seed in relation to color of spikelets and panicle branches at harvest. *Crop Science*, **18**, 683–684.

Phaneendranath, B. R. & Funk, C. R. (1981). Effect of storage conditions on viability, after-ripening and induction of secondary dormancy of Kentucky bluegrass seed. *Journal of Seed Technology*, **6**, 9–22.

Pharis, R. P. & King, R. W. (1985). Gibberellins and reproductive development in seed plants. In *Annual Review of Plant Physiology*, vol. 36, ed. R. W. Briggs, R. L. Jones & V. Walbot, pp. 517–568. Palo Alto, California, USA: Annual Reviews Inc.

Piacco, R. (1940). La germinazione dei semi di *Panicum crusgalli* e *Panicum phillopogon*. *Risicoltura*, **30**, 101–113.

Piech, J. (1970). Inheritance of seed dormancy stage duration in winter wheat (*Triticum aestivum* L.). *Genetica Polonica*, **11**, 227–235.

Pieczur, E. A. (1952). Effect of tissue cultures of maize endosperm on the growth of excised

maize embryos. *Nature (London)*, **170**, 241–242.
Pierpoint, M. & Jensen, L. (1958). Activation of germination of Highland bentgrass by infra-red lamp. *Proceedings. Association of Official Seed Analysts*, **48**, 75–80.
Pili, E. C. (1968). Dormancy of rice seeds – its causes and methods of breaking. *Philippine Journal of Plant Industry*, **33**, 127–136.
Pinnell, E. L. (1949). Genetic and environmental factors affecting corn seed germination at low temperatures. *Agronomy Journal*, **41**, 562–568.
Pinthus, M. J. & Rosenblum, J. (1961). Germination and emergence of sorghum at low temperatures. *Crop Science*, **1**, 293–296.
Plarre, W. K. F. (1975). Selection of high sprouting disposition and seed dormancy in rye. *Hodowla Roslin Aklimatyzacja i Nasiennictwo*, **19**, 595–602.
Pollard, F. (1982). Light induced dormancy in *Bromus sterilis*. *Journal of Applied Ecology*, **19**, 563–568.
Pollhamer, E. (1960). Experiments on the duration of seed dormancy in different varieties of winter and spring barley. *Novenytermeles*, **9**, 2313–2346.
Pollock, J. R. A. (1956). Studies in barley and malt. VIII. Survey of the dormancy of barleys in Britain in 1955. *Journal of the Institute of Brewing*, **62**, 331–333.
Pollock, J. R. A., Kirsop, B. H. & Essery, R. E. (1955). Dormancy in barley. *Proceedings. 5th European Brewery Convention, Baden Baden*, Abstr., **1955**, 203–211.
Pollock, J. R. A. & Pool, A. A. (1962). Studies in barley and malt. XIX. Further observations on water-sensitivity in barley. *Journal of the Institute of Brewing*, **68**, 427–431.
Pont, J. W. (1935). *Physiological Studies with Seeds of* Andropogon sorghum *Brot*. Union of South Africa, Department of Agriculture, 64pp. Amsterdam: Swets and Zeitlinger.
Popay, A. I. (1973). Germination and dormancy in the seeds of certain East African weed species. *Proceedings. 4th Asian-Pacific Weed Science Conference*, **1**, 77–81.
(1975). Laboratory germination of barley grass. *Proceedings. 28th New Zealand Weed and Pest Control Conference*, pp. 7–11.
(1981). Germination of five annual species of barley grass. *Journal of Applied Ecology*, **18**, 547–558.
Pope, M. N. & Brown, E. (1943).Induced vivipary in three varieties of barley possessing extreme dormancy. *Journal of the American Society of Agronomy*, **35**, 161–163.
Popova, D. (1979). Effect of light at constant and variable temperature conditions on the germination of green amaranth (*Amaranthus retroflexus* (L.) R. et S.) and barnyard grass (*Echinochloa crus-galli* L.) seeds. *Plant Science*, **16**, 39–48.
Povilaitis, B. (1956). Dormancy studies with seeds of various weed species. *Proceedings. International Seed Testing Association*, **21**, 88–111.
Prasad, V. N., Gupta,V. N. P. & Bajracharya, D. (1983). Alleviation by gibberellic acid and kinetin of the inhibition of seed germination in maize *Zea mays* under submerged conditions. *Annals of Botany*, **52**, 649–652.
Prentice, L. J. (1981). Observations on the germination time on range grasses. *Association of Official Seed Analysts News Letter*, **55**, 59.
Probert, R. J. (1981). The promotive effects of a mould, *Penicillium funiculosum* Thom. on the germination of *Oryzopsis miliacea* (L.) Asch. and Schw. *Annals of Botany*, **48**, 85–88.
Probert, R. J., Smith, R. D. & Birch, P. (1985a). Germination responses to light and temperatures in European populations of *Dactylis glomerata* L. I. Variability in relation to origin. *New Phytologist*, **99**, 305–316.
(1985b). Germination responses to light and alternating temperatures in European populations of *Dactylis glomerata* L. II. The genetics and environmental components of germination. *New Phytologist*, **99**, 317–322.
(1985c). Germination responses to light and alternating temperatures in European populations of *Dactylis glomerata* L. III. The role of the outer covering structures. *New Phytologist*, **100**, 447–455.

(1985d). Germination responses to light and alternating temperatures in European populations of *Dactylis glomerata* L. IV. The effects of storage. *New Phytologist*, **101**, 521–530.

Purseglove, J. W. (1972). *Tropical Crops. Monocotyledons.* Burnt Mill, England: Longman Group Ltd.

Purvis, O. N. (1948). Studies in vernalisation. XI. The effect of date of sowing and of excising the embryo upon the response of Petkus winter rye to different periods of vernalisation treatment. *Annals of Botany*, **12**, 183–206.

Quail, P. H. (1968). *A study of the biology and control of wild oats (Avena fatua* L. and *A. ludoviciana* L. Dur.). Ph.D. Thesis, Faculty of Agriculture, Sydney University, Australia. 266pp.

Quail, P. H. & Carter, O. G. (1968). Survival and seasonal germination of seeds of *Avena fatua* and *A. ludoviciana. Australian Journal of Agricultural Research*, **19**, 721–729.

(1969). Dormancy in seeds of *Avena ludoviciana* and *A. fatua. Australian Journal of Agricultural Research*, **20**, 1–11.

Quick, W. A. & Hsiao, A. I. (1983). The role of phosphorus in wild oat seed dormancy. In *Wild Oat Symposium '83*, ed. A. E. Smith, pp. 161–172. Regina, Saskatchewan, Canada: Canadian Plains Research Centre.

Rademacher, B. & Kiewnick, L. (1964). Uber den Einfluss einer mineralischen und organischen Dungung auf die Periodizitat de Keimung sowie die Lebensdauer der Damen des Flughafers (*Avena fatua* L.). *Zeitschrift für Acker- und Pflanzenbau*, **119**, 369–385.

Radley, M. (1967). Site of production of gibberellin-like substances in germinating barley embryos. *Planta (Berlin)*, **75**, 164–171.

Ragavan, K. (1960). Potential vivipary in *Saccharum spontaneum* and hybrid sugarcane. *Science and Culture*, **26**, 129–130.

Raju, M. V. S. (1983). Awn anatomy and its relation to germinability of wild oat caryopses. In *Wild Oat Symposium '83*, ed. A. E. Smith, pp. 153–159. Regina, Saskatchewan,Canada: Canadian Plains Research Centre.

(1984). Studies on the inflorescence of wild oats (*Avena fatua* L.). Morphology and anatomy of the awn in relation to its movements. *Canadian Journal of Botany*, **62**, 2237–2247.

Raju, M. V. S. & Barton, R. (1984). On dislodging caryopses of wild oats. *Botanical Magazine (Tokyo)*, **127-130**.

Raju, M. V. S., Hsiao, A. I. & McIntyre, G. I. (1986). Seed dormancy in *Avena fatua*. III. The effect of mechanical injury on the growth and development of the root and scutellum. *Botanical Gazette*, **147**, 4443–4452.

(1988). Seed dormancy in *Avena fatua*. IV. Further observations on the effect of mechanical injury on water uptake and germination in different pure lines. *Botanical Gazette*, **149**, 419–426.

Raju, M. V. S., Jones, G. J. & Ledingham, G. F. (1985). Floret anthesis and pollination in wild oats (*Avena fatua*). *Canadian Journal of Botany*, **63**, 2187–2195.

Raju, M. V. S. & Ramaswamy, S. N. (1983). Studies on the inflorescence of wild oats (*Avena fatua* L.). *Canadian Journal of Botany*, **61**, 74–78.

Ramakrishnan, P. S. (1960). Ecology of *Echinochloa colonum* Link. *Proceedings. Indian Academy of Science, B*, **52**, 73–90.

Ramakrishnan, P. S. & Khoshla, A. K. (1971). Seed dormancy in *Digitaria adscendens* M. and *Echinochloa colonum* with particular reference to covering structures. *Tropical Ecology*, **12**, 112–122.

Rana, O. P. S. & Maherchandani, N. (1982). Studies on the dormancy of cereal seeds as affected by gamma radiation. *Indian Journal of Plant Physiology*, **25**, 324–329.

Rao, D. V. & Raju, M. V. S. (1985). Radial elongation of the epidermal cells of scutellum during caryopsis germination of wild oats (*Avena fatua*). *Canadian Journal of Botany*, **63**, 1789–1793.

Rao, U. P. (1985). Breeding of rice varieties for rainfed upland areas. *Indian Journal of Genetics and Plant Breeding*, **44**,42–48.

Rauber, R. (1984a). Dormancy in winter barley (*Hordeum vulgare* L.) – Influence of temperature during seed development and germination test. *Landwirtschaftliche Forschung*, **37**, 102–110.

(1984b). Influence of temperature before and after early dough stage on dormancy of winter barley (*Hordeum vulgare* L.). *Landwirtschaftliche Forschung*, **37**, 218–222.

(1985). Experiments on primary dormancy of naked barley (*Hordeum vulgare* var. Hexastichon, *Hordeum vulgare* var. Nudum). *Angewandte Botanik*, **59**, 261–266.

(1987). Secondary dormancy in winter barley (*Hordeum vulgare* L.) – Effect of temperature during imbibition. *Landwirtschaftliche Forschung*, **40**, 131–140.

Rauber, R. & Isselstein, J. (1985). Studies on the thermosensitivity and water-sensitivity during germination of freshly ripened winter barley (*Hordeum vulgare*). *Angewandte Botanik*, **59**, 267–277.

Ray, C. B. & Stewart, R. T. (1937). Germination of seeds from certain species of *Paspalum*. *Journal of the American Society of Agronomy*, **29**, 548–554.

Raza, S. H. (1977). Effect of temperature pre-treatment on germination of seeds of *Pennisetum typhoides* va. HB1. *Indian Journal of Agricultural Research*, **11**, 241–242.

Reddy, L. V., Metzger, R. J. & Ching, T. M. (1985). Effect of temperature on seed dormancy of wheat. *Crop Science*, **25**, 455–458.

Rejowski, A. (1971). Effect of gibberellic acid on ribonuclease activity in the embryos and endosperm of spring barley grain in the post-harvest dormancy period. *Acta Societatis Botanicorum Poloniae*, **40**, 423–430.

Renard, C. & Capelle, P. (1976). Seed germination in ruzizi grass (*Brachiaria ruziziensis* (Germain and Evrard)). *Australian Journal of Botany*, **24**, 437–446.

Renard, H. A. (1960). Contribution to the study of the influence of gibberellin on the dormancy of certain grains. *Annales de Physiologie Vegetale (Bruxelles)*, **2**, 99–107.

Richards, P. W. (1966). *The Tropical Rainforest. An Ecological Study*, London, Ibadan, New York: Cambridge University Press.

Richardson, S. G. (1979). Factors influencing the development of primary dormancy in wild oat seeds. *Canadian Journal Of Plant Science*, **59**, 777–784

Rijkslandbouwhogeschool Gent. (1960). *Wild Oats:* Avena fatua *and* Avena strigosa. Gent: Beknopt Versl. Centr. Onkruidonerz. pp. 45–47.

Rines, H. W., Stuthman, D. D., Briggle, L. W., Youngs, V. L., Jedlinski, A., Smith, D. H., Webster, J. A. & Rothman, P. G. (1980). Collection and evaluation of *Avena fatua* for use in oat improvement. *Crop Science*, **20**, 63–68.

Ringlund, K. (1987). Pre-harvest sprouting in barley. In *4th International Symposium. Preharvest Sprouting in Cereals*, ed. D. J. Mares, pp. 15–23. Boulder, Color. USA: Westview Press.

Rizk, T. Y., Fayed, M. T. & El-Deepah, H. R. (1978). *Effect of some promoters on weed seed germination*. Research Bulletin. Faculty of Agriculture, Ain Shamo University, No. 818. 30pp.

Robbins, W. A. & Porter, R. H. (1946). Germinability of sorghum and soybean exposed to low temperatures. *Journal of the American Society of Agronomy*, **38**, 905–913.

Roberts, E. H. (1961). Dormancy in rice seed. 2. The influence of covering structures. *Journal of Experimental Botany*, **12**, 430–445.

(1962). Dormancy in rice seed. 3. The influence of temperature, moisture, and gaseous environment. *Journal of Experimental Botany*, **13**, 75–94.

(1963a). An investigation of inter-varietal differences in dormancy and viability of rice seed. *Annals of Botany*, **27**, 365–369.

(1963b). The effects of some organic growth substances and organic nutrients on dormancy in rice seed. *Physiologia Plantarum*, **16**, 745–755,

(1964). A survey of the effects of chemical treatments on dormancy in rice seed. *Physiologia Plantarum*, **17**, 30–43.
(1965). Dormancy in rice seed. IV. Varietal responses to storage and germination temperatures. *Journal of Experimental Botany*, **16**, 341–349.
(1967). Factors affecting dormancy in rice seed. In *Physiologie, Oekologie und Biochemie der Keimung*, vol. 1, ed. H. Boriss, pp. 359–365. Greifswald: Ernst-Moritz-Arndt-Universität.
(1969). Seed dormancy and oxidation processes. *Symposium. Society for Experimental Biology*, **23**, 161–192.
(1972). *Viability of Seeds*. London: Chapman and Hall.
(1973). Oxidative processes and the control of seed germination. In *Seed Ecology*, ed. W. Heydecker, pp. 189–218. London: Butterworths.
Roberts, E. H. & Benjamin, S. K. (1979). The interaction of light, nitrate and alternating temperature on the germination of *Chenopodium album, Capsella bursa pastoris* and *Poa annua* before and after chilling. *Seed Science and Technology*, **7**, 379–392.
Roberts, E. H. & Ellis, R. H. (1982). Physiological, ultrastructural and metabolic aspects of seed viability. In *The Physiology and Biochemistry of Seed Development, Dormancy and Germination*, ed. A. A. Khan, pp. 465–485. Amsterdam: Elsevier Biomedical Press.
Roberts, H. A. (1986). Persistence of seeds of some grass species in cultivated soil. *Grass Forage Science*, **41**, 273–276.
Roberts, H. A. & Lockett, P. M. (1978). Seed dormancy and field emergence in *Solanum nigrum* L. *Weed Research*, **18**, 231–241.
Rogler, G. A. (1960). Relation of seed dormancy of green needlegrass (*Stipa viridula* Trin.) to age and treatment. *Agronomy Journal*, **52**, 467–469.
Rost, T. L. (1972). Ultrastructure and physiology of protein bodies and lipids from hydrated dormant and non-dormant embryos of *Setaria lutescens*. *American Journal of Botany*, **59**, 607–616.
(1975). The morphology of germination in *Setaria lutescens* (Gramineae): The effects of covering structures and chemical inhibitors on dormant and non-dormant florets. *Annals of Botany*, **39**, 21–30.
Rostrup, O. (1897/98). Germination of *Poa palustris*. *Dansk Frøkontrol Beretning*, **1897/98**, 32–34.
Roy, M. & Gupta, K. (1976). Effect of temperature and promoters on germination of some commonly cultivated varieties of rice. *Geobios (Johdpur)*, **3**, 210–212.
Ruediger, K. R. (1982). Natural inhibitors of germination and growth. 1. Development of a quantitative biotest and application upon extracts from husks of *Avena sativa*. *Zeitschrift für Naturforschung. Section C. Biosciences*, **37**, 793–801.
Ruediger, W. & Lohaus, E. (1985). Germination and growth inhibitors as allelochemicals. *Abstracts of Papers. American Chemical Society*, AGFD No. 203.
Rumpho, M. E. & Kennedy, R. A. (1983). Activity of the pentose phosphate and glycolytic pathways during anaerobic germination of *Echinochloa crus-galli* (barnyard grass) seeds. *Journal of Experimental Botany*, **34**, 893–902.
Ruszkowski, M. & Piech, J. (1969). Seed dormancy inheritance in crosses of winter wheat, *Triticum aestivum* L. *Genetica Polonica*, **10**, 97.
Safaralieva, R. A., Alekperova, S. I. & Mekhtizade, R. M. (1977). The dynamics of auxin inhibitor activity in wheat seeds in the process of emergence from the dormant state. *Biologicheskie Nauki (Moscow)*, **20**, 90–93.
Saini, H. S., Bassi, P. K. & Spencer, M. S. (1986). Interactions among ethephon, nitrate, and after-ripening in the release of dormancy of wild oat (*Avena fatua*) seed. *Weed Science*, **34**, 43–47.
Saini, H. S., Bassi, P. K., Goudey, J. S. & Spencer, M. S. (1986/87). Germination stimulants for more effective weed control. *Agriculture and Forestry Bulletin, University of Alberta*, **9**, 10–12.

Sampson, D. R. & Burrows, V. D. (1972). Influence of photoperiod, short day vernalisation and cold vernalisation on days to heading in *Avena* species and cultivars. *Canadian Journal of Plant Science*, **52**, 471–482.

Sandhu, A. S. & Husain, A. (1961). Effect of seed treatment with gibberellic acid on germination and growth of bajra (*Pennisetum typhoides*). *Indian Journal of Agronomy*, **5**, 269–72.

Sano, Y. (1980). Adaptive strategies compared between the di ploid and tetra ploid forms of *Oryza punctata*. *Botanical Magazine (Tokyo)*, **93**, 171–180.

Sautter, E. H. (1962). Germination of switchgrass (*Panicum virgatum*). *Journal of Range Management*, **15**, 108–110.

Sawhney, R., Hsiao, A. I. & Quick, W. A. (1984). Temperature control of germination and its possible role in the survival of a non-dormant population of *Avena fatua*. *Physiologia Plantarum*, **61**, 331–336.

(1986). The influence of diffused light and temperature on seed germination of three genetically non-dormant lines of wild oats (*Avena fatua*) and its adaptive significance. *Canadian Journal of Botany*, **64**, 1910–1915.

Sawhney, R., Quick, W. A. & Hsiao, A. I. (1985). The effect of temperature during parental vegetative growth on seed germination of wild oats (*Avena fatua* L.). *Annals of Botany*, **55**, 25–28.

Sawhney, R. & Naylor, J. M. (1979). Dormancy studies in seed of *Avena fatua*. 9. Demonstration of genetic variability affecting the response to temperature during seed development. *Canadian Journal of Botany*, **57**, 59–63.

(1980). Dormancy studies in seed of *Avena fatua*. 12. Influence of temperature on germination behaviour of nondormant families. *Canadian Journal of Botany*, **58**, 578–581.

Schafer, D. E. & Chilcote, D. O. (1970). Factors influencing persistence and depletion in buried seed populations. II. The effects of soil temperature and moisture. *Crop Science*, **10**, 342–345.

Schleip, H. (1938). Untersuchungen uber die Auswuchsfestigkeit bei Weizen. *Landwirtschaftliches Jahrbuch*, **86**, 795–822.

Schmidt, B. (1969). On the influence of temperature on the course of germination of some important lawn grasses. *Saatgut Wirtschaftliches*, **21**, 584–589.

Schonfeld, M. A. & Chancellor, R. J. (1983). Factors influencing seed movement and dormancy in grass seeds. *Grass Forage Science*, **38**, 243–250.

Schooler, A. B. (1960). The effect of gibrel and gibberellic acid (K salt) in embryo culture media for *Hordeum vulgare*. *Agronomy Journal*, **52**, 411.

Schroeder, H. (1911). Uber die selektiv Permeable Hulle des Weizenkornes. *Flora oder Allgemeine Botanische Zeitung (Jena)*, **105**(N. F. 2), 186–208.

Schultz, Q. E. & Kinch, R. C. (1976). The effect of temperature, light and growth promoters on seed dormancy in western wheat grass seed (*Agropyron smithii*). *Journal of Seed Technology*, **1**, 79–85.

Schwendiman, A. & Shands, H. L. (1943). Delayed germination or seed dormancy in Vicland oats. *Journal of the American Society of Agronomy*, **35**, 681–688.

Seely, C. I. (1976). Variability in wild oat types and its implications. *Washington Weed Conference*, pp. 29–30. Moscow, Idaho, USA: University of Idaho.

Selman, M. (1970). The population dynamics of *Avena fatua* (wild oats) in continuous spring barley. Desirable frequency of spraying with tri-allate. *Proceedings. 10th British Weed Control Conference*, 1176–1184, Oxford: Weed Research Organisation.

Semple, A. T. (1970). *Grassland Improvement*. London: Leonard Hill Books.

Sendulsky, T., Filguerias, T. S. & Burman, A. G. (1986). Fruits, embryos and seedlings. In *Grass Systematics and Evolution*, ed. T. R. Soderstrom, K. W. Hilu, C. S. Campbell & M.E. Barkworth, pp. 31–36. Washington, D.C.: Smithsonian Institute.

Sexsmith, J. J. (1969). Dormancy of wild oat seed produced under various temperature and moisture conditions. *Weed Science*, **17**, 405–407.

Sexsmith, J. J. & Pittman, U. J. (1963). Effect of nitrogen fertilisers on germination and stand of wild oats. *Weeds*, **11**, 99–101.

Sgambetti-Araujo, L. (1978). *Germination and dormancy studies in* Poa annua *L.* Ph.D. Thesis. University of California, Riverside, California, USA. 151pp.

Sharir, A. (1971). Germination temperatures of dormant and non-dormant Rhodes grass seed. *Proceedings. International Seed Testing Association*, **36**, 109–113.

Sharma, M. P. (1979). Wild oat – a billion dollar problem. *Weeds Today*, **10**, 5–6.

Sharma, M. P., McBeath, D. K. & Vanden Born, W. H. (1976). Studies on the biology of wild oats. 1. Dormancy, germination and emergence. *Canadian Journal of Plant Science*, **56**, 611–618.

(1977). Studies on the biology of wild oats. II. Growth. *Canadian Journal of Plant Science*, **57**, 811–817.

Shetty, S. V. R. (1976). Weeds and weed management in sorghum, pearl millet, chick pea and pigeon pea. *Action for Food Production. AFPO Training Centre, CPPTI*, Lectures. Rajendranagar, Hyderabad, India.

Shimizu, M. (1959). Relation of the enclosing structure of the *Digitaria adscendens* spikelet to its germination. II. The role of the enclosing structure (glumes and shell coat) of crab grass (*D. adscendens*) and rice spikelets in the absorption of water and water soluble substances. *Proceedings. Crop Science Society of Japan*, **28**, 239–243.

Shimizu, N. (1979). Dormancy and germination of seeds in grasses of *Panicum* species. I. Light-temperature response in germination and dormancy-breaking effect of metabolic inhibitors. *Sochi Shikenjo Kenkyu Hokoku*, pp. 94–101.

Shimizu, N. & Mochizuki, N. (1978). Studies on dormancy and germination of seed in finger millet (*Eleusine coracana*). 2. Varietal differences of degree of dormancy and germination behaviour. *Bulletin. National Grassland Institute of Japan*, **12**, 76–91.

Shimizu, N. & Tajima, K. (1979). Studies on dormancy and germination of seed in finger millet (*Eleusine coracana*). 1. Effect of light, temperature and their interaction on germination. *Journal. Japanese Society of Grassland Science*, **24**, 289–295.

Shimizu, N., Tajima, K. & Ogata, R. (1970). Studies on the germination at low temperatures in tropical grass seed. 1. Promotion of germination at low temperature in seeds of *Chloris* spp. and *Eragrostis* spp. *Bulletin. National Grassland Research Institute*, **11**, 47–56.

Shimizu, N., Takahashi, H. & Tajima, K. (1974). Effect of some artificial treatments and various chemicals as oxidase inhibitors on breaking seed dormancy in *Echinochloa crus-galli* var. Caudata. *Nippon Sochi Gakkai-shi*, **1974**, 173–180.

Shimizu, N. & Ueki, K. (1972). Breaking of dormancy in *Echinochloa crus-galli* var. Oryzicola (barnyard grass) seed. III. Change of the dormancy breaking effect of various compounds, especially phenolic ones, concerned with the oxidation-reduction system during the dormancy stage. *Nippon Sakumotsu Gakkai Kiji*, **41**, 488–495.

Shuster, L. & Gifford, R. H. (1962). Changes in 3'-nucleotidase during the germination of wheat embryos. *Archives of Biochemistry and Biophysics*, **96**, 534–540.

Sikder, H. P. (1967). Dormancy of paddy seeds in relation to different seed treatments. *Experimental Agriculture*, **3**, 249–255.

(1973). Dormancy of paddy seeds in relation to different moisture contents of grains and dates of harvests. *Journal of Agricultural Science*, **81**, 159–163.

(1974). Dormancy of paddy seeds of a non-dormant variety. *Journal of Experimental Botany*, **25**, 176–179.

(1983). Effects of cropping seasons and temperatures on the dormancy of rice (*Oryza sativa*) seeds. *Experimental Agriculture*, **19**, 349–354.

Simmonds, J. A. (1971) *Oxidative metabolism and dormancy of caryopses of Avena fatua*. Ph.D. Thesis. University of Saskatchewan, Saskatoon, Canada. 132pp.

Simmonds, J. A. & Simpson, G. M. (1971). Increased participation of pentose phosphate pathway in response to after-ripening and gibberellic acid treatment in caryopses of

Avena fatua. Canadian Journal of Botany, **49**, 1833–1840.

(1972). Regulation of the Krebs cycle and pentose phosphate pathway activities in the control of dormancy of *Avena fatua. Canadian Journal of Botany*, **50**, 1041–1048.

Simpson, G. M. (1965). Dormancy studies in seed of *Avena fatua* 4. The role of gibberellin in embryo dormancy. *Canadian Journal of Botany*, **43**, 793–816.

(1966a). A study of germination in the seed of wild rice (*Zizania aquatica*). *Canadian Journal of Botany*, **44**, 1–9.

(1966b). The suppression by (2–chloroethyl) trimethylammonium chloride of synthesis of a gibberellin-like substance by embryos of *Avena fatua. Canadian Journal of Botany*, **44**, 115–116.

(1978). Metabolic regulation of dormancy in seeds – a case history of the wild oat (*Avena fatua*). In *Dormancy and Developmental Arrest*, ed. M. Clutter, pp. 167–220. New York: Academic Press.

(1981). *Water Stress on Plants*. New York: Praeger Publishers.

(1983). A review of dormancy in wild oats and the lesson it contains for today. In *Wild Oat Symposium '83*, vol, 2, ed. A. E. Smith, pp. 3–20. Regina, Saskatchewan, Canada: Canadian Plains Research Centre.

(1987). *Bibliography of Seed Dormancy in Grasses*. 1405 refs. Saskatoon, Saskatchewan: Department of Crop Science and Plant Ecology, University of Saskachewan.

(1988). *Seed Dormancy in the Wild Oat (*Avena fatua *L.). An annotated bibiography 1850–1987*. 864 refs. Saskatoon, Saskatchewan, Canada: Department of Crop Science and Plant Ecology, University of Saskatchewan.

Simpson, G. M. & Naylor, J. M. (1962). Dormancy studies in seed of *Avena fatua*. 3. A relationship between maltase, amylase and gibberellin. *Canadian Journal of Botany*, **40**, 1659–1673.

(1963). On the mechanism of action of gibberellic acid action in germination. *Abstract. Canadian Society of Plant Physiologists, Annual Meeting*, 4, p. 5.

Sims, R. C. (1959). Germination of barley: Effects of varying water contents upon the initiation and maintenance of growth. *Journal of the Institute of Brewing*, **65**, 46–50.

Singh, A., Datta, D. D. & Singh, D. (1971). Laboratory germination findings on bajra seed (*Pennisetum typhoides*). *Proceedings. International Seed Testing Association*, **36**, 105–107.

Sinyagin, I. I. & Teper, E. N. (1967). The effect of fertilizer on the germination of weed seeds. *Doklady Vsesoyuznoi Akademii Sel'khoz Nauk*, **1**, 2–4.

Sircar, S. M. & Lahiri, A. N. (1956). Studies on the physiology of rice. XII. Culture of excised embryos in relation to endosperm auxin and other growth factors. *Proceedings. National Institute of Science, India part B*, **22**, 212–225.

Skinnes, H. (1984). Genetic investigations on seed dormancy in barley. *Hereditas*, **101**, 285.

Smith, A. L. (1972). Factors influencing germination of *Scolochloa festucacea* caryopses. *Canadian Journal of Botany*, **50**, 2085–2092.

Smith, C. J. (1971). Seed dormancy in sabi panicum. *Proceedings. International Seed Testing Association*, **36**, 81–97.

Smith, D. F. (1968). The growth of barley grass (*Hordeum leporinum*) in annual pasture. 1. Germination and establishment in comparison with other annual pasture species. *Australian Journal of Experimental Agriculture and Animal Husbandry*, **8**, 478–483.

Smith, R. L. (1979). Seed dormancy in *Panicum maximum. Tropical Agriculture*, **56**, 233–240.

Solange, M., Cruz, D. & Takaki, M. (1983). Dormancy and germination of seeds of *Chloris orthonothon. Seed Science and Technology*, **11**, 323–330.

Somody, C. N., Nalewaja, J. D. & Miller, S. D. (1981). Morphology characteristics and dormancy of 1200 wild oat selections. *Proceedings. North Central Weed Control Conference*, **36**, 34.

(1984a). The response of wild oat (*Avena fatua*) and *Avena sterilis* accessions to photoperiod and temperature. *Weed Science*, **32**, 206–213.

(1984b). Wild oat (*Avena fatua*) seed environment and germination. *Weed Science*, **32**, 502–507.

(1984c). Wild oat (*Avena fatua*) and *Avena sterilis* morphological characteristics and response to herbicide. *Weed Science*, **32**, 353–359.

Soriano, A. (1961). Germination of *Stipa neaei* in relation to imbibition and moisture level. *Arid Zone Research. UNESCO. Plant Water Relationships in Arid and Semi-arid Conditions*, **17**, 261–264.

Stanway, V. (1959). Germination of *Sorghum vulgare* Pers. at alternating temperature of 20–30°C and 20–35°C. *Proceedings. Association of Official Seed Analysts*, **49**, 84–87.

(1971). Laboratory germination of giant foxtail, *Setaria faberii* Herrm., at different stages of maturity. *Proceedings. Association of Official Seed Analysts*, **61**, 85–90.

Steinbauer, G. P. & Grigsby, B. A. (1957). Field and laboratory studies on the dormancy and germination of the seeds of chess (*Bromus secalinus* L.) and downy brome-grass (*Bromus tectorum* L.). *Weeds*, **5**, 1–4.

Stinson, R. H. & Peterson, R. L. (1979). On sowing wild oats. *Canadian Journal of Botany*, **57**, 1292–1295.

Stoskopf, N. C. (1985). *Cereal Grain Crops*. Reston, Virginia, USA: Reston Publishing Co.

Stoyanova, S. S., Kostov, K. & Algelova, S. (1984). Influence of storage temperature on the post-harvest dormancy period of *Festuca arundinacea. Rasteniev'd Nauki*, **21**, 99–107.

Strand, E. (1965). Studies on seed dormancy in barley. *Science Reports. Agricultural College of Norway*, **44**, 1–23.

Striegel, W. L. & Boldt, P. F. (1981). Germination and emergence characteristics of wild proso millet. *Proceedings. North Central Weed Control Conference*, **36**, 22.

Stryckers, A. & Pattou, (1963). Biology and survey of wild oats in Belgium. *Mededelingen van de Landbouwhogeschool (Gent)*, **28**, 1063–1086.

Stubbendieck, J. (1974). Effect of pH on germination of three grass species. *Journal of Range Management*, **27**, 78–79.

Stubbendieck, J. & McCully, W. G. (1976). Effect of temperature and photoperiod on germination and survival of sand bluestem. *Journal of Range Management*, **29**, 206–208.

Sugawara, S. (1959). Studies on the germination capacity of upland rice seeds in relation to the location on the flower panicle. *Bulletin. Faculty of Agriculture Niigata University*, **11**, 9–22.

Sumner, D. C. & Cobb, R. D. (1962). Post-harvest dormancy of Coronado side-oats Grama as affected by storage temperature and germination inhibitors. *Crop Science*, **2**, 321–325.

Sung, S-J, S., Hale, M. G., Leather, G. R. & Hurtt, W. (1983). The mechanism of barnyard grass (*Echinochloa crus-galli*) seed dormancy. *Plant Physiology*, **72**, Supp. p. 53.

Sung, S-J, S., Leather, G. R. & Hale, M. G. (1987). Development and germination of barnyard grass (*Echinochloa crus-galli*) seeds. *Weed Science* , **35**, 211–215.

Suss, A. & Bachthaler, G. (1968). Preliminary experiments on gamma-irradiation of weed seeds. *Proceedings. 9th British Weed Control Conference*, pp. 20–24.

Suzuki, S. (1987). Seed dormancy induced at moderate temperature and broken at both high and low temperatures in *Eragrostis ferruginea. Abstracts. International Botanical Congress*, **17**, 137.

Svare, C. W. (1960). The effects of various oxygen levels on germination and early development of wild rice. *Report. Minnesota Department of Conservation, Game and Fish*, No. 3.

Symons, S. J., Angold, R. E., Black, M. & Chapman, J. M. (1983). Changes in the growth capacity of the developing wheat embryo. *Journal of Experimental Botany*, **34**, 1541–1550.

Symons, S. J., Naylor, J. M., Simpson, G. M. & Adkins, S. W. (1986). Secondary dormancy in *Avena fatua*: induction and characteristics in genetically pure dormant lines. *Physiologia Plantarum*, **68**, 27–33.

Symons, S. J., Simpson, G. M. & Adkins, S. W. (1987). Secondary dormancy in *Avena fatua*: Effect of temperature and after-ripening. *Physiologia Plantarum*, **70**, 419–426.

Szekeres, F. (1982). Effect of light, range of temperature and photoperiodic induction on germinating seeds of *Avena fatua* L. *Acta Phytopathologica. Academiae Scientiarum Hungaricae*, **17**, 280–290.

Takabayashi, M. & Nakayama, K. (1981). The seasonal change in seed dormancy of main upland weeds. *Weed Research. (Japan)*, **26**, 249–253.

Takagi, Y., Samoto, S., Kishikawa, H. & Motomura, T. (1986). Effect of hydrogen peroxide on dormancy breaking in rice seed. *Saga Daigakku Nogakubu Iho*, **1986**, 55–59.

Takahashi, H. (1978). Seed germination, ecology and cultivation of barnyard grass after Italian ryegrass in unflooded paddy fields. *Japan Agriculture Research Quarterly*, **12**, 44–48.

Takahashi, N. (1955). Effect of varietal difference on the velocity of germination of rice seeds. *Tohoku Daigaku Nogaku Kenkyusho Iho*, **7**, 1–12.

(1962). Physicogenetical studies on germination of rice seeds with special reference to its genetical factors. *Tohoku University, Institute of Agricultural Research Bulletin*, **14**, 1–87.

(1984). Seed germination and seedling growth. In *Biology of Rice*, ed. S. Tsunodo & N. Takahashi, pp. 71–88. Tokyo, Japan: Japan Scientific Societies Press.

(1985). Inhibitory effects of oxygen on the germination of *Oryza sativa* seeds. *Annals of Botany*, **55**, 597–600.

Takahashi, N., Kato, T. & Tsunagawa, M. (1976). Mechanisms of dormancy in rice seeds. II. New growth inhibitors momilactone-A and -B isolated from the hulls of rice seed. *Ikushugaku Zasshi*, **26**, 91–98.

Takahashi, N. & Miyoshi, K. (1985). Inhibitory effects of oxygen on seed germination as a specific trait of Japonica rice (*Oryza sativa*). *Japanese Journal of Breeding*, **35**, 383–389.

Takahashi, N, & Oka, H. (1958). Observations on natural populations of Formosan wild rice particularly on variations in seed dormancy. *National Institute of Genetics (Mishima) Annual Report*, **8**, 58–59.

(1959). A preliminary report of dormancy in wild rice seed. *Tohoku University. Institute of Agricultural Research Report*, **10**, 81–85.

Tal, M. (1977). Abscisic acid and germination in *Avena sterilis*. *Israel Journal of Botany*, **26**, 100–103.

Tang, W. T. & Chiang, S. M. (1955). Studies on the dormancy of rice seed. *Memoirs. National Taiwan University, College of Agriculture*, **4**, 1–7.

Tao, K-L. J. (1982). Improving the germination of Johnsongrass seeds. *Journal of Seed Technology*, **7**, 1–9.

Taylor, J. S. & Simpson, G. M. (1980). Endogenous hormones in after-ripening wild oat (*Avena fatua*) seeds. *Canadian Journal of Botany*, **58**, 1016–1024.

Taylorson, R. B. (1970). Changes in dormancy and viability of weed seeds in soils. *Weed Science*, **18**, 265.

(1972). Phytochrome controlled changes in dormancy and germination of buried weed seeds. *Weed Science*, **20**, 417–422.

(1975). Inhibition of pre-chill-induced dark germination in *Sorghum halapense* (L.) Pers. seeds by phytochrome transformations. *Plant Physiology*, **55**, 1093–1097.

(1979a). Control of fall panicum (*Panicum dichotomiflorum*) seed dormancy by light. *Proceedings. Northeast Weed Science Society*, **33**, 330.

(1979b). Release of volatiles during accelerated after-ripening of seeds. *Seed Science and Technology*, **7**, 369–378.

(1980). Aspects of dormancy in fall panicum (*Panicum dichotomiflorum*). *Weed Science*, **28**, 64–67.

(1982). Anaesthetic effects on secondary dormancy and phytochrome responses in *Setaria faberii* seeds. *Plant Physiology*, **70**, 882–886.

(1986). Water stress-induced germination of giant foxtail (*Setaria faberii*) seeds. *Weed Science*, **34**, 871–875.

(1988). Anaesthetic enhancement of *Echinochloa crus-galli* (L.) Beauv. seed germination: possible membrane involvement. *Journal of Experimental Botany*, **39**, 50–58.

Taylorson, R. B. & Borthwick, H. A. (1969). Light filtration by foliar canopies; significance for light-controlled weed seed germination. *Weed Science*, **17**, 48–51.

Taylorson, R. B. & Brown, M. M. (1977). Accelerated after-ripening for overcoming seed dormancy in grass weeds. *Weed Science*, **25**, 473–476.

Taylorson, R. B. & Hendricks, S. B. (1979a). Overcoming dormancy in seeds with ethanol and other anaesthetics. *Planta (Berlin)*, **145**, 507–510.

(1979b). Effects of ethanol and other anaesthetics on seed dormancy. *Plant Physiology*, **63**, 68.

(1981). Anaesthetic release of seed dormancy – an overview. *Israel Journal of Botany*, **29**, 273–280.

Taylorson, R. B. & McWhorter, C. G. (1969). Seed dormancy and germination in ecotypes of Johnson grass. *Weed Science*, **17**, 359–361

Tester, W. C.. & McCormick, G. (1954). Germination of Johnson grass; results of tests made by the Arkansas State Plant Board. *Proceedings. Association of Official Seed Analysts*, **44**, 96–99.

Thanos, C. A. & Mitrakos, K. (1979). Phytochrome-mediated germination control of maize caryopses. *Planta (Berlin)*, **146**, 415–417.

Thill, D. C., Schirman, R. D. & Appleby, A. P. (1980). Influence of after-ripening temperature and endogenous rhythms on downy brome (*Bromus tectorum*) germination. *Weed Science*, **28**, 321–323.

Thomas, P. E. L. & Allison, J. C. S. (1975). Seed dormancy and germination in *Rottboellia exaltata*. *Journal of Agricultural Science*, **85**, 129–134.

Thornton, M. L. (1966a). Seed dormancy in tall wheatgrass (Agropyron elongatum). *Proceedings. Association of Official Seed Analysts*, **56**, 116–119.

(1966b). Seed dormancy in buffalo grass (*Buchloe dactyloides*). *Proceedings. Association of Official Seed Analysts*, **56**, 120–123.

Thornton, M. L. & Thornton, B. J. (1962). Firm seed and longevity of blue grama (*Bouteloua gracilis*). *Proceedings. Association of Official Seed Analysts*, **52**, 112–115.

Thornton, N. C. (1945). Importance of oxygen supply in secondary dormancy and its relation to the inhibiting mechanism regulating dormancy. *Contributions. Boyce Thompson Institute*, **13**, 487–500.

Thurston, J. M. (1951a). A comparison of the growth of wild and of cultivated oats in manganese-deficient soils. *Annals of Applied Biology*, **38**, 289–302.

(1951b). Some experiments and field observations on the germination of wild oat (*Avena fatua* and *A. ludoviciana*) seeds in soil and the emergence of seedlings. *Annals of Applied Biology*, **38**, 812–832.

(1952). Biology of wild oats: germination and dormancy. *Report. Rothamsted Experimental Station*, **1952**, 67–68.

(1957a). Weed studies: wild oats. *Report. Rothamsted Experimental Station*, **1957**, 92–93.

(1957b). Morphological and physiological variation in wild oats (*Avena fatua* L. and *Avena ludoviciana* Dur.) and in hybrids between wild and cultivated oats. *Journal of Agricultural Science (Cambridge)*, **49**, 259–274.

(1960). Weed studies: wild Oats. *Report. Rothamsted Experimental Station*, **1960**, 102–103.

(1961). Weed studies: wild Oats. *Report. Rothamsted Experimental Station*, **1961**, 81–83.

(1962a). The effect of competition from cereal crops on the germination and growth of *Avena fatua* L. in a naturally infested field. *Weed Research*, **2**, 192–207.

(1962b). Biology and control of wild oats. *Report. Rothamsted Experimental Station*, **1962**, 236–253.

(1963). Geographical variation with species: effect of photoperiod on germination. *Report. Rothamsted Experimental Station*, **1963**, 90–91.

(1965). Weed studies. Wild oats (*Avena fatua* and *Avena ludoviciana*). *Report. Rothamsted Experimental Station*, **1965**, 102–103.

Thurston, J. M. & Phillipson, A. (1976). Distribution. In *Wild Oats in World Agriculture*, ed. D. Price Jones, pp. 19–64. London, England: Agricultural Research Council.

Tibelius, A. C. & Klinck, H. (1986). Development and yield of oat (*Avena sativa*) plants from primary and secondary seeds. *Canadian Journal of Plant Science*, **66**, 299–306.

Tilsner, H. R. & Upadhyaya, M. K. (1985). Induction and release of secondary dormancy in genetically pure lines of *Avena fatua*. *Physiologia Plantarum*, **64**, 377–382.

(1987). Action of respiratory inhibitors on seed germination and oxygen uptake in *Avena fatua* L. *Annals of Botany*, **59**, 477–482.

Tingey, D. C. (1961). Longevity of seeds of wild oats (*Avena fatua*), winter rye and wheat in cultivated soil. *Weeds*, **9**, 607–611.

Tinnin, R. O. & Muller, C. H. (1971). The allelopathic potential of *Avena fatua*: influence on herb distribution. *Bulletin. Torrey Botanical Club*, **98**, 243–250.

(1972). The allelopathic influence of *Avena fatua*: the allelopathic mechanism. *Bulletin. Torrey Botanical Club*, **99**, 287–292.

Tischler, C. R. & Young, B. A. (1983). Effects of chemical and physical treatments on germination of freshly harvested Klein grass (*Panicum coloratum*) seed. *Crop Science*, **23**, 789–792.

(1987). Development and characteristics of a Kleingrass population with reduced post-harvest seed dormancy. *Crop Science*, **27**, 1238–1241.

Tokuhisa, J. G. & Quail, P. H. (1987). The levels of two distinct species of phytochrome are regulated differently during germination in *Avena sativa*. *Planta (Berlin)*, **172**, 371–377.

Tomar, J. B. (1984). Genetics of grain dormancy in rice (*Oryza sativa* L.). *Genetica Agraria*, **38**, 443–446.

Toole, E. H. (1961). The effect of light and other variables on the control of seed germination. *Proceedings. International Seed Testing Association*, **26**, 659–673.

Toole, E. H. & Coffman, F. A. (1940). Variations in the dormancy of the seeds of the wild oat, *Avena fatua*. *Journal of the American Society of Agronomy*, **32**, 631–638.

Toole, E. H. & Toole, V. K. (1940). Germination of seeds of goosegrass, *Eleusine indica*. *Journal of the American Society of Agronomy*, **32**, 320–321.

Toole, V. K. (1939). Germination of the seed of poverty grass, *Danthonia spicata*. *Journal of the American Society of Agronomy*, **31**, 954–965.

(1940a). Germination of seed of vine-mesquite, *Panicum obtusum*, and plains bristle-grass, *Setaria macrostachya*. *Journal of the American Society of Agronomy*, **32**, 503–512.

(1940b). The germination of seed of *Oryzopsis hymenoides*. *Journal of the American Society of Agronomy*, **32**, 33–41.

(1941). Factors affecting the germination of various dropseed grasses (*Sporobolus* spp.). *Journal of Agricultural Research*, **62**, 691–715.

(1976). Light and temperature control of germination in *Agropyron smithii* seeds. *Plant and Cell Physiology*, **17**, 1263–1272.

(1968). Light responses of *Eragrostis curvula* seed. *Proceedings. International Seed Testing Association*, **33**, 515–530.

Toole, V. K. & Borthwick, H. A. (1971). Effect of light, temperature and their interactions on germination of seeds of Kentucky bluegrass (*Poa pratensis* L.). *Journal American Society of Horticultural Science*, **96**, 301–304.

Toole, V. K. & Koch, E. J. (1977). Light and temperature controls of dormancy and germination in bentgrass seeds. *Crop Science*, **17**, 806–811.

Topornina, N. A. (1958). New data on seed germination of *Avena fatua* and *Setaria glauca*. *Agrobiologiya*, **3**, 149–151.

Trewavas, A. J. (1987). Timing and memory processes in seed embryo dormancy – a conceptual paradigm for plant development questions. *BioEssays*, **6**, 87–92.

Tseng, S. (1964). Breaking dormancy of rice seed with carbon dioxide. *Proceedings. International Seed Testing Association*, **29**, 445–450.

Twentyma, J. D. (1974). Environmental control of dormancy and germination in seeds of *Cenchrus longispinus* (Hack.) Fern. *Weed Research*, **14**, 1–11.

Ujiihra, K. (1982). Studies on the preharvest sprouting of grain sorghum. *Bulletin. Chukogu National Agricultural Experiment Station, A*, **30**, 1–33.

Ulali, D. L., Barker, M. B. & Dumlao, R. C. (1960). A preliminary study of the cancellation of dormancy period of rice seeds. *Philippine Agriculturist*, **44**, 279–289.

Uotani, K., Umezu, T., Teruhiko, M., Meguro, M., Tuzimura, K. *et al.* (1972). Germination inhibitor in dormant rice seed. Isolation of vanillin from the active fraction. *Tohoku Journal of Agricultural Research*, **23**, 58–63.

Upadhyaya, M. K. (1986). Effects of salicylhydroxamate on respiration, seed germination and seedling growth in *Avena fatua*. *Physiologia Plantarum*, **67**, 43–48.

(1987). Inhibition of gibberellic acid-induced starch mobilization by salicylhydroxamic acid in *Avena fatua* L. seed. *Physiologia Plantarum*, **59**, 265–268.

Upadhyaya, M. K., Hsiao, A. I. & Bonsor, M. E. (1986). Gibberellin A_3-like action of substituted phthalimides on *Avena fatua* seed. *Abstract. Annual Meeting, American Society of Plant Physiologists*, June 8–12, 1986. Baton Rouge, USA.

Upadhyaya, M. K., Naylor, J. M. & Simpson, G. M. (1982a). The physiological basis of seed dormancy in *Avena fatua*. I. Action of the respiratory inhibitors sodium azide and salicylhydroxamic acid. *Physiologia Plantarum*, **54**, 419–424.

(1982b). Co-adaptation of seed dormancy and hormonal dependence of α-amylase production in endosperm segments of *Avena fatua*. *Canadian Journal of Botany*, **60**, 1142–1147.

(1983). The physiological basis of seed dormancy in *Avena fatua*. II. On the involvement of alternative respiration in the stimulation of germination by sodium azide. *Physiologia Plantarum*, **58**, 119–123.

Upadhyaya, M. K., Simpson, G. M. & Naylor, J. M. (1981). Levels of glucose-6–phosphate and 6–phosphogluconate dehydrogenase in the embryos and endosperms of some dormant and non-dormant lines of *Avena fatua* during germination. *Canadian Journal of Botany*, **59**, 1640–1646.

Urion, E. & Chapon, L. (1956). Contribution a l'etude de la dormance de l'orge. *Brasserie*, **11**, 151–162, 181–190.

Van Staden, J. & Hendry, N. S. (1985). An evaluation of the problem of volunteer ryegrass (*Lolium mulitflorum*) in seed production. *South African Journal of Plant and Soil*, **2**, 157–160.

Vanderzee, D. & Kennedy, R. A. (1981). Germination and seedling growth in *Echinochloa crus-galli* var. oryzicola under anoxic conditions; structural aspects. *American Journal of Botany*, **68**, 1269–1277.

Varshney, K. A. & Baijal, B. D. (1978). Synergistic and antagonistic behaviour of some growth regulators on germination of seeds of *Pennisetum pedicellatum*. *Comparative Physiology and Ecology*, **3**, 178–180.

Varty, K., Arreguin, L. B., Gomez, T. M., Lopez, T. P. J. & Gomez, L. M. A. (1983). Effects of abscisic acid and ethylene on the gibberellic acid-induced synthesis of α-amylase by isolated wheat aleurone layers. *Plant Physiology*, **73**, 692–697.

Veeraraja Urs, Y. S. (1987). Dormancy in some early and medium duration varieties. *International Rice Research Newsletter*, **12**, 5.

Vegis, A. (1964). Dormancy in higher plants. In *Annual Review of Plant Physiology*, vol. 15, ed. R. W. Briggs, pp.185–224. Palo Alto, California, USA: Annual Reviews Inc.

Venugopal, K. & Krishnamurthy, K. (1972). Dormancy studies in barley varieties. *Mysore Journal of Agricultural Science*, **6**, 113–117.

Voderberg, K. (1965). The physiology of the germination of wild oats. *Archiv für Pflanzenschutz*, **1**, 49–66.

Voigt, P. W. (1973). Induced seed dormancy in weeping lovegrass (*Eragrostis curvula*). *Crop Science*, **13**, 76–79.

Von Bertalanffy, L. (1968). *General Systems Theory*. New York, USA: George Braziller.

Von Guttenberg, H. & Wiedow, H. L. (1952). Uber den Wirkstoffbedarf isolierter Hafer-embryonen. *Planta (Berlin)*, **41**, 145–166

Von Prante, G. (1971). Contribution to the systematics of wild oats, (*Avena fatua* L.). *Zeitschrift für Pflanzenkrankheiten*, **11/12**, 675–694.

Wagenvoort, W. A. & Opstal, N. A. V. (1979). The effect of constant and alternating temperatures, rinsing, stratification and fertiliser on germination of some weeds. *Scientific Horticulture*, **10**, 15–20.

Walker, R. G. (1934). The inimical effects of pre-soaking on the seeds of oats. *Welsh Journal of Agriculture*, **10**, 278–284.

Walker-Simmons, M. (1987). ABA levels and sensitivity in developing wheat embryos of sprouting resistant and susceptible cultivars. *Plant Physiology*, **84**, 61–66.

Walker-Simmons, M. & Sesing, J. D. (1987). Effect of temperature on abscisic acid sensitivity and levels in the developing wheat embryo. *Plant Physiology*, **83**, Supp. p. 38.

Walton, D. C. (1980/81). Does abscisic acid play a role in seed germination? *Israel Journal of Botany*, **29**, 168–180.

Wang, W. Z., Chen, J. & Liu, E. J. (1986). Study on relations between seed coat and dormancy of *Pinus koraiensis* seeds. *Northeast Forestry Institute*, **2**, 83–86.

Watanabe, Y. (1981). Ecological studies on seed germination and emergence of some summer annual weeds of Hokkaido. *Weed Research (Japan)*, **26**, 193–199.

Watanabe, Y. & Hirokawa, F. (1979). Ecological studies on the germination and emergence of annual weeds. 1. Effect of temperatures on the dormancy breaking in seeds of *Chenopodium album*, *Echinochloa crus-galli* var. praticola and *Polygonum lapathifolium*. *Weed Research (Japan)*, **17**, 24–28.

Watkins, F. B. (1969). Techniques used in laboratory studies on dormancy of wild oats. *Australian Weed Research Newsletter*, **13**, 41–42.

Watt, L. A. & Whalley, R. D. B. (1982). Establishment of small-seeded perennial grasses on black clay soils in North-western New South Wales. *Australian Journal of Botany*, **30**, 611–623.

Weber, R. P. & Simpson, G. M. (1967). Influence of water on wild rice (*Zizania aquatica*) grown in a prairie soil. *Canadian Journal of Plant Science*, **47**, 657–663.

Weidner, S. (1984). Studies on the mechanism which prevents germination of unripe Triticale caryopses. *Acta Societatis Botanicorum Poloniae*, **53**, 325–338.

Weidner, S., Makowski, W., Sojka, E. & Rejowski, A. (1984). The role of zeatin and gibberellic acid in breaking of the abscisic acid-induced dormancy in Triticale caryopses. *Acta Societatis Botanicorum Poloniae*, **53**, 339–351.

Weinberg, G. W. (1975). *An Introduction to General Systems Thinking*. London: John Wiley and Sons.

Wellington, P. S. (1956). Studies on the germination of cereals. 1. The germination of wheat grains in the ear during development, ripening and after-ripening. *Annals of Botany*, **20**, 105–120.

(1964). Studies on the germination of cereals. 5. The dormancy of barley grains during ripening. *Annals of Botany*, **28**, 113–126.

Wellington P. S. & Durham, V. M. (1961). Studies on the germination of cereals. 4. The oxygen requirement for germination of wheat grains during maturation. *Annals of Botany*, **25**, 197–205.

Wells, P. V. (1959). Ecological significance of red light sensitivity in germination of tobacco seed. *Science*, **129**, 41–42.

Werker, W. A. (1981). Seed dormancy as explained by the anatomy of embryo envelopes.

Israel Journal of Botany, **29**, 22–44.

Westra, R. N. & Loomis, W. E. (1966). Seed dormancy in *Uniola paniculata*. *American Journal of Botany*, **53**, 407–411.

Wheeler, W. A. (1950). *Forage and Pasture Crops*. Toronto, London & New York: D. Van Nostrand Co. Inc.

Whiteman, P. C. & Mendra, K. (1982). Effects of storage and seed treatments on germination of *Brachiaria decumbens*. *Seed Science and Technology*, **10**, 233–242.

Whittington, W. J., Hillman, J., Gatenby, S. M., Hooper, B. E. & White, J. C. (1970). Light and temperature effects on the germination of wild oats. *Heredity*, **25**, 641–650.

Whyte, L. L., Wilson, A. G. & Wilson, D. (1969). Hierarchical structures. New York: American Elsevier.

Wiberg, H. (1959). Viability of wild oat kernels in a compost. *Vaxtodling*, **10**, 54–57.

Wiberg, H. & Kolk, H. (1960). Effect of gibberellin on germination of seeds. *Proceedings. International Seed Testing Association*, **25**, 440–445.

Wiesner, L. E. & Grabe, D. F. (1972). Effect of temperature preconditioning and cultivar on ryegrass (*Lolium* sp.) seed dormancy. *Crop Science* , **12**, 760–764.

Wiesner, L. E. & Kinch, R. C. (1964). Seed dormancy in green needlegrass. *Agronomy Journal*, **56**, 371–373.

Williams, E. D. (1968). Preliminary studies of germination and seedling behaviour in *Agropyron repens* (L.) Beauv. and *Agrostis gigantea* Roth. *Proceedings. 9th British Weed Control Conference*, **1**, 119–124.

(1973). Seed germination of *Agrostis gigantea* Roth. *Weed Research*, **13**, 310–324.

(1978). Germination and longevity of seeds of *Agropyron repens* and *Agrostis gigantea* in soil in relation to different cultivation regimes. *Weed Research*, **18**, 129–138.

(1983a). Effects of temperature fluctuation, red and far-red light and nitrate on seed germination of five grasses. *Journal of Applied Ecology*, **20**, 923–935.

(1983b). Germinability and enforced dormancy in seeds of species of indigenous grassland. *Annals of Applied Biology* **102**, 557–566.

Williams, G. C. & Thurston, J. M. (1964). The effect of temperature in a sack-drier on survival of insects (*Oryzaephilus surinamensis* (L.) (Col., Silvanidae)) and weed seeds (*Avena fatua* L. and *A. ludoviciana* Dur.). *Annals of Applied Biology*, **53**, 29–32.

Wilson, B. J. (1979). The costs of wild oats (*Avena* spp.) in Australian wheat production. *Proceedings. 7th Asian-Pacific Weed Society Conference*, pp. 441–444.

Wilson, R. D. (1973). Characterization of the dormancy of the seed of wild cane (*Sorghum bicolor* (L.) Moench.). *Dissertation Abstracts International. Series B*, **33**, 5099.

Wright, L. N. (1973). Seed dormancy, germination environment, and seed structure of Lehmann lovegrass, *Eragrostis lehmanniana* Nees. *Crop Science*, **13**, 432–435.

Wright, L. N. & Baltensperger, A. A. (1964). Influence of temperature, light radiation, and chemical treatment on laboratory germination of black gramagrass, *Bouteloua eriopoda* Torr. *Crop Science*, **4**, 168–171.

Wright, W. G. & Kinch, R. C. (1959). Dormancy in *Sorghum vulgare* Pers. *Proceedings. Association of Official Seed Analysts*, **52**, 169–177.

Wu, L. (1978). Seed dormancy of a Taiwan wild rice population and its potential for rice breeding. *Botanical Bulletin. Academia Sinica*, **19**, 1–12.

(1980). The variation of salt tolerance and seed dormancy among natural populations of *Cynodon dactylon*. In *Abstracts. Second International Congress of Systematic and Evolutionary Biology*, p. 400, Vancouver: University of British Columbia, Canada.

Wu, L., Till-Bottraud, I. & Torres, A. (1987). Genetic differentiation in temperature-enforced seed dormancy among golf course populations of *Poa Annua* L. *New Phytologist*, **107**, 623–631.

Wurzburger, J. & Koller, D. (1973). Onset of seed dormancy in *Aegilops kotschyi* Boiss. – its experimental modification. *New Phytologist*, **72**, 1057–1061.

(1976). Differential aspects of parental photothermal environment on development of dormancy in caryopses of *Aegilops kotschyi*. *Journal of Experimental Botany*, **27**, 43–48.

Wurzburger, J. & Leshem, Y. (1967). Gibberellin and hull controlled inhibition of germination in *Aegilops kotschyi* Boiss. *Israel Journal of Botany*, **16**, 181–186.

(1969). Physiological action of the germination inhibitor in the husk of *Aegilops kotschyi* Boiss. *New Phytologist*, **68**, 337–341.

Wurzburger, J., Leshem, Y. & Koller, D. (1974). The role of gibberellin and the hulls in the control of germination in *Aegilops kotschyi* caryopses. *Canadian Journal of Botany*, **52**, 1597–1601.

(1976). Correlative aspects of imposition of dormancy in caryopses of *Aegilops kotschyi*. *Plant Physiology*, **57**, 670–671.

Wverson, E. H. & Hart, R. B. (1961). Varietal variation for dormancy in mature wheat. *Michigan Agricultural Experiment Station. Quarterly Bulletin*, **43**, 820–829.

Yabuki, K. & Miyagawa, I. (1958). Studies on the effects of diurnal variation of temperature upon the germination of seeds. II. The germination of barley and vegetables. *Journal of Agricultural Meteorology (Japan). Society of Agricultural Meteorology*, **13**, 105–109.

Yamasue, Y., Sudo, K. & Ueki, K. (1977). Physiological studies on seed dormancy of barnyard grass (*Echinochloa crus-galli* Beauv. var. Oryzicola Ohwi.). *Proceedings. 6th Asian-Pacific Weed Science Society Conference, Indonesia*, **1**, 42–51.

Yasue, T. (1973). Effect of moisture content of seeds and soaking in water on breaking dormancy in Indica rice. *Research Bulletin. Faculty of Agriculture, Gifu University*, **34**, 1–10.

Yokum,H. C., Jutras, M. W. & Peters, R. A. (1961). Preliminary investigations of a germination and growth inhibitor produced by yellow foxtail (*Setaria glauca* L. Beauv.). *Proceedings. Northeast Weed Control Conference*, **15**, 341–349.

Yomo, H. (1958). Sterilization of barley seeds and formation of amylase by separated embryos and endosperms. *Hakko Kyokaishi*, **16**, 444–448.

Young, J. A. & Evans, R. A. (1980). Germination of desert needlegrass. *Journal of Seed Technology*, **5**, 40–46.

(1984). Germination of seeds of cultivars Paloma and Nezpar Indian rice grass. *Journal of Range Management*, **37**, 19–21.

Young, J. A., Evans, R. A. & Easi, D. A. (1985). Enhancing germination of Indian rice grass (*Oryzopsis hymenoides*) seeds with sulfuric-acid. *Agronomy Journal*, **77**, 203–206.

Young, J. A., Kay, B. L. & Evans, R. A. (1977). Accelerating the germination of common Bermudagrass for hydroseeding. *Agronomy Journal*, **69**, 115–119.

Yumoto, S., Shimamoto, Y. & Tsuda, C. (1980). Studies on ecotype variations among natural populations of timothy (*Phleum pratense* L.). I. Variation in germination characteristics. *Journal of the Japanese Society of Grassland Science*, **26**, 243–250.

Zade, A. (1909). *Der Flughafer*. Inaug. Dissertation, pp. 1–48. Jena: Frommannsche Hofbuchdruckerei (Hermann Pohle).

(1912). Der Flughafer (*Avena fatua*). *Arbeiten. Deutsche Landwirtschafts-Gesellschaft (Berlin)*, **229**, 99pp.

Zee, S-Y., Chan, H-Y. & Ma, C. Y. (1984). Localization of prolamine in oat scutellum cells before and after seed germination. *Journal of Plant Physiology*, **116**, 91–94.

Zemenova, A. (1975). A contribution to the study of dormancy of wild oat (*Avena fatua* L.). *Acta Universitatis Agriculturae (Brno). Facultas Agriculturae*, **23**, 957–966.

Zieber, N. K. & Brink, R. A. (1951). The stimulative effects of *Hordeum* endosperms on the growth of immature plant embryos *in vitro*. *American Journal of Botany*, **38**, 253–256.

Zimmerman, H. (1978). Untersuchungen zur Extraktion von Indolyl-3–essigsaure-Protein aus Hafer. *Zeitschrift für Pflanzenphysiologie*, **89**, 115–118.

Zimmerman, H., Siegert, C. & Karl, R. (1976). Transformations of the phytohormone IAA during the germination of *Avena sativa* . *Zeitschrift für Pflanzenphysiologie*, **80**, 225–235.

Zorner, P. S. (1981). *Seed dormancy and germination of* Avena fatua *L. and* Kochia scoparia *(L.) Scop*. Ph.D. Thesis. Colorado State University. USA.

Zwar, J. A. & Hooley, R. (1986). Hormonal regulation of α-amylase gene transcription in wild oat (*Avena fatua*) aleurone protoplasts. *Plant Physiology*, **80**, 459–463.

Index

Index

Index

Index

Index